Quantum Computing

A pathway to quantum logic design

Quantum Computing

A pathway to quantum logic design

Hafiz Md Hasan Babu
Department of Computer Science and Engineering, University of Dhaka, Dhaka-1000, Bangladesh

IOP Publishing, Bristol, UK

ISBN 978-0-7503-2747-3 (ebook)
ISBN 978-0-7503-2745-9 (print)
ISBN 978-0-7503-2748-0 (myPrint)
ISBN 978-0-7503-2746-6 (mobi)

DOI 10.1088/978-0-7503-2747-3

Version: 20200501

IOP ebooks

British Library Cataloguing-in-Publication Data: A catalogue record for this book is available from the British Library.

Published by IOP Publishing, wholly owned by The Institute of Physics, London

IOP Publishing, Temple Circus, Temple Way, Bristol, BS1 6HG, UK

US Office: IOP Publishing, Inc., 190 North Independence Mall West, Suite 601, Philadelphia, PA 19106, USA

Cover image: An artistic interpretation of a 4-bit quantum ripple carry adder inspired by and adapted from imagery in Lisa N J and Babu H M H, “A compact realization of an n-bit quantum carry skip adder circuit with optimal delay,” 2014 NASA/ESA Conference on Adaptive Hardware and Systems (AHS), Leicester, 2014, pp 270–7 with permission of the IEEE.

To my beloved parents and also to my beloved wife, daughter, and son, who made it possible for me to write this book.

Contents

Preface

With the advent of nanotechnology, quantum computing is increasingly playing a large role in developing more compact and lower power consumption computers. The use of the properties of quantum mechanics has increased the speed certain computations dramatically. Quantum computing is one of the most promising nanotechnologies for designing modern compact computers. The quantum computer is an entirely different concept than the traditional computer and it does not use binary logic. Quantum computers can assess more than one probability at a time through something called quantum entanglement.

The switching and memory units of a computing system are becoming as small as an atom, and switching techniques have reached their peak of development. Therefore, it is necessary to discover an entirely new way of thinking to build computing machines. Recently, advances in mathematics, materials science, and computer engineering have transformed quantum computing machines from theory to practice.

This quantum computing book covers basic quantum computing, fault-tolerant quantum computing, and quantum dot cellular automata (QCA). The basic quantum computing part of this book comprises designs for different quantum circuits, including the adder, multiplier, divider, encoder, decoder, barrel shifter, comparator, and arithmetic logic unit. These quantum circuits are the building blocks of a quantum computer. Readers will obtain an understanding of quantum computing from the basic to the advanced level, which will help them to design new quantum circuits.

The discovery of quantum error correction has greatly improved the long-term prospects for quantum computing technology. Encoded quantum information can thus be protected from the errors that arise due to uncontrolled interactions with the environment, or due to imperfect implementations of quantum logical operations. The fault-tolerant quantum computing part of this book consists of the designs for different fault-tolerant quantum circuits, such as the adder, multiplier, and divider. Fault tolerance is a very complex part of making a quantum circuit. It is necessary to develop fault-tolerant circuits which allow the qubits to be protected from quantum errors produced by inadequate control of environmental interactions. Readers will obtain a clear understanding of fault-tolerant quantum circuits and will be able to design their own fault-tolerant circuits.

QCA are a new computational paradigm which encode binary information by charge configuration within a cell instead of the conventional current switches. As there is no current flow, the Coulombic interaction is sufficient for computation. This revolutionary paradigm provides a possible solution for transistorless computation at the nanoscale. Readers will obtain knowledge on QCA and the design processes of different quantum circuits using QCA design software. They will be acquainted with the costs of QCA circuits, which include cell area, delay, kink energy, and power.

As a whole, this book is a great resource for quantum computing researchers and students alike. There are very few books written in this field. This book will satisfy the interest of beginner to advanced level readers.

Acknowledgments

I would like to express my sincerest gratitude and particular appreciation to the various researchers in the field of quantum computing. The contents in this quantum computing book have been compiled from a wide variety of research works, where the researchers are pioneers in their respective fields. All the research articles related to the contents are listed at the end of each chapter.

I am grateful to my parents and family members for their endless support. Most of all, I want to thank my wife Mrs Sitara Roshan, my daughter Ms Fariha Tasnim, and my son Md Tahsin Hasan for their invaluable cooperation in completing this book.

Finally, I am also grateful to all of those who have provided their support and valuable time to finish this book.

Author biography

Hafiz Md Hasan Babu

Dr Hafiz Md Hasan Babu is currently working as the Pro-Vice-Chancellor of the National University, Bangladesh. He is on deputation from the Department of Computer Science and Engineering, University of Dhaka, Bangladesh. He is also a Professor and the former Chairman of the same department. Due to his excellent academic and administrative capabilities, he also served as the Professor and founder Chairman of the Department of Robotics and Mechatronics Engineering, University of Dhaka, Bangladesh. He served as a World Bank Senior Consultant and General Manager of the Information Technology & Management Information System Departments of Janata Bank Limited, Bangladesh. Dr Hasan Babu was the World Bank Resident Information Technology Expert of the Supreme Court Project Implementation Committee, Supreme Court of Bangladesh. He was also the Information Technology Consultant of the Health Economics Unit and Ministry of Health and Family Welfare in the project 'SSK (Shasthyo Shurokhsha Karmasuchi) and Social Health Protection Scheme' under the direct supervision and funding of the German Financial Cooperation through KfW.

Professor Dr Hafiz Md Hasan Babu received his MSc degree in Computer Science and Engineering from the Brno University of Technology, Czech Republic, in 1992 under a Czech Government Scholarship. He also obtained a Japanese Government Scholarship to pursue his PhD from the Kyushu Institute of Technology, Japan, in 2000. He also received a Deutscher Akademischer Austauschdienst (DAAD) Fellowship from the Federal Republic of Germany.

Professor Dr Hasan is a very eminent researcher. He has received the best paper award at three reputed international conferences. In recognition of his valuable contributions in the field of computer science and engineering, he received the Bangladesh Academy of Sciences Dr M O Ghani Memorial Gold Medal Award for the year 2015, which is one of the most prestigious research awards in Bangladesh. He was also awarded the University Grants Commission (UGC) of Bangladesh Gold Medal Award 2017 for his outstanding research contributions to computer science and engineering. He has written more than one hundred research articles published in reputable international journals (*IET Computers & Digital Techniques*, *IET Circuits and Systems*, *IEEE Transactions on Instrumentation and Measurement*, *IEEE Transactions on VLSI Systems*, *IEEE Transactions on Computers*, *Journal of Microelectronics* (Elsevier), *Journal of Systems Architecture* (Elsevier), *Journal of Quantum Information Processing* (Springer), etc) and at international conferences. According to Google Scholar, Professor Hafiz has received around 960 citations with an h-index of 14 and an i10-index of 21. He is a regular reviewer for reputable international journals and international conferences. He has presented invited talks and chaired scientific sessions or worked as a member of the organizing committee or international advisory board at many international conferences held in different countries. For his excellent research record, he has also been appointed as the Associate Editor of *Computers and Digital Techniques*, published by the Institution of Engineering and Technology of the United Kingdom.

Professor Dr Hasan was appointed as a member of the Prime Minister's ICT Task Force Committee, Government of the People's Republic of Bangladesh, on recognition of his national- and international-level contributions to engineering sciences. He is currently the President of the Bangladesh Computer Society and also the President of the Internet Society, Bangladesh Chapter. He has been recently appointed as a part-time member of Bangladesh Accreditation Council to ensure the quality of higher education in Bangladesh.

Part I

Quantum logic

An overview of quantum circuits

Quantum computation and quantum information are the study of the information processing tasks that can be accomplished using quantum mechanical systems. Quantum computation is an application of quantum mechanics, where the evolution of a quantum system is described by a quantum algorithm. This observation suggests that quantum computing can be used as an introduction to quantum mechanics, because quantum computation and quantum information offer an excellent conceptual laboratory for understanding the basic concepts and unique aspects of quantum mechanics.

The advancements in higher-level integration and fabrication processes have resulted in better logic circuits and energy loss has also been dramatically reduced over the last few decades. This trend of reduction of heat in computation also has its physical limit. It is easily understandable that in logic computation every bit of information loss generates $kT\ln 2$ joules of heat energy, where k is Boltzmann's constant of 1.38×10^{-23} J K^{-1} and T is the absolute temperature of the environment. At room temperature, the dissipating heat is around 2.9×10^{-21} J. The energy loss limit is also important as it is likely that the growth of heat generation as a cause of loss will be noticeable in the future.

Quantum computing is an emerging technology that has the potential to change the perspectives and applications of computing in general. A wide range of applications can be enabled, from faster algorithmic solutions of still-difficult classical problems to theoretically more secure communication protocols. A quantum computer uses the quantum mechanical effects of particles or particle-like systems, and a major similarity between quantum and classical computers is that both are abstracted as information processing machines. Whereas a classical computer operates on classical digital information, a quantum computer processes quantum information, which shares similarities to analog signals. One of the central differences between the two types of information is that classical information is more fault-tolerant compared to its quantum counterpart.

It is possible to visualize a quantum circuit as a series of gates acting on the qubits which are abstracted as wires. The wires do not have a direct physical representation, but are associated with a temporal axis. The inputs are on the left of the diagram, the outputs on the right, and in each time step a quantum gate is applied to a qubit's state abstracted by the wire. It is standard to assume that the inputs are being initialized in the computational basis $|0\rangle$, and it will be indicated if the case otherwise.

Reliable quantum circuits are the result of designing circuits that operate directly on encoded quantum information, but the circuits' reliability is also increased by supplemental redundancies, such as sub-circuit repetitions. Reliable quantum circuits have not been widely used, and one of the major obstacles is their vast associated resource overhead, however, recent quantum computing architectures show promising scalabilities. Consequently, the number of particles used for computing can be increased more easily and classical control hardware (inherent for quantum computation) is also more reliable.

The number of output bits is relatively small compared to the number of input bits in most computing tasks. For example, in a decision problem, the output is only one bit (yes or no) and the input can be as large as desired. However, computational tasks in quantum computing require that all of the information encoded in the input should be preserved in the output. One might expect to obtain further speed-ups by adding instructions to allow computation of a quantum circuit.

Quantum communication and computation study information transmission and processing as physical phenomena that follow the laws of quantum mechanics. Considering quantum mechanics introduces new possibilities, such as private communication with quantum cryptography or efficient factoring algorithms. Most quantum information protocols and algorithms can be explained as a sequence of transformations applied to a known initial state and a final measurement stage. The intermediate evolution is usually the key to the procedure. This state evolution can be studied from different perspectives.

This part of the book starts with some background and preliminary studies on quantum logic and some characteristics, as well as the evolution of quantum computers, which are discussed in chapter 1. Many of the quantum gates are included in multiple chapters. The basic definitions of the different efficiency parameters of quantum logic, which are needed to design quantum circuits, are described in chapter 2. Chapter 3 presents the characteristics and design procedure of a quantum bit string comparator. The designs for quantum full-adder and quantum subtractor circuits are shown in chapter 4. In this chapter, the design techniques for a quantum full-adder as well as a quantum half-subtractor and full-subtractor are described. The designs for a quantum multiplexer and quantum demultiplexer are discussed in chapter 5. In this chapter, a quantum 2-to-1 multiplexer, quantum 4-to-1 multiplexer, and quantum 2^n-to-1 multiplexer are shown. In addition, a quantum 1-to-2 demultiplexer, quantum 1-to-4 demultiplexer, and quantum 1-to-2^n demultiplexer are also provided in this chapter. Chapter 6 presents the design for a quantum adder circuit which is optimized in terms of quantum cost. The four-bit quantum carry skip adder and eight-bit quantum carry skip adder circuits are described in this chapter. The design for a quantum multiplier accumulator is described in chapter 7. The reduction of garbage outputs and ancillary inputs of quantum circuits is also described in this chapter. Chapter 8 presents the quantum divider circuit. A tree-based quantum division technique is also described in this chapter.

The design procedure and analysis of the properties of the quantum binary coded decimal (BCD) priority encoder circuit are introduced in chapter 9. Chapter 10 describes the design procedure and analysis of the properties of a quantum decoder circuit. The quantum square root circuit is described in chapter 11. In this chapter, the design of a quantum adder/subtractor circuit is also shown. The designs for a quantum SR latch, quantum D latch, quantum T latch, and quantum J-K latch are discussed in chapter 12. The quantum asynchronous and quantum synchronous counters are also introduced in this chapter. Chapter 13 presents the design of a quantum controlled ternary barrel shifter circuit. In this chapter, quantum ternary Peres gates and quantum ternary modified Fredkin gates are used. The construction procedure for quantum random access memory is shown in chapter 14. The construction procedure for a quantum memory unit is also given in this chapter. Three approaches for designing a quantum arithmetic logic unit are described in chapter 15. Finally, several real-life applications of quantum computing technology are given in chapter 16.

IOP Publishing

Quantum Computing
A pathway to quantum logic design
Hafiz Md Hasan Babu

Chapter 1

Quantum logic

In the age of developing nanotechnology, quantum computing can play an incredibly important role in developing more compact and lower power consumption computers. The main appeal of quantum logic originates in its reflection of the physical law of energy conservation, in which the creation or deletion of energy is impossible, and only transformation from one form to another is possible. Hence, the fundamental law of energy preservation is incorporated into the logic design of circuits and systems by the quantum logic. The motivation to implement circuits and systems using quantum computing is the fact that, theoretically, the internal computations in quantum logic systems consume no power.

1.1 Overview

Quantum computing is a technology which consumes less power and for which the design is compact. In irreversible logic, energy dissipation is a common phenomenon since every bit loss causes energy loss in the irreversible operation. Quantum logic circuits are necessarily reversible and hence there is no dissipation of energy while processing a bit in quantum computation.

Quantum technology is one of the most promising nanotechnologies which are useful for designing modern circuits. Logic design with quantum logic is of great interest in recent technologies which allow scaling to atomistic dimensions. In this particular logic design approach, quantum cells are arranged in a particular fashion to define the logic. A classical gate cannot handle the superposition of states represented by a qubit (discussed in section 2.1 of chapter 2). Thus, this forms a special case of quantum device.

Quantum registers, which are necessary for the implementation of a quantum electronics device, combine n qubits to form larger Hilbert spaces H_n using the tensor product ($\otimes$) operator to form

$$|\Psi\rangle = |\Psi_1\rangle \otimes \ |\Psi_2\rangle \otimes \cdots \otimes |\Psi_n\rangle = \sum_{i=0}^{n} |\alpha_i|i\rangle;$$

where $\alpha_i \in C$, $|\Psi_i\rangle$ represents a qubit and $\sum_{i=0}^{n} |\alpha_i|^2 = 1$. High-speed multiplication has always been a fundamental requirement of high-performance processors and systems. In quantum signal processing (QSP) applications, multiplication is one of the most utilized arithmetic operations. Improving multiplier design directly benefits the high-performance embedded processors and QSP applications used in consumer and industrial electronic products. Moreover, quantum information processing (QIP) is a high-impact research area in quantum information science to construct a quantum computer. The main goal of QIP is to harness the fundamental laws of quantum mechanics to dramatically improve all aspects (e.g. acquisition, transmission, and processing) of information processing as well as enhance the performance of quantum computers.

There are several tasks for which a quantum computer will be useful. The first, which is mentioned most frequently, is that quantum computers will be able to read secret messages communicated over the Internet using current technologies such as Rivest–Shamir–Adleman (RSA), Diffie–Hellman, and other cryptographic protocols; these protocols are based on the difficulty of number theoretic problems such as factoring and discrete logarithms. In addition, quantum computers are useful for scientists conducting virtual experiments and searching huge amounts of data.

1.2 Motivations towards quantum computing

Quantum logic is a great achievement for very-large-scale integrated (VLSI) circuit design, and can work faster than classical logic circuits. Quantum circuits are used to build quantum super computers and can solve the complex problems in polynomial time. Quantum algorithms are used to implement quantum circuits.

Quantum bits have the important property of superposition, which means that the values of quantum bits can stay in more than one position at the same time, which is impossible in the case of classical logic designs. Reversible computation supports binary (0,1) values, whereas quantum logic can hold multiple values (binary, ternary, quaternary, etc). This multi-value support makes quantum circuits more compact and efficient with optimal delay. Unlike the two fixed values 0 or 1 of classical logic, quantum bits can take the values of a linear combination of 0 and 1. Quantum circuits are inherently reversible and there is no dissipation of energy in these circuits. Thus these circuits prevent the loss of information, since energy dissipation causes bit or information loss. The above mentioned properties of quantum logic circuits have motivated researchers to design circuit components using quantum logic.

1.3 The relationship between reversible and quantum logic

Reversible logic circuits have one-to-one mapping between inputs and outputs. In other words, if the number of outputs in a logic circuit is equal to the number of inputs, and any input pattern may map to a unique output pattern, it is called a reversible logic circuit.

All the reversible circuits can be represented by quantum logic gates. Quantum circuits maintain the rules of reversible logic. Quantum logic circuits must have a one-to-one relationship between the input and output vectors like reversible logic. Quantum logic gates have unique unitary matrices which are also present in reversible logic. However, there are also differences between these two types of logic. Unlike in quantum logic, the superposition of bits and multiple values is not possible in reversible logic. Thus both similarities and differences exist between reversible and quantum logic.

Reversible circuits have broad applications in nanocomputing. Nanoelectronics engineering that would enable device scaling down to molecular levels will almost surely imply a cellular architecture with near-neighbor connectivity. The scheme which has been developed to physically realize such a concept is called the quantum-dot cellular automata (QCA). QCA have drawn a lot of attention for their very small feature size and ultra-low power consumption, which make them one candidate for substituting CMOS technology.

1.4 Quantum computers

In 1959, Richard Feynman noted that 'as electronic components begin to reach microscopic scales, effects projected by quantum mechanics occur'. He advised that it might be used in the design of more powerful computers. In particular, quantum researchers hope to bind the superposition. In the quantum mechanical world, objects do not necessarily have visibly defined states. A traditional digital computer works with binary digits that can be in one of two states, denoted as 0 and 1; thus, for example, a four-bit computer register can hold any one of 16 (2^4) possible numbers. However, a quantum bit (qubit) exists in a wavelike superposition of values from 0 to 1; thus, for example, a four-qubit computer register can hold 16 different numbers concurrently. In theory, a quantum computer can therefore control many values in parallel, so that a 30-qubit quantum computer would be equivalent to a digital computer proficient at performing 10 trillion floating-point operations per second.

During the 1980s and 1990s, the theory of quantum computers advanced considerably beyond Feynman's early assumptions. In 1985, David Deutsch designated the structure of quantum logic gates for a universal quantum computer. In 1994, Peter Shor established an algorithm to factor numbers with a quantum computer that would involve six qubits (although more qubits would be essential for factoring large numbers in a rational time).

In 1998, Isaac Chuang, Neil Gershenfeld, and Mark Kubinec produced the first quantum computer (two-qubit) that could be burdened with data and output a result. Although their system was rational for only a few nanoseconds and slight from the perspective of solving meaningful problems, it verified the principles of quantum computation. This type of quantum computer can be prolonged by using molecules with more individually addressable elements. In March 2000, Emanuel

Knill, Raymond Laflamme, and Rudy Martinez reported that they had generated a seven-qubit quantum computer. However, many researchers are skeptical about extending magnetic techniques much beyond 10 to 15 qubits because of diminishing consistency among the elements.

Quantum computers established on semiconductor technology are yet another opportunity. In a collective approach, a discrete number of free electrons (qubits) reside within incredibly small sections known as quantum dots. Although relying on decoherence, such quantum computers are constructed on well-established, solid-state procedures and offer the prospect of readily applying integrated circuit 'scaling' technology. In addition, large groups of identical quantum dots could theoretically be contrived on a single silicon chip. The chip controls an external magnetic field that controls the electron spin states, while neighboring electrons are weakly coupled through quantum mechanical effects. An array of covered wire electrodes allows individual quantum dots, which have been discussed.

1.5 The working principles of quantum computers

The huge amount of processing power created by computer designers has not yet been able to satisfy the desire for speed and computing ability. In 1947, Howard Aiken claimed that only six electronic digital computers would satisfy the computing needs of the United States. Others have made similar inaccurate predictions about the amount of computing power that would satisfy the ever-increasing needs of technology. Of course, Aiken did not take into account the large amounts of data produced by scientific research, the explosion of personal computers, or the appearance of the Internet, which have further driven the need for computing power.

Scientists have already constructed basic quantum computers that can carry out certain calculations. Quantum computers will harness the power of atoms and molecules to achieve memory and processing activities. Quantum computers have the potential to achieve certain controls significantly faster than any silicon-based computer. In this book, the reader will learn the design mechanisms of a quantum computer and just how they will be adapted for the next era of computing.

It is not necessary to go back too far to find the origins of quantum computing, as it was first conceived of less than 30 years ago. On the other hand, classical computers have been used, with many difficulties, for a comparatively long time. Paul Benioff is credited with being the first to apply quantum theory to computers, in 1981, when he posited creating a quantum Turing machine.

1.6 The evolution of quantum computers

Quantum computing is still in its emerging stages. There is a long way to go before a usably running quantum computer can be built, let alone brought to the market. However, progress in this new technology is occurring regularly and no chronological record can ever be complete. What follows is a brief timeline clarifying key areas of progress in quantum computing. Much of the technical development has been achieved in this century while most of the primary theoretical perspectives were laid down in the late twentieth century.

In 1980, Paul Benioff was the first to design a computer which operated under quantum mechanical principles. His idea of a quantum computer was based on Alan Turing's famous paper tape computer, described in his article published in 1936. In 1981, Feynman proved that it was impossible to simulate quantum systems on a classical computer. His argument hinged on Bell's theorem, written in 1964. In 1985, David Deutsch published a report describing the world's first universal quantum computer. He showed how such a quantum machine could reproduce any realizable physical system. Enthusiasm for creating the first quantum computer was stimulated by Peter Shor's algorithm in 1994. Shor described a method for factorizing large integers. This had serious implications for cryptography, which relies on this operation being difficult in order to keep codes secure. Shor's algorithm searched for periodicities in long integer-sequences of repeated digits. It used the quantum principles of superposition to scour for periodicities in the astonishingly fast time of a few minutes. To perform this same computation on a classical computer would take longer than the age of the Universe. In 1996, Lov Grover used quantum mechanics to solve an old unstructured search problem. For example, if someone wants to match a large database of names with a long list of telephone numbers, a classical computer could only solve this problem by querying each name with a telephone number until it found the right one, which is not fewer than $O(N)$ evalutions of the function. Grover's quantum algorithm, however, produces the output value using only $O(\sqrt{N})$ evaluations of the function.

In 2000, the first working five-qubit nuclear magnetic resonance (NMR) computer was put through its paces at the Technical University of Munich. Shortly after, the Los Alamos National Laboratory surpassed this feat with a working seven-qubit NMR quantum computer.

The year 2001 is known for the demonstration of Shor's algorithm. A team at the IBM Almaden Research Center in California succeeded in factorizing the integer 15 into 5 and 3. They used a thimbleful of a bespoke liquid containing billions of molecules. The molecules were constructed from five fluoride and two carbon atoms, each with their own nuclear spin state. The molecules worked as a seven-qubit quantum computer when pulsed with electromagnetic waves and monitored using NMR.

In 2006 researchers offered a new functioning standard by monitoring a 12-qubit quantum system with only minimal decoherence. NMR quantum information processors were used to decrypt the computation. These quantum controls led to the hope that higher-qubit quantum computers might be available one day. The same year, scientists took a step closer to the building of a quantum gate, the quantum representation of a mathematical rule. Also in 2006, scholars created molecules of quantum-dot pairs. These have great potential for quantum computers, in particular if more complex elements can be created.

This book focuses on how quantum gates and quantum representations of different circuits are to be designed to make a complete quantum computer. Readers will be motivated to design such circuits and to make quantum representations of different sequential and logical circuits of their own with different examples and designs of the circuit.

1.7 Why pursue quantum computing?

Despite the most remarkable wave of technological inventions, there are definite computational problems that the digital revolution still cannot seem to solve, even though some of these computational problems could be solved by scientific advances. Although conventional computers have doubled in processing speed and power nearly every two years for decades, they still do not appear to be fast eough to solve these enduring problems. In the long run, to solve the world's most tenacious computing problems competently, we will have to turn to an utterly new and more capable machine: the quantum computer.

Finally, the dissimilarity between a classical computer and a quantum computer is not like the difference between an old car and a new one. Instead, it is like the difference between a horse and a hawk: while one can run, the other can fly. Classical computers and quantum computers certainly have that difference. In this chapter we have taken a careful look at where the key differences lie and have taken a profound plunge into what makes quantum computers unique.

1.8 Summary

This chapter mainly introduces researchers to quantum logic and simply delineates why future generations will choose quantum computing instead of conventional computers. In this chapter, the relation between reversible and quantum logic, the salient principles of quantum computers, their evolution, and necessity are briefly described.

Further reading

[1] Bonsor K and Strickland J 2000 How quantum computers work *How Stuff Works* https://computer.howstuffworks.com/quantum-computer.htm (Accessed: 4 December 2018)

[2] Wikipedia Quantum cellular automaton https://en.wikipedia.org/wiki/Quantum_cellular_automaton (Accessed: 4 December 2018)

[3] Holton W C Quantum computer *Encyclopædia Britannica* https://www.britannica.com/technology/quantum-computer (Accessed: 4 December 2018)

[4] Timeline of quantum computers *Quantum Computing 101* http://quantumly.com/timeline-of-quantum-computing-history-of-quantum-computers/dates.html (Accessed: 4 December 2018)

[5] Akbar E P A, Haghparast M and Navi K 2011 Novel design of a fast reversible Wallace sign multiplier circuit in nanotechnology *Microelectron. J.* **42** 973–81

[6] Barenco A, Bennett C H, Cleve R, DiVincenzo D P, Margolus N, Shor P, Sleator T, Smolin J A and Weinfurter H 1995 Elementary gates for quantum computation *Phys. Rev.* A **52** 3457

[7] Cai X-D, Wu D, Su Z-E, Chen M-C, Wang X-L, Li L, Liu N-L, Lu C-Y and Pan J-W 2015 Entanglement-based machine learning on a quantum computer *Phys. Rev. Lett.* **114** 110504

[8] Hudson R L 1990 An introduction to Hilbert space and quantum logic, by D. W. Cohen. pp 148. DM79. 1989. ISBN 3-540-96870-9 (Springer) *Mathematical Gazette* **74** 418–9

[9] Dibbo S V, Babu H M H and Jamal L 2016 An efficient design technique of a quantum divider circuit *IEEE Int. Symp. on Circuits and Systems (ISCAS)* (Piscataway, NJ: IEEE) pp 2102–5

[10] Yu Kitaev A 1997 Quantum computations: algorithms and error correction *Russ. Math. Surv.* **52** 1191–249

[11] Landaurer R 1961 Irreversibility and heat generation in the computational process *IBM J. Res. Dev.* **5** 183–91

[12] Li D-F, Wang R-J, Zhang F-L, Deng F-H and Baagyere E 2015 Quantum information splitting of arbitrary two-qubit state by using four-qubit cluster state and bell-state *Quantum Inf. Process.* **14** 1103–16

[13] Nielsen M A and Chuang I 2002 *Quantum Computation and Quantum Information* 10th edn (Leiden: Cambridge University Press)

[14] Schubert M and Rana F 2006 Analysis of terahertz surface emitting quantum-cascade lasers *IEEE J. Quantum Electron.* **42** 257–65

[15] Tóth G, Timler J and Lent C S 1998 Quantum computing with quantum-dot cellular automata using coherence vector formalism *Proc. of the IEEE Int. Workshop on Computational Electronics (IWCE-6)*

Chapter 2

Basic definitions of quantum logic

Quantum computing was advanced as a technological attempt to build a propositional structure that would permit relating the events of interest in quantum mechanics. Quantum computing seeks to replace the Boolean structure which, although appropriate for addressing classical physics, is insufficient for representing atomic elements. The scientific structure of the propositional linguistics of classical systems is a power set, partly ordered by set enclosure, with a pair of operations that denote conjunction and disjunction.

During the progress of quantum computing, numerous lines of study have tried to address quantum mechanics from a logical viewpoint. This book offers a map of these multiple methods in order to familiarize the reader with the very different approaches and problems deliberated in the quantum computing literature. When possible, redundant theories, algorithms, and examples, are avoided in order to provide an instinctive understanding of the ideas before designing or presenting the associated mathematics.

The procedure of a two-bit 'controlled-NOT' quantum logic gate has been designed which, in concurrence with simple single-bit operations, forms a common quantum logic gate for quantum computation. The two quantum bits are placed in the internal and external degrees of choice of a single trapped element which is first laser cooled to the zero-point energy. Decoherence properties are acknowledged for the operation, and the prospect of expanding the system to more qubits seems favorable.

2.1 The quantum bit

A quantum bit (qubit) is typically derived from the state of a two-level quantum system, such as the ground and excited states of an atom or the vertical and horizontal polarizations of a single photon. A qubit, represented by $|A\rangle$, is the basic unit of information in a quantum computer, which can hold two states, 0 or 1, simultaneously or at different times. A qubit can also be a superposition (both states

doi:10.1088/978-0-7503-2747-3ch2

at the same time) of these two states, i.e. a linear combination of the binary values $|1\rangle$ and $|0\rangle$ ($\alpha|0\rangle_i + \beta|1\rangle_i$, where α and β are the probabilities of being the $|1\rangle$ and $|0\rangle$ states, respectively), whereas classical bits or binary bits are in one of two possible states, labeled 1 and 0.

2.2 The quantum gate

A quantum gate is a basic quantum circuit operating on a small number of qubits. Previously, various quantum gates with different functionalities have been designed. Among them, the NOT, CNOT, controlled-V, and controlled-V^+ gates represent an important class of quantum gates. These gates are shown in figure 2.1. In this figure, the control, target, and contact qubits are represented by the ·, ⊗, and || symbols, respectively. In a quantum gate the number of outputs must be equal to the number of inputs.

2.2.1 The quantum Feynman gate

The quantum Feynman gate is a 2 × 2 quantum gate which implements the logical functions of $P = A$ and $Q = A \oplus B$, and is illustrated in figure 2.2. The quantum Feynman gate can be used for copying a bit. When B is set to zero, then the output vector will be $P = A$ and $Q = A$, which ensures copying of the input A.

2.2.2 The quantum Tofolli gate

The quantum Tofolli gate is a 3 × 3 gate in which A, B, and C are input vectors and the output vectors are $P = A$, $Q = B$, and $R = AB \oplus C$ logical functions. The quantum Tofolli gate can implement the 'AND' operation when C is set to zero. The quantum Tofolli gate is presented in figure 2.3. The Tofolli gate is an important quantum gate and is widely applied in the construction of quantum circuits.

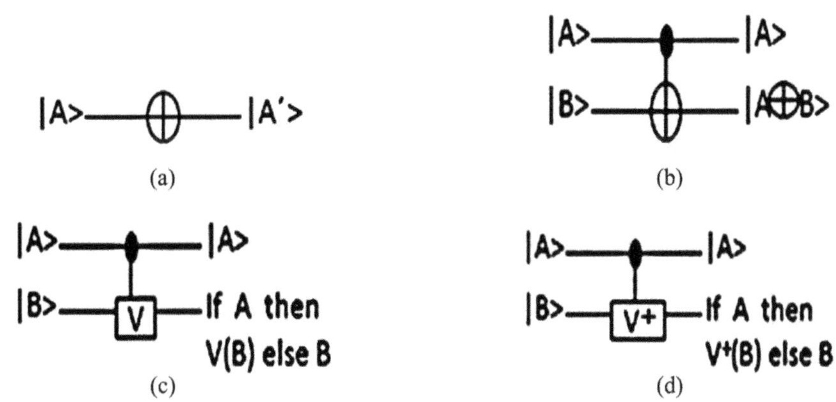

Figure 2.1. Basic quantum gates. (a) Quantum NOT gate. (b) Quantum CNOT gate. (c) Controlled-V gate. (d) Controlled-V^+ gate.

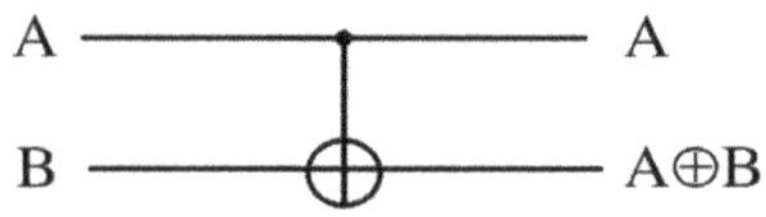

Figure 2.2. The quantum Feynman gate.

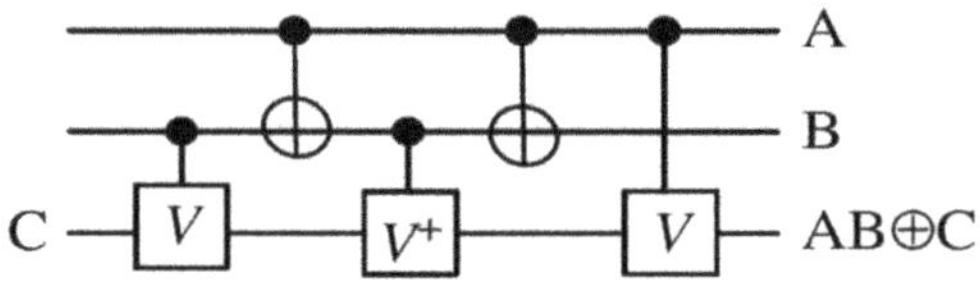

Figure 2.3. The quantum Tofolli gate.

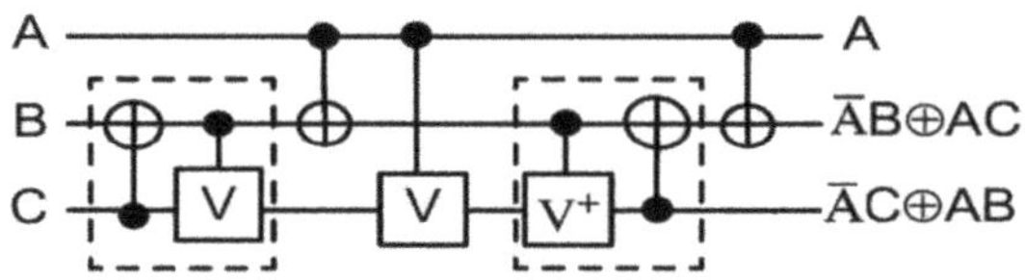

Figure 2.4. The quantum Fredkin gate.

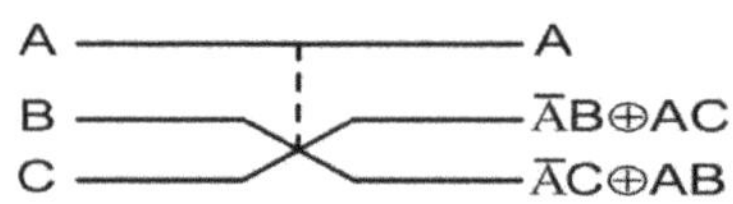

Figure 2.5. Swapping of the quantum Fredkin gate.

2.2.3 The quantum Fredkin gate

The quantum Fredkin gate is illustrated in figure 2.4, and is a 3×3 quantum gate in which A, B, and C are input vectors and the output vectors are the $P = A$, $Q = \bar{A}B \oplus AC$, and $R = \bar{A}B \oplus AB$ logical functions. One of the important applications of the quantum Fredkin gate is swapping. The input bit A is used as a control bit which decides whether the other two bits will swap or not, which is illustrated in figure 2.5. It shows that the quantum Fredkin gate can swap the values when A is set to 1.

2.3 Garbage outputs

In a quantum gate the number of outputs must be equal to the number of inputs. As a consequence, there are usually some outputs which are not required further. They are called garbage outputs. Each garbage output incurs a heavy price. In figure 2.1, the output $|A\rangle$ is the garbage output for the CNOT quantum gate.

2.4 Constant inputs

Constant inputs are the inputs which are added to a function to make the one-to-one mapping between inputs and outputs. For example, to perform the adder operation

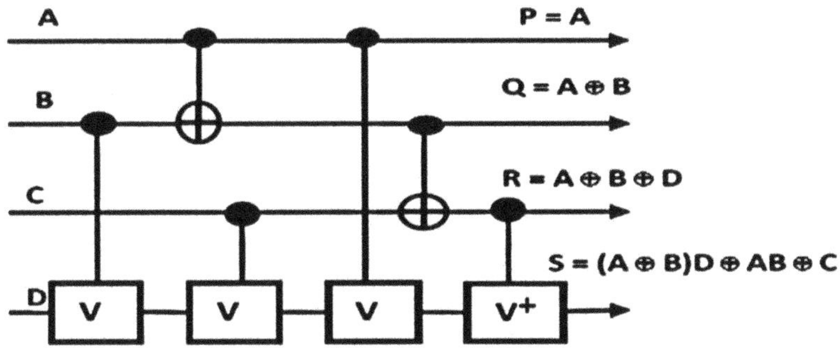

Figure 2.6. Quantum realization of the DPG circuit.

using a double Peres gate (DPG) quantum circuit, the C input bit in figure 2.6 has to remain 0 and this input is called the constant input.

2.5 Area

The area of a gate is defined by the circuit size. This size varies according to the number of quantum gates of the circuit. As the basic quantum gates are fabricated with quantum dots with size ranging from several to tens of nanometers (10^{-9} m) in diameter, the size of basic quantum gates ranges from 50 Å–300 Å. The angstrom (Å) is a unit equal to 10^{-10} m (one ten-billionth of a meter) or 0.1 nm. Its symbol is the Swedish letter Å. Quantum circuits can be implemented with the basic quantum gates and the quantum cost of a gate depends on the number of basic quantum gates needed to implement it. Thus the area of a gate can be defined as follows: $A = N_q \times S_q$, where A = area, N_q = the number of quantum gates, and S_q = the size of basic quantum gates. According to the circuit size, the area of the quantum Toffoli gate is ((50 × 5) Å – (300 × 5) Å) = (250 Å – 1500 Å), where the number of quantum gates of the quantum Toffoli gate is five.

2.6 Power

The power of a gate is defined by the energy consumed. The energy of basic quantum gates is 142.3 meV (micro-electronvolts). Quantum circuits can be implemented with basic quantum gates and the quantum cost of a gate depends on the number of basic quantum gates needed to implement it. Thus the power of a gate can be defined as follows, for example: the energy of the Toffoli gate is (5 × 142.3) meV = 711.5 meV, where the number of quantum gates of the quantum Toffoli gate is five.

2.7 Delay

The delay represents the critical delay of the circuit. In delay calculations, the logical depth is used as the measure of the delay. The delays of all 1 × 1 and 2 × 2 quantum gates are taken as the unit delay, designated by Δ. Any quantum gate can be designed from 1 × 1 and 2 × 2 quantum gates, such as the CNOT, controlled-V, or controlled-V^+ quantum gates. Thus, the delay of a quantum gate can be computed by calculating its logical depth when it is designed using smaller 1 × 1 and 2 × 2 quantum gates. Each 2×2 quantum gate in the logic depth contributes to 1Δ delay. For example, the quantum Toffoli gate requires a 5Δ delay, as shown in figure 2.3.

2.8 Depth

The depth of a quantum circuit is the maximum number of stages or slices where each stage or slice represents a quantum gate or a number of quantum gates along the same vertical line.

A quantum circuit is constructed using different quantum gates. These quantum gates are placed in different input lines. To find the depth of any quantum circuit, it is necessary to divide it into some slices. There may be more than one quantum gate in any slice. The maximum number of slices is considered as the depth of that quantum circuit. For example, consider the Thapliyal Ranganathan (TR) gate-based half subtractor, shown in figure 2.7. This quantum circuit needs four quantum gates to be implemented. Now, to find the depth of this circuit, it will be divided into stages according to the quantum gates. The vertical lines are used to split the circuit into slices. In the first slice there is a controlled-V^+ gate, in the second slice there is a CNOT gate, and in the third and fourth slices there are controlled-V gates. This circuit has a maximum of four slices, as numbered in the figure. Thus it can be said that the depth of the circuit is 4.

2.9 Quantum cost

The quantum cost of a quantum circuit is the number of basic quantum gates in the circuit. Quantum cost is an important measure of the performance of a quantum circuit. The quantum cost of the basic quantum gates, such as the NOT, CNOT, controlled-V, and controlled-V^+ gates, is considered to be 1. The quantum circuit in figure 2.7 consists of four basic quantum gates and thus the quantum cost of this circuit is 4.

2.10 Quantum gate calculation complexity

The quantum gate calculation complexity refers to the number of quantum gates (NOT, CNOT, controlled-V, and controlled-V^+) used to synthesize a given circuit, with ρ being the NOT quantum gate calculation complexity, σ the CNOT quantum gate calculation complexity, and Ω the controlled-V (controlled-V^+) gate calculation

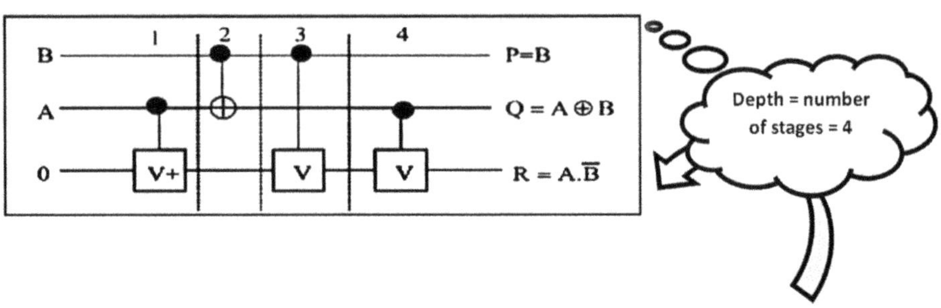

Figure 2.7. Illustration of the depth of a quantum circuit.

complexity. For example, the DPG quantum circuit has two CNOT quantum gates and four controlled-V (controlled-V^{+}) gates. Therefore, the quantum gate calculation complexity of the DPG quantum circuit is $2\sigma + 4\Omega$, which is depicted in figure 2.6.

2.11 Summary

Quantum gates and quantum networks offer a very useful language for constructing any quantum computer or (basically the same) quantum multi-particle circuits. Now, the question is whether it it possible to build quantum logic gates or not?

Single-qubit quantum gates are viewed as comparatively easy to implement. For example, a classic quantum optical realization uses elements as qubits and switches their states with laser light pulses of appropriately selected frequency, strength, and duration; any recommended superposition of two selected logical states can be prepared this way.

Research into quantum computation and all of its possible variations has become vigorously active and any comprehensive review of the field cannot help but be obsolete as soon as it is written. Here, only some of the very basic knowledge has been provided, hoping that this will serve as a good starting point to enter into the field.

Further reading

[1] Oxford quantum http://oxfordquantum.org/ (Accessed: 5 December 2018)

[2] Balandin A A and Wang K L 1999 Implementation of quantum controlled-not gates using asymmetric semiconductor quantum dots *Quantum Computing and Quantum Communications* (Berlin: Springer) pp 460–7

[3] Landaurer R 1961 Irreversibility and heat generation in the computational process *IBM J. Res. Dev.* **5** 183–91

[4] Li X, Steel D, Gammon D and Sham L J 2004 Quantum information processing based on optically driven semiconductor quantum dots *Opt. Photon. News* **15** 38–43

[5] Mohammadi M and Eshghi M 2009 On figures of merit in reversible and quantum logic designs *Quantum Inf. Process.* **8** 297–318

[6] Monroe C, Meekhof D M, King B E, Itano W M and Wineland D J 1995 Demonstration of a fundamental quantum logic gate *Phys. Rev. Lett.* **75** 4714

[7] Nielsen M A and Chuang I L 2000 *Quantum Computation and Quantum Information* 10th edn (Leiden: Cambridge University Press)

[8] Shende V V, Prasad A K, Markov I L and Hayes J P 2003 Synthesis of reversible logic circuits *IEEE Trans. Comput.-Aided Des. Integr. Circuits Syst.* **22** 710–22

[9] Toffoli T 1980 Reversible computing *International Colloquium on Automata, Languages, and Programming* (Berlin: Springer) pp 632–44

[10] Zhou R-G, Li Y-C and Zhang M-Q 2014 Novel designs for fault tolerant reversible binary coded decimal adders *Int. J. Electron.* **101** 1336–56

Chapter 3

The quantum bit string comparator

The quantum bit string comparator enables the implementation of a quantum algorithm using conditional statements, a fundamental structure for designing algorithms. This widens the number of applications where quantum algorithms can be used and at the same time it brings quantum programmers close to some of the successful techniques used in classical computation based on comparisons. For example, constructing a database composed of only prime numbers is possible using the quantum bit string comparator.

3.1 Characteristics of a comparator

A comparator, as suggested by the name, compares a signal on one input of an op-amp, as shown in figure 3.1, with a known voltage called the reference voltage on the other input. The comparator is nothing other than an open loop op-amp with two analog inputs (differential input) and one digital output (signal ended output). The op-amp has a very large gain when used in an open loop. Hence the output may be in the positive or negative saturation voltage depending upon which input is larger. An op-amp is perfectly suited to comparator applications because of its high input impedance and large open loop gain.

The important characteristics of a comparator are:

1. Speed of operation.
2. Accuracy.
3. Compatibility of output.

3.2 The magnitude comparator

A magnitude comparator is a logic circuit that first compares the sizes of A and B and then determines the result among $A > B$, $A < B$, and $A = B$. When the two numbers in the comparator circuit are two one-bit numbers, the result will be only one bit from 0 and 1. Thus the circuit is called a one-bit magnitude comparator, which is a basis of comparison of the two numbers of n bits. A quantum bit string

doi:10.1088/978-0-7503-2747-3ch3

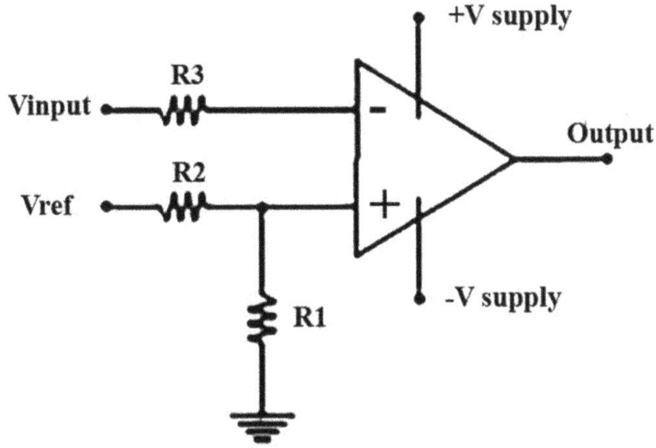

Figure 3.1. An open loop op-amp.

comparator is designed for the quantum bit string comparator circuit. In this chapter two quantum states are identified by providing a comparison status, such as equality or larger or smaller, after performing a comparison between these states. In addition, this chapter shows that the quantum bit string comparator enables us to implement conditional statements in quantum computation, which is a basic structure for designing a comparison algorithm. However, the design requires a huge number of quantum gates, garbage outputs, and constant inputs. Moreover, the design does not show the area and power requirements of the circuit. A quantum comparator circuit is designed for a quantum comparator circuit using quantum dot cellular automata. The design requires a huge number of quantum gates and garbage outputs.

An ASIC implementation of a low power area efficient folded binary comparator circuit was invented in 2014. This comparator consists of a pre-computation unit and an encoder block. The basic principle is to group the binary inputs into digit sets. The digit sets are sent to the pre-computation unit starting from the most significant digit (MSD) to check for equality, and the computations in the pre-computation unit are stopped at the first digit set which produces a 1 as output. The corresponding digit set is then sent to the carry look-ahead (CLA) encoder block to find the greater of the two inputs. However, as this is an irreversible comparator circuit, this circuit has huge dynamic power dissipation.

3.3 The design of a quantum comparator

There are two important properties to define a quantum gate which are as follows.

Property 3.1. Each quantum gate has an unique unitary matrix. A complex square matrix U is unitary only if

$$U^*U = UU^* = I,$$

where I is the identity matrix and U^* is the conjugate transpose of U.

Property 3.2. A quantum state is represented by a state vector in a Hilbert space over complex numbers.

The matrices for the controlled-E and controlled-E$^+$ gates are

$$M_{\text{controlled-E}} = \begin{pmatrix} -i/2 & -i/2 \\ -i/2 & i/2 \end{pmatrix};\ M_{\text{controlled-E}^+} = \begin{pmatrix} i/2 & i/2 \\ i/2 & -i/2 \end{pmatrix}.$$

The conjugate transpose of $M_{\text{controlled-E}}$ is $M_{\text{controlled-E}^+}$ and vice versa. After multiplying $M_{\text{controlled-E}}$ and $M_{\text{controlled-E}^+}$, we obtain an identity matrix which is

$$M_{\text{controlled-E}} \times M_{\text{controlled-E}^+} = \begin{pmatrix} -i/2 & -i/2 \\ -i/2 & i/2 \end{pmatrix} \times \begin{pmatrix} i/2 & i/2 \\ i/2 & -i/2 \end{pmatrix}$$
$$= \begin{pmatrix} 1 & 0 \\ 0 & 1 \end{pmatrix} = I.$$

The controlled-E and controlled-E$^+$ gates have a unique matrix which is unitary. Figures 3.2(a) and (b) show the diagrams of the controlled-E and controlled-E$^+$ gates.

The matrices for the XN (M_{XN}) gate and its conjugate transpose (M_{XN^*}) are

$$M_{XN} = \begin{pmatrix} 0 & 1 & 0 & 0 \\ 1 & 0 & 0 & 0 \\ 0 & 0 & 1 & 0 \\ 0 & 0 & 0 & 1 \end{pmatrix}\quad M_{XN^*} = \begin{pmatrix} 0 & 1 & 0 & 0 \\ 1 & 0 & 0 & 0 \\ 0 & 0 & 1 & 0 \\ 0 & 0 & 0 & 1 \end{pmatrix}.$$

After multiplying M_{XN} and M_{XN^*}, we obtain an identity matrix which is

$$M_{XN} \times M_{XN^*} = \begin{pmatrix} 0 & 1 & 0 & 0 \\ 1 & 0 & 0 & 0 \\ 0 & 0 & 1 & 0 \\ 0 & 0 & 0 & 1 \end{pmatrix} \times \begin{pmatrix} 0 & 1 & 0 & 0 \\ 1 & 0 & 0 & 0 \\ 0 & 0 & 1 & 0 \\ 0 & 0 & 0 & 1 \end{pmatrix} = \begin{pmatrix} 1 & 0 & 0 & 0 \\ 0 & 1 & 0 & 0 \\ 0 & 0 & 1 & 0 \\ 0 & 0 & 0 & 1 \end{pmatrix} = I.$$

Thus the XN gate has a unique matrix which is unitary. Figure 3.3 shows the diagram of the XN gate.

Figure 3.2. Controlled gate. (a) Controlled-E$^+$ gate. (b) Controlled-E gate.

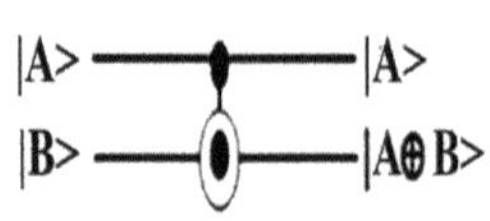

Figure 3.3. The XN gate.

Algorithm 3.1 describes the computational process for the designed method for the proof of the time complexity. The technique of comparison of the comparator has four basic steps. First, sort the two numbers and also calculate their middle bit position. Second, find Ex-OR of the middle bit position and decide based on that value whether to go to the left half or to the right half of the numbers. Third, repeat the second step after entering into the left or right half and also calculate Ex-OR of the other bits at the same time. Finally, take a decision if any value of Ex-OR is $|1\rangle$ as to which number is greater, otherwise carry out Ex-NOR to find whether they are equal or not.

Property 3.3. Algorithm for the comparison technique for an n quantum bit string comparator which requires a time complexity of $O(2(2 + \log n))$.

Proof. Let n be the number of qubits of two strings. To sort two strings of n qubits in parallel using merge sorting, the total time complexity in the best case is $O(\log n)$. Each string is divided into half ($n/2$) of the original string when the comparison technique works in each time. Thus it reduces the working space by half. In the worst case, the comparison technique operations need to be executed on both the left and right halves of the two n-qubit strings, where the time required for one half is the run time ($T(|n/2|)$) for half ($n/2$) of the n-qubit string using the comparison technique, and the time required for the other half is the same time which is treated as the time for copying the same operations in the other half of the string. In contrast, in the best case only the left/right half operations need to be executed. The positions of the qubit strings are stored in two different arrays. When comparing two qubits, the two arrays containing the qubit positions need to be accessed in parallel and the complexity of an array access is $O(1)$. According to the algorithm, Ex-OR of midpoint of the sorted strings is performed and depending on the result, the left/right half of the unsorted strings is used for comparison. Hence we need to access the two unsorted array of strings again and this also has time complexities. Thus the recurrence of the comparison technique in the best case can be specified as

$$T(n) = T(|n/2|) + 1,$$

where $T(n)$ denotes the run time of the comparison technique, $T(|n/2|)$ is the time required to perform an operation on the left/right half of the n-qubit strings, n is the length of the qubit strings, and a constant 1 (one) is the time for midpoint qubit operation of the qubit string.

Algorithm 3.1 Algorithm of the comparison technique.

INPUT: INPUT: Two n-qubit strings $|A\rangle = |A_n\rangle|A_{n-1}\rangle \cdots |A_0\rangle$ and $|B\rangle = |B_n\rangle|B_{n-1}\rangle \cdots |B_0\rangle$
OUTPUT: $|A\rangle > |B\rangle$ or $|A\rangle < |B\rangle$ or $|A\rangle = |B\rangle$
1: Begin
2: $|A[i]\rangle$ = array containing n qubits of the first string
3: $|B[i]\rangle$ = array containing n qubits of the second string
4: $|SA[i]\rangle$ = sorted $|A[i]\rangle$

5: $|SB[i]\rangle$ = sorted $|B[i]\rangle$
6: n = size of array
7: mid=$|\frac{n}{2}|$
8: $Ex-OR[\text{mid}] = |SA[\text{mid}]\rangle \oplus |SB[\text{mid}]\rangle$
9: If .Ex-OR[mid] = 0 then
10: Find Ex-OR values of the left half of $|SA[i]\rangle$ and $|SB[i]\rangle$ s.t.i < mid
11: If Ex-OR of the left side $|SA[i]\rangle$ and $|SB[i]\rangle$ is 0 then
12: Find Ex-OR values of the left half of $|A[i]\rangle$ and $|B[i]\rangle$ s.t.i < mid
13: If Ex-OR of any position is 1 and $|A[i]\rangle > |B[i]\rangle$, then
14: $|A\rangle > |B\rangle$
15: Else If Ex-OR of any position is 1 and $|A[i]\rangle < |B[i]\rangle$, then
16: $|B\rangle > |A\rangle$
17: Else goto Step 19
18: End If
19: Repeat Steps 12 to 16 for right side, i.e. mid < i < = n
20: Else goto Step 26
21: End If
22: End If
23: Else
24: Repeat Steps 10–20 for the right half of the numbers $|SA[i]\rangle$ and $|SB[i]\rangle$ s.t.i > mid
25: End Else
26: If Ex-NOR of both halves of $|A[i]\rangle$ and $|B[i]\rangle$ is 1, where $|SA[\text{mid}]\rangle \oplus |SB[\text{mid}]\rangle = 0$, then
27: $|A\rangle = |B\rangle$
28: End If
29: End

Guess

It is assumed that the solution of the recurrence is $T(n) = O(\log n)$, i.e. it is true iff $T(n) \leqslant c(\log(n))$, where $c > 0$ is a constant.

Proof by substitution. Assume that this bound holds for all positive $m < n$ in the particular $m = [n/2]$, where n is the number of qubits and m is a constant term. It yields that

$$T(n/2) \leqslant c(\log(|n/2|)).$$

By substituting into the recurrence, we obtain

$$\begin{aligned} T(n) &\leqslant c(\log(|n/2|)) + 1 \\ &= c(\log(n)) - c(\log(2)) + 1 \\ &= c(\log(n)) - c + 1 \\ &\leqslant c(\log(n)) \text{ as long as } c \geqslant 1. \end{aligned}$$

Thus, $T(n) \leqslant c(\log(n))$, that is, $T(n) = O(\log n)$. Therefore, the total time complexity of the comparison algorithm in the best case is sorting time + comparison time + array of position access time + array of unsorted qubits access time

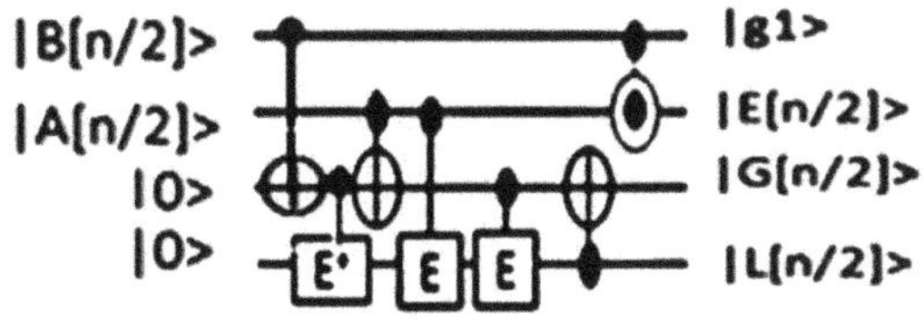

Figure 3.4. The MQC circuit. Reproduced with permission from [2]. Copyright 2017 IEEE.

$$
\begin{aligned}
&= O(\log n) + O(\log n) + O(2) + O(2) \\
&= O(2(2 + \log n)).
\end{aligned}
$$

Thus this completes the proof and property 3.3 is true.

The design of a quantum comparator consists of two circuits: the midpoint qubit comparison (MQC) circuit and the rest qubit comparison (RQC) circuit. First, the most significant quantum bits using the MQC circuit are compared. Second, the rest of the qubits using the RQC circuit are compared, where each qubit comparison is performed by one RQC circuit. In figure 3.4, the MQC circuit consists of two controlled-E gates, one controlled-E^+ gate, three CNOT gates, and one *XN* gate. This circuit generates the following three outputs as presented in equation (3.1) and produces one garbage output which is $|g_1\rangle$:

$$
\begin{aligned}
&|E_{n/2}\rangle = |A_{n/2}\rangle \odot |B_{n/2}\rangle \\
&|G_{n/2}\rangle = |A_{n/2}\ \overline{B_{n/2}} \\
&|L_{n/2}\rangle = |\ \overline{A_{n/2}} B_{n/2}\rangle .
\end{aligned}
\tag{3.1}
$$

In figure 3.5, the RQC circuit consists of four controlled-E gates, three controlled-E^+ gates, one *XN* gate, and eight CNOT gates. This circuit generates three outputs, as presented in equation (3.2). It also produces two garbage outputs which are $|g_1\rangle$ and $|g_2\rangle$:

$$
\begin{aligned}
&|AGB_{n/2-1}\rangle = |G_{n/2}\rangle + |E_{n/2}\rangle .\ |A_{n/2-1}\ \overline{B_{n/2-1}}\rangle \\
&|ALB_{n/2-1}\rangle = |L_{n/2}|E_{n/2}\rangle .\ |\ \overline{A_{n/2-1}} B_{n/2-1}\rangle \\
&|AEB_{n/2-1}\rangle = |AGB_{n/2-1}\rangle \odot |ALB_{n/2-1}\rangle .
\end{aligned}
\tag{3.2}
$$

3.3.1 Example

To construct a two-qubit quantum comparator circuit, one MQC circuit and one RQC circuit are needed to perform the greater than, less than, and equality operations. In figure 3.6, the detailed quantum circuit for a two-qubit comparator has been shown. The midpoint is the qubits in position 1. At first, the midpoint qubits $|A_1\rangle$ and $|B_1\rangle$ are applied in the MQC circuit to generate the following outputs as

$$
\begin{aligned}
&|E_1\rangle = |A_1\rangle \odot |B_1\rangle \\
&|G_1\rangle = |A_1 \overline{B_1}\rangle\ 3 \\
&|L_1 = |\overline{A_1}\, B_1\rangle .
\end{aligned}
\tag{3.3}
$$

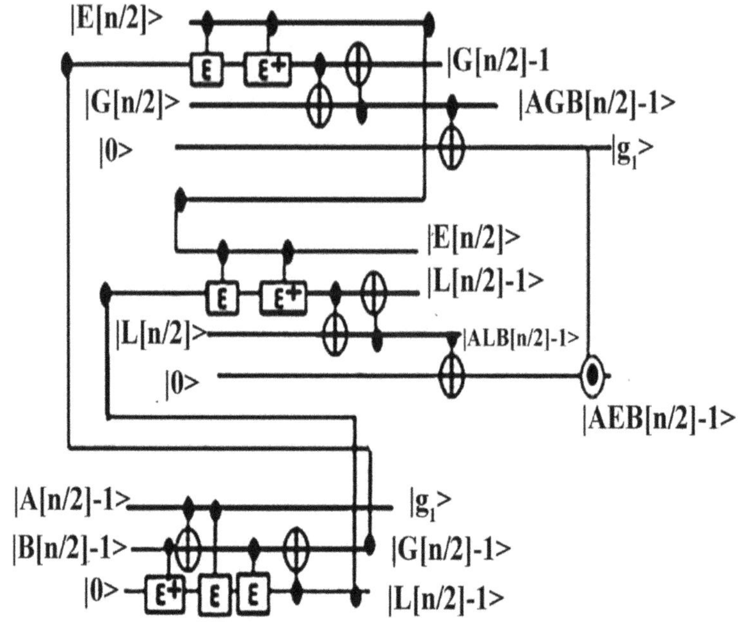

Figure 3.5. The RQC circuit. Reproduced with permission from [2]. Copyright 2017 IEEE.

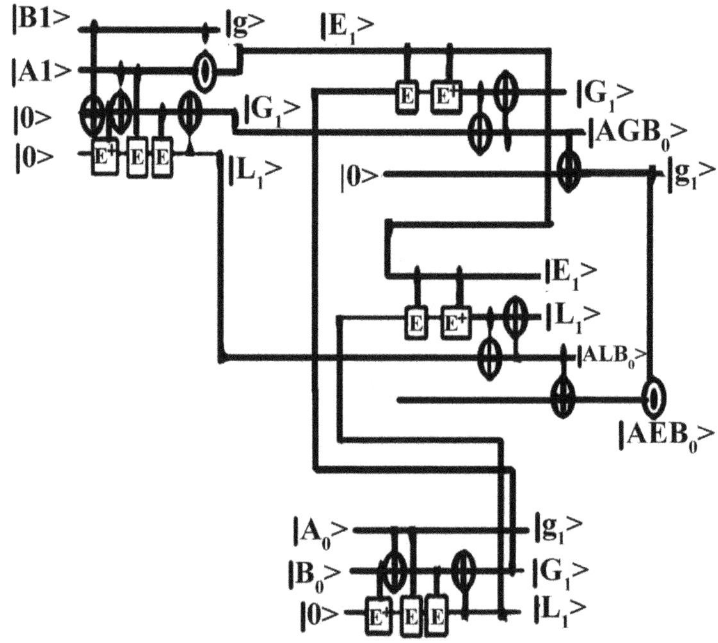

Figure 3.6. The two-qubit quantum comparator circuit. Reproduced with permission from [2]. Copyright 2017 IEEE.

Then the outputs of the MQC circuit ($|E_1\rangle$, $|G_1\rangle$, $|L_1\rangle$) and the rest of the qubits ($|A_0\rangle$, $|B_0\rangle$) are used as inputs to the next circuit which is the RQC circuit. This circuit produces the final outputs

$$\begin{aligned}|AGB_0\rangle &= |G_1\rangle + |E_1\rangle.\ |A_0(B_0)\rangle\\ |ALB_0\rangle &= |L_1\rangle + |E_1\rangle.\ |(A_0)B_0\rangle\\ |AEB_0\rangle &= |AGB_0\rangle \odot |ALB_0\rangle.\end{aligned} \tag{3.4}$$

Similarly, an n-qubit quantum comparator circuit can be constructed using one MQC circuit and maximum $(n - 1)$ RQC circuits or minimum $((n/2) - 1)$ RQC to perform the greater than, less than, and equality operations. In figure 3.7, the detailed quantum circuit for an n-qubit quantum comparator is shown.

3.4 Summary

This chapter presents the design methodology for a noble quantum n-qubit comparator using a fast comparison technique. The time complexity of the comparison technique is $O(\log n)$ where n is the number of qubits. The comparator was constructed in two steps. First, bit comparisons for greater than and less than operations were performed in parallel. Then bit comparison for the equality operation was performed. The quantum comparator circuit can be used in designing different quantum circuits, such as quantum processing units, complex arithmetic circuits, and communication systems. Thus the design which has been shown in this chapter should help the reader understand the next three chapters better. In addition, the concept of designing a quantum bit string comparator will help in designing the quantum subtraction, multiplier, divider, and adder circuits.

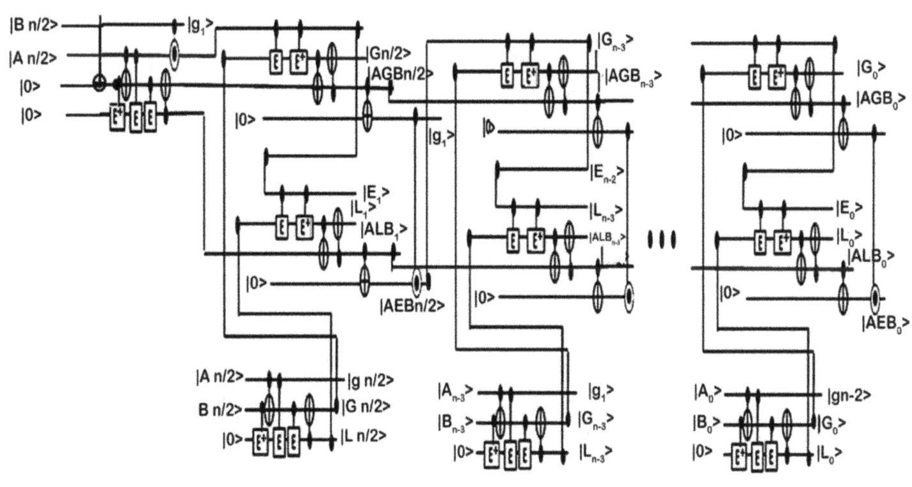

Figure 3.7. The n-qubit quantum comparator circuit. Reproduced with permission from [2]. Copyright 2017 IEEE.

Further reading

[1] Barwad R 2015 The characteristics of comparator *Polytechnic Hub* http://www.polytechnichub.com/characteristics-comparator/ (Accessed: 5 December 2018)

[2] Babu H M H, Jamal L, Dibbo S V and Biswas A K 2017 Area and delay efficient design of a quantum bit string comparator *2017 IEEE Computer Society Annual Symp. on VLSI* (Piscataway, NJ: IEEE) pp 51–6

[3] Das J C and De D 2016 Reversible comparator design using quantum dot-cellular automata *IETE J. Res.* **62** 323–30

[4] Dibbo S V, Babu H M H and Jamal L 2016 An efficient design technique of a quantum divider circuit *IEEE Int. Symp. on Circuits and Systems (ISCAS)* (Piscataway, NJ: IEEE) pp 2102–5

[5] Nielsen M A and Chuang I L 2000 *Quantum Computation and Quantum Information* 10th edn (Leiden: Cambridge University Press)

[6] Parida D 2010 A novel high speed CMOS comparator with low power dissipation and low offset *PhD Thesis*

[7] Phaneendra P S, Vudadha C, Sreehari V and Srinivas M B 2014 An optimized design of reversible quantum comparator *27th Int. Conf. on VLSI Design and 13th Int. Conf. on Embedded Syst.* (Piscataway, NJ: IEEE) pp 557–62

[8] Saravanakumar N, NirmalKumar A, Nandhakumar A and KanyaKumari G E 2013 ASIC implementation of low power area efficient folded binary comparator *Int. J. Sci. Eng. Technol.* **5** 4582

Chapter 4

The quantum adder and subtractor

In circuit design the adder and subtractor are the circuits that are capable of adding or subtracting numbers. In this chapter quantum adder, namely half-adder and full-adder, circuits and subtractor, namely half-subtractor and full-subtractor, circuits are presented with their quantum representations.

4.1 The quantum adder

An adder is a circuit in electronics that implements the addition of numbers. In many computers and other types of processors, adders are used to calculate addition and similar operations in the arithmetic logic unit (ALU) and also in other parts of processors. These can be constructed for many numerical representations such as excess-3 or binary coded decimal. Adders are classified into two types: half-adder and full-adder. The half-adder circuit has two inputs, A and B, which add two input digits and generate a carry and a sum. The full-adder circuit has three inputs, A, B, and C, which add the three input numbers and generate a carry and a sum.

Table 4.1 is the truth table of a half-adder. From this table, we can obtain the half-adder circuit's outputs which are

$$\text{Sum} = A \oplus B$$
$$\text{Carry} = AB.$$

The output equations of the half-adder can be mapped with a quantum Peres gate, as shown in figure 4.1, and the quantum half-adder can be obtained by putting $C = 0$, which is illustrated in figure 4.2. The quantum cost of the quantum half-adder is 4 and the delay is 4Δ.

4.1.1 The quantum full-adder

Table 4.2 is the truth table of a full-adder. From this table we can obtain the output of the full-adder circuit:

Table 4.1. The truth table of the half-adder.

A	*B*	Carry	Sum
0	0	0	0
0	1	0	1
1	0	0	1
1	1	1	0

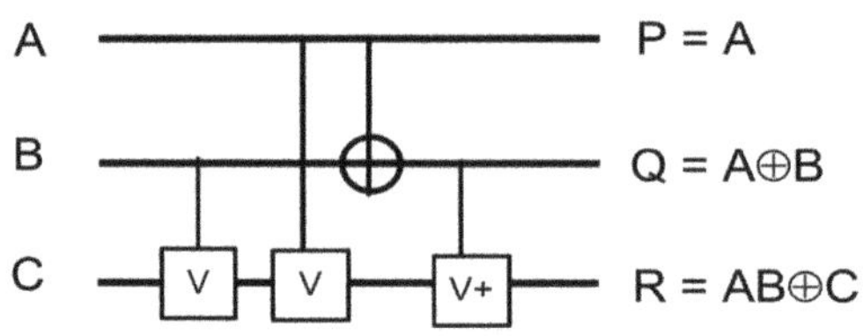

Figure 4.1. The quantum Peres gate.

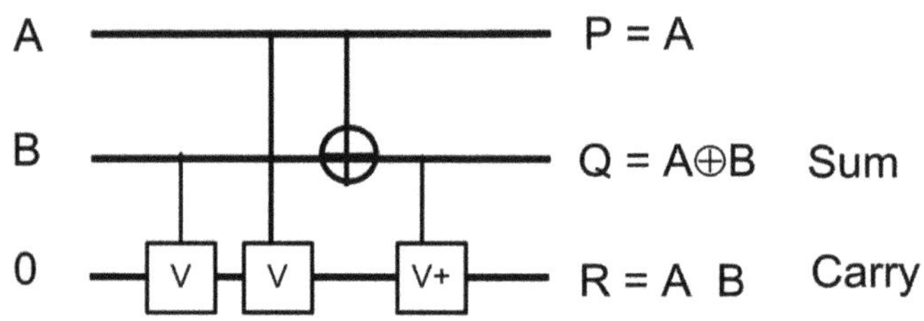

Figure 4.2. The quantum Peres gate as a quantum half-adder.

Table 4.2. The truth table of the full-adder.

A	*B*	C_{in}	Carry	Sum
0	0	0	0	0
0	0	1	0	1
0	1	0	0	1
0	1	1	1	0
1	0	0	0	1
1	0	1	1	0
1	1	0	1	0
1	1	1	1	1

$$\text{Sum} = A \oplus B \oplus C_{\text{in}}$$
$$\text{Carry} = (A \oplus B)C_{\text{in}} \oplus AB.$$

The output equations of the full-adder can be mapped with the quantum modified Thapliyal Srinivas gate (MTSG), as shown in figure 4.3, and the quantum full-adder

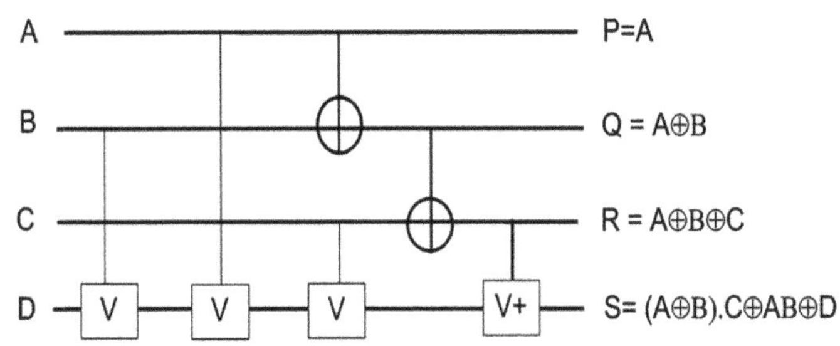

Figure 4.3. The quantum MTSG gate.

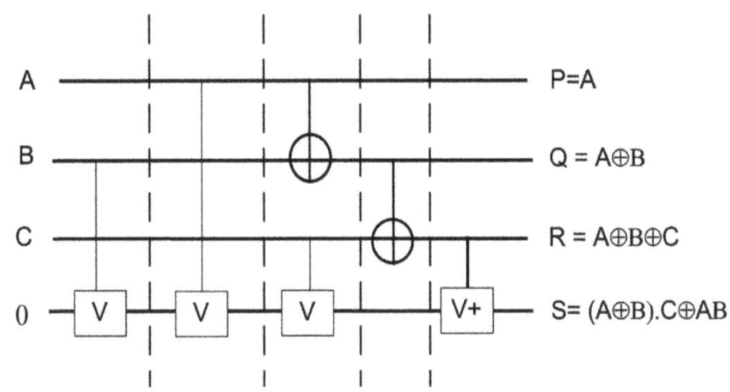

Figure 4.4. The quantum MTSG gate as a quantum full-adder.

can be obtained by putting $D = 0$, which is shown in figure 4.4. The quantum cost of the quantum full-adder is 6 and the delay is 5Δ.

4.2 The quantum subtractor

A subtractor is also an important logic component in circuit design. Subtractors are classified into two types: half-subtractors and full-subtractors. The half-subtractor circuit has two inputs A and B where the half-subtractor performs the $A - B$ operation.

4.2.1 The quantum half-subtractor

Table 4.3 is the truth table of a half-subtractor. From this table we can obtain the half-subtractor circuit:

$$\text{Difference} = A \oplus B$$
$$\text{Borrow} = \bar{A}B.$$

The output equations of the half-subtractor can be mapped with a quantum Thapliyal Ranganathan (TR) gate, as shown in figure 4.5, and the quantum half-subtractor can be obtained by putting $C = 0$, which is shown in figure 4.6. The quantum cost of the quantum half-subtractor is 4 and the delay is 4Δ.

Table 4.3. The truth table of the half-subtractor.

A	*B*	Borrow	Difference
0	0	0	0
0	1	1	1
1	0	0	1
1	1	0	0

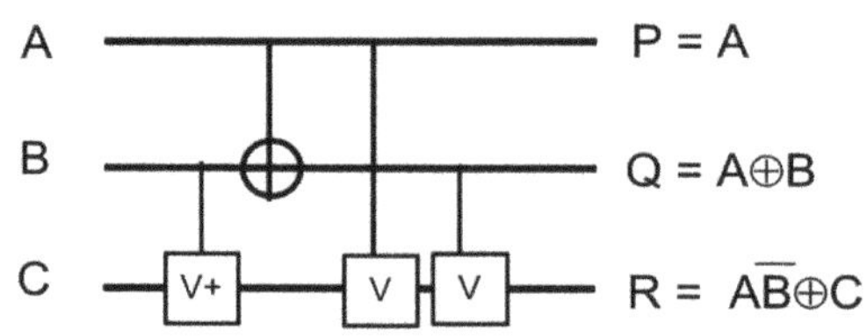

Figure 4.5. The quantum TR gate.

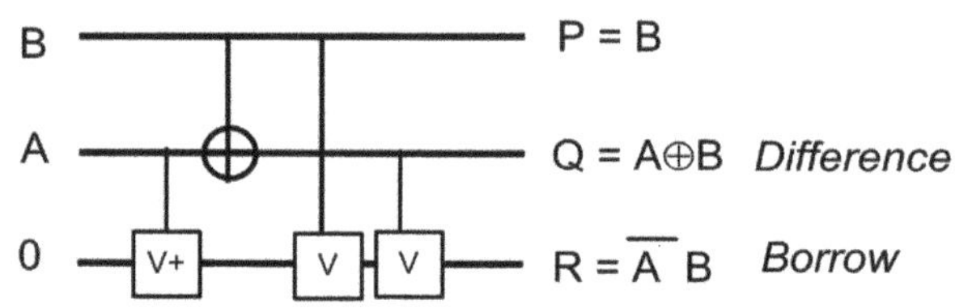

Figure 4.6. The quantum half-subtractor.

Table 4.4. The truth table of full-subtractor.

A	*B*	C_{in}	Borrow	Difference
0	0	0	0	0
0	0	1	1	1
0	1	0	1	1
0	1	1	1	0
1	0	0	0	1
1	0	1	0	0
1	1	0	0	0
1	1	1	1	1

4.2.2 The quantum full-subtractor

The full-subtractor circuit has three inputs A, B, and C_{in}, which realize the operation $Y = A - B - C$. Table 4.4 is the truth table of a full-subtractor. From this table we can obtain the output of the full-subtractor circuit:

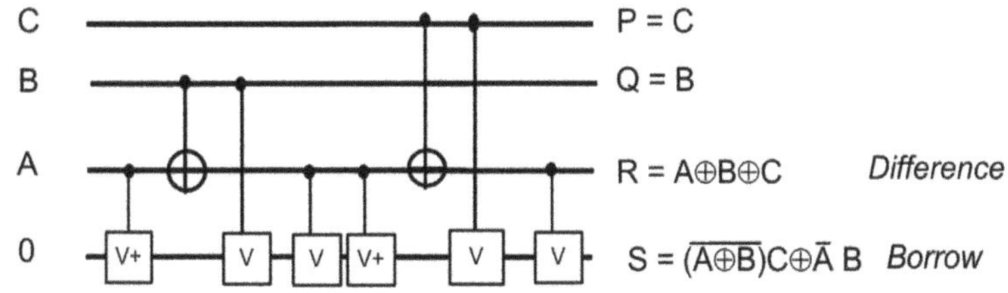

Figure 4.7. The quantum full-subtractor by cascading two quantum TR gates.

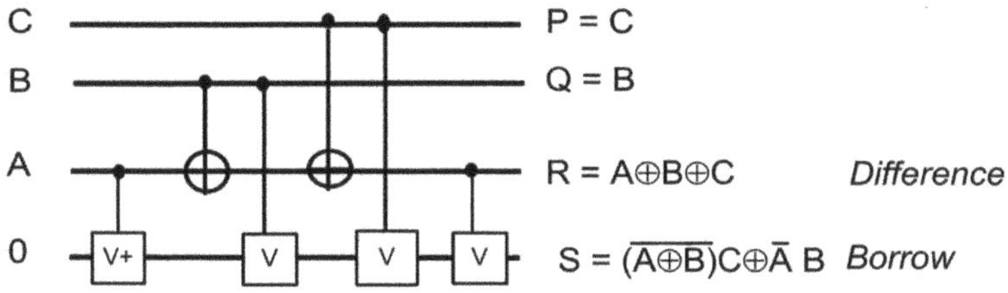

Figure 4.8. The optimized quantum full-subtractor.

$$\text{Difference} = A \oplus B \oplus C_{\text{in}}$$
$$\text{Borrow} = (\overline{A \oplus B})C_{\text{in}} \oplus AB.$$

A quantum full-subtractor can be designed using two quantum TR gates, shown in figure 4.7. Figure 4.8 is the optimized version of the full-subtractor. The quantum cost of a quantum full-subtractor is 6 and the delay is 4Δ.

4.3 Summary

This chapter presents the quantum full-adder and full-subtractor from the half-adder and half-subtractor circuits. The quantum costs of the adder and subtractor circuits are discussed for better understanding of the circuits.

Further reading

[1] Babu H M H, Islam M R, Chowdhury S M A and Chowdhury A R 2004 Synthesis of a full-adder circuit using reversible logic *Proc. 17th Int. Conf. on VLSI Design (Mumbai, 2004)* pp 757–60

[2] Babu H M H, Islam R, Chowdhury A R and Chowdhury S M A 2003 On the realization of reversible full-adder circuit *Int. Conf. on Computer and Information Technology* pp 880–3

[3] Babu H M H, Islam R, Chowdhury A R and Chowdhury S M A 2003 Reversible logic synthesis for minimization of full-adder circuit *Proc. Euromicro Symp. on Digital System Design* pp 50–4

[4] Babu H M H, Jamal L and Saleheen N 2013 An efficient approach for designing a reversible fault tolerant *n*-bit carry look-ahead adder *IEEE Int. SOC Conf.* pp 98–103

[5] Cheng K-W and Tseng C-C 2002 Quantum full adder and subtractor *Electron. Lett.* **38** 1343–4

[6] Cuccaro S A, Draper T G, Kutin S A and Moulton D P 2004 A new quantum ripple-carry addition circuit (arXiv: quant-ph/0410184)

[7] Khan M H A and Perkowski M A 2007 Quantum ternary parallel adder/subtractor with partially-look-ahead carry *J. Syst. Archit.* **53** 453–64

[8] Monfared A T and Haghparast M 2016 Design of new quantum/reversible ternary subtractor circuits *J. Circuit. Syst. Comp.* **25** 1650014

[9] Murali K V R M, Sinha N, Mahesh T S, Levitt M H, Ramanathan K V and Kumar A 2002 Quantum-information processing by nuclear magnetic resonance: experimental implementation of half-adder and subtractor operations using an oriented spin-7/2 system *Phys. Rev.* A **66** 22313

[10] Takahashi Y 2009 Quantum arithmetic circuits: a survey *IEICE Trans. Fundam. Electron. Commun. Comput. Sci.* **92** 1276–83

[11] Takahashi Y, Tani S and Kunihiro N 2010 Quantum addition circuits and unbounded fan-out *Quantum Inf. Comput.* **10** 872–90

Chapter 5

The quantum multiplexer and demultiplexer

In large-scale computing systems, it is necessary for a single line to carry two or more signals. One signal can be placed on one line at a time. However, the device which will allow us to select and place the signal we choose on a common line, is called a multiplexer (MUX). The purpose of a MUX is to select the input of any n inputs and feed that to one output line. The function of a demultiplexer (DEMUX) is the opposite to the function of the MUX.

5.1 The quantum multiplexer

This section presents the design of the quantum multiplexer. The multiplexer causes the transmission of a large number of information units over a smaller number of channels. Architecturally a digital multiplexer is a logic circuit that puts one of several inputs to a single output. A set of selected inputs controls the inputs. Generally a multiplexer has m inputs and n selected inputs where $m = 2^n$.

5.1.1 The quantum 2-to-1 multiplexer

A 2-to-1 multiplexer is the smallest unit of architecture of a quantum multiplexer. The characteristic function of a multiplexer is $s_0'I_0 + s_0I_1$. A quantum Fredkin gate, as shown in figures 5.1(a) and (b), can be used as a 2-to-1 quantum multiplexer as it can map the characteristic function of a multiplexer.

Let I_0 and I_1 be the inputs and S_0 be the selected input of a 2-to-1 multiplexer. When $S_0 = 0$, then input I_0 transmits to the output Y and when $S_0 = 1$, then input I_1 transmits to the output Y. Figure 5.2 shows the architecture of a quantum 2-to-1 multiplexer using a quantum Fredkin gate. The quantum cost and delay of this quantum 2-to-1 multiplexer are 5 and 5Δ, respectively. Moreover, the number of garbage outputs is two.

doi:10.1088/978-0-7503-2747-3ch5

5.1.2 The quantum 4-to-1 multiplexer

The quantum 4-to-1 multiplexer has four inputs, two select lines, and one output. Figure 5.3 illustrates the design of a quantum 4-to-1 multiplexer, where I_0, I_1, I_2, and I_3 are the inputs, and S_0 and S_1 are the select lines. The bit combination of select lines controls the function of a 4-to-1 multiplexer, as presented in table 5.1. Three quantum Fredkin gates are used in this design. Thus the quantum cost of the quantum 4-to-1 multiplexer is 15 and the delay is 15Δ in the logic circuits, respectively; whereas the number of garbage outputs is five.

5.1.3 The quantum 2^n-to-1 multiplexer

Figure 5.4 shows the design of a quantum 8-to-1 multiplexer. As the consequence of the design of quantum multiplexers, a 2^n-to-1 multiplexer can be constructed using two 2^{n-1}-to-1 quantum multiplexers and one 2-to-1 quantum multiplexer. Figure 5.5

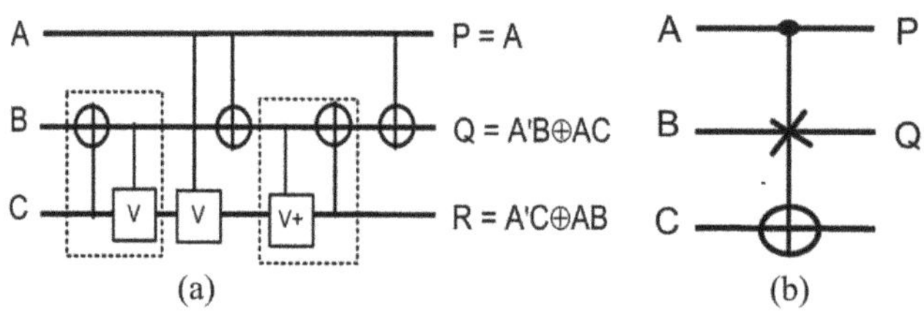

Figure 5.1. The quantum Fredkin gate. (a) Quantum circuit of a quantum Fredkin gate. (b) Symbol of a quantum Fredkin gate.

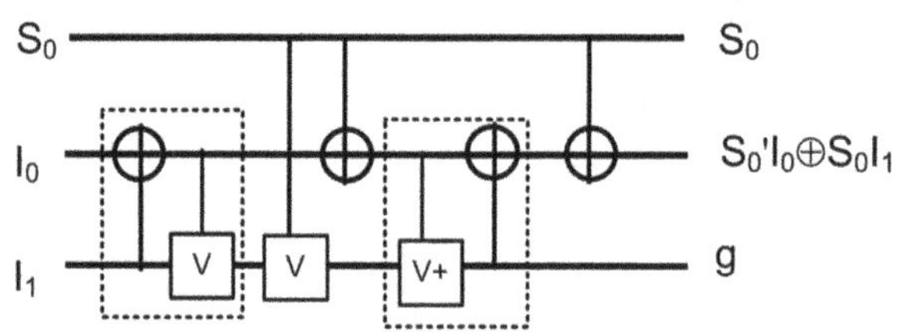

Figure 5.2. The quantum Fredkin gate as a quantum 2-to-1 multiplexer.

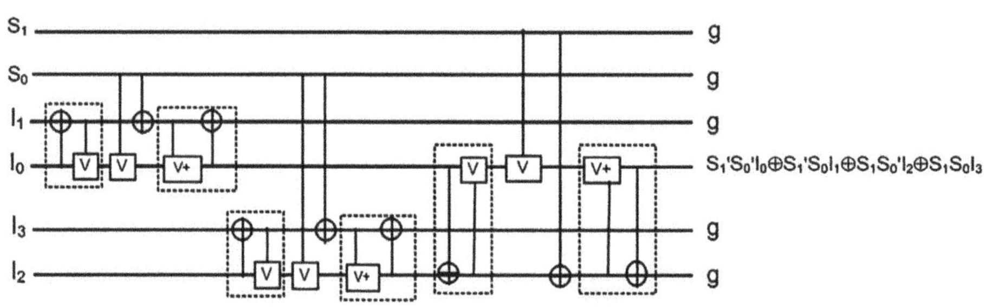

Figure 5.3. The quantum 4-to-1 multiplexer.

Table 5.1. Function of S_0 and S_1 select lines.

S_0	S_1	Output(O)
0	0	I_0
0	1	I_1
1	0	I_2
1	1	I_3

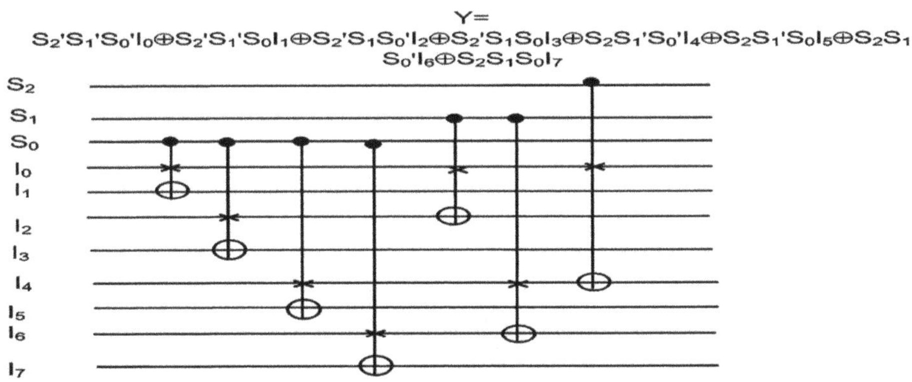

Figure 5.4. The quantum 8-to-1 multiplexer.

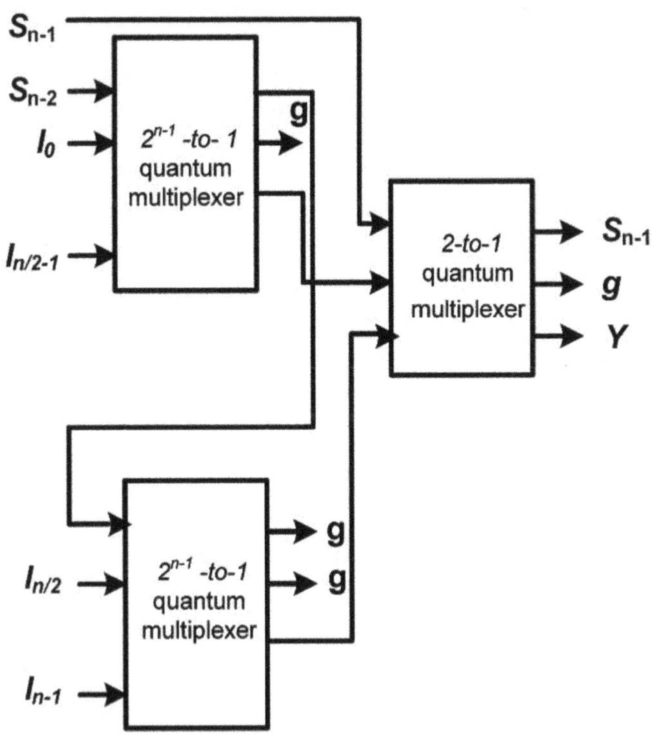

Figure 5.5. Block diagram of 2^n-to-1 multiplexers.

presents the 2^n-to-1 multiplexer, and the properties of the 2^n-to-1 quantum multiplexer are given in property 5.1.

Property 5.1. A quantum 2^n-to-1 multiplexer can be designed with a $2^n - 1$ gate which produces $2^n + n - 1$ garbage outputs. It also requires a $5(2^n - 1)$ quantum cost and a delay of $5(2^n - 1)\Delta$, where n denotes the number of selection lines and Δ denotes the unit delay.

5.2 The quantum demultiplexer

This section presents the design of the quantum demultiplexer. A demultiplexer (or DEMUX) is a device that takes a single input line and routes it to one of several digital output lines. A demultiplexer of 2^n outputs has n select lines which are used to select the output line to which to send the input. A demultiplexer is also called a data distributor. The demultiplexer can be used to implement general purpose logic. By setting the input to true, the DEMUX behaves as a decoder. The reverse of a multiplexer is the demultiplexer.

5.2.1 The quantum 1-to-2 demultiplexer

A 1-to-2 demultiplexer is the smallest unit of the architecture of a quantum demultiplexer. The characteristic function of a 1-to-2 demultiplexer is $s_0'D$ s_0D on the different output line, as shown in table 5.2. A quantum Fredkin gate can be used as a 1-to-2 quantum demultiplexer as it can map the characteristic functions of a demultiplexer.

Let, D be the inputs and S_0 the select input of a 1-to-2 demultiplexer. When $S_0 = 0$, then D input is transmitted to the second output Y_0 and when $S_0 = 1$, then the D input is transmitted to the third output Y_1. Figure 5.6 shows the architecture of a quantum 1-to-2 demultiplexer using a quantum Fredkin gate. The quantum cost and delay of this quantum 1-to-2 demultiplexer are 5 and 5Δ,respectively. Moreover, the number of garbage outputs is one.

5.2.2 The quantum 1-to-4 demultiplexer

The quantum 1-to-4 demultiplexer has two select lines, one data input, and four outputs. Figure 5.7 shows the design of a quantum 1-to-4 demultiplexer where Y_0, Y_1, Y_2, and Y_3 are the outputs, and S_0 and S_1 are the select lines. The bit combination of select lines controls the function of the 1-to-4 demultiplexer, as shown in table 5.3. Three quantum Fredkin gates are used in this design. Thus the quantum cost of the

Table 5.2. Truth table of a 1-to-2 demultiplexer.

S	Y_1	Y_0
0	0	D
1	D	0

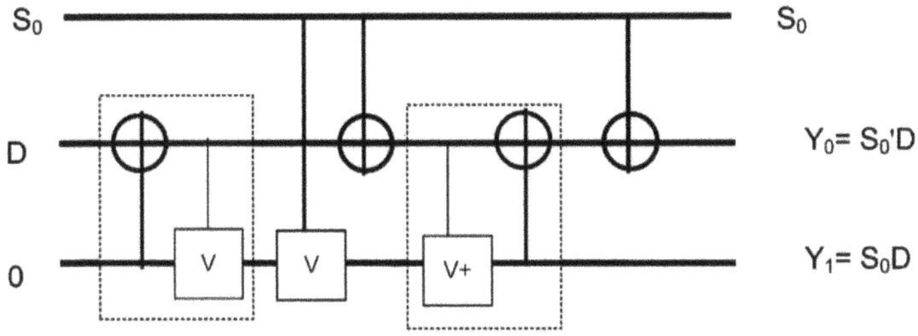

Figure 5.6. The quantum Fredkin gate as a quantum 1-to-2 demultiplexer.

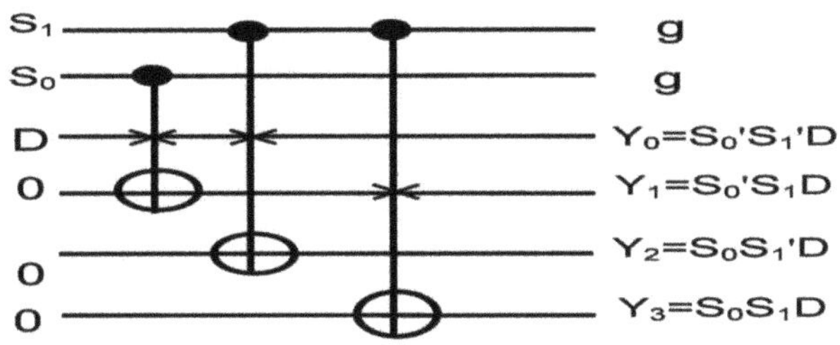

Figure 5.7. The quantum 1-to-4 demultiplexer.

Table 5.3. The truth table of a 1-to-4 demultiplexer.

S_1	S_0	Y_3	Y_2	Y_1	Y_0
0	0	0	0	0	D
0	1	0	0	D	0
1	0	0	D	0	0
1	1	D	0	0	0

quantum 1-to-4 demultiplexer is 15 and the delay of the quantum 1-to-4 demultiplexer is 15Δ, while the number of garbage outputs is two.

5.2.3 The quantum 1-to-2^n demultiplexer

Figure 5.8 shows the design of a quantum 1-to-8 demultiplexer. As a consequence of the design of the quantum demultiplexer, a 1-to-2^n quantum demultiplexer can be constructed using a 1-to-2^{n-1} quantum demultiplexer and 2^{n-1} quantum Fredkin gates, which is shown in figure 5.9. The properties of the 1-to-2^n quantum demultiplexer are given in property 5.2.

Property 5.2. A quantum 1-to-2^n demultiplexer can be designed with $2^n - 1$ gates which produce n garbage outputs. It also requires a $5(2^n - 1)$ quantum cost and a delay of $5(2^n - 1)\Delta$, where n denotes the number of selection inputs and Δ denotes the unit delay.

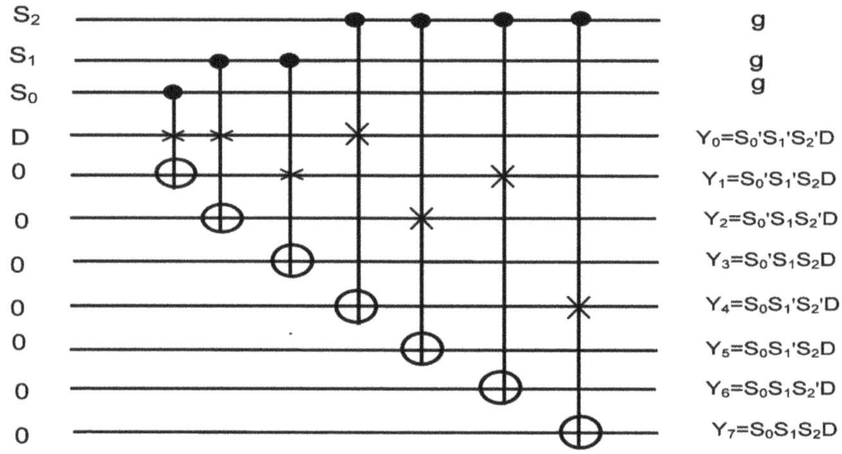

Figure 5.8. The quantum 1-to-8 demultiplexer.

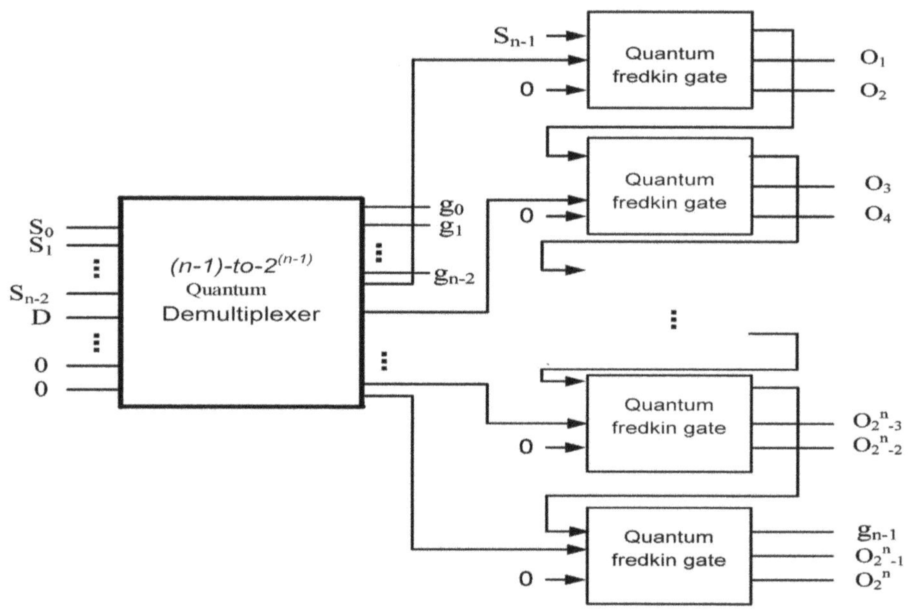

Figure 5.9. Block diagram of a 1-to-2^n demultiplexer.

5.3 Summary

In this chapter the quantum multiplexer (MUX) and demultiplexer (DEMUC) are presented and explained with the quantum circuit representation. In the multiplexer section, a 2-to-1 MUX, 4-to-1 MUX, and 8-to-1 MUX are described and generalized to the 2^n-to-1 multiplexer. In the demultiplexer section, a 1-to-2 DEMUX, 1-to-4 DEMUX, and 1-to-8 DEMUX are described and generalized to the 1-to-2^n demultiplexer. Moreover, the quantum cost and delay of the multiplexers and demultiplexers are provided.

Further reading

[1] Haghparast M and Monfared A T 2017 Novel quaternary quantum decoder, multiplexer and demultiplexer circuits *Int. J. Theor. Phys.* **56** 1694–707

[2] Khan M H A 2008 Reversible realization of quaternary decoder, multiplexer, and demultiplexer circuits *38th Int. Symp. on Multiple Valued Logic* pp 208–13

[3] Mardiris V A and Karafyllidis I G 2010 Design and simulation of modular 2^n to 1 quantum-dot cellular automata (QCA) multiplexers *Int. J. Circuit Theory Appl.* **38** 771–85

Chapter 6

Quantum adder circuits

Several types of adders are used in computing systems, such as carry skip adders (CSAs), binary coded decimal (BCD) adders, and so on. Among them the most common is the ripple carry adder, in which full-adders are connected in series and carry is propagated through all the stages and hence requires more carry propagation time to generate the carry output. Carry look ahead adders are the fastest of all the adders since the carry output is generated in parallel computation, but this requires a large number of gates. The CSA is the most promising adder which presents a compromise between hardware and performance compared to both adders mentioned above. In the full-adder operation, if either of the inputs is a logical one, then the cell will propagate the carry input to the carry output. Hence the *i*th full-adder carry input C_i will propagate to its carry output, C_{i+1}, when $P_i = X_i \oplus Y_i$.

In computers, numbers are stored in straight binary format. Due to the inherent characteristics of floating-point numbers and limitations on storage formats, not all floating-point numbers can be represented with the desired precision. Thus computing in decimal format is gaining popularity as losses due to precision can be avoided in this format. However, hardware support for binary arithmetic allows it to be performed faster than decimal arithmetic. Faster hardware for decimal floating-point arithmetic is also imminent as it has importance in financial and Internet-based applications. Thus faster circuits for BCD numbers will have a great impact as they are likely to be incorporated in more complex circuits such as future mathematical processors.

In this chapter the quantum CSA (QCSA) and BCD adder circuits are presented in sections 6.1 and 6.5, respectively.

6.1 The carry skip adder

Adders are widely used in generic computers because they are very important for adding data in the processor. The simplest binary adder is the ripple carry adder. It is

doi:10.1088/978-0-7503-2747-3ch6

easy to understand and implement. A more complex binary adder is the CSA. It is widely used due to its superior performance compared to the ripple carry adder. In this chapter an n-bit QCSA is presented which has fewer quantum gates. It also produces fewer garbage outputs and requires less area, power, and delay. With the help of these properties the efficiency of the quantum logic synthesis of an n-bit CSA is also proved in this chapter.

Adders are one of the most common elements in circuit design since addition is a fundamental operation for any kind of digital system. However, traditional ripple carry adders have some defects such as low computing efficiency and a long delay. When generalizing and designing a large adder, the ripple carry adder results in a large number of quantum gates, garbage outputs, and constant inputs, and high hardware complexity. The reason behind this lies in the carry propagation. If the carry propagation is not performed in an effective way, a huge summation network is required to design a large-scale adder. In order to overcome the disadvantages mentioned above, a new kind of adder called the CSA is used which has a shorter delay.

6.2 The quantum comparison circuit

The quantum comparison circuit (QCC) which is depicted in figure 6.1 is constructed with 11 inputs comprising three constant inputs. This quantum circuit consists of four CNOT gates and three quantum Peres gates. The circuit performs the following function:

$$X = (A_3 \oplus B_3) \cdot (A_2 \oplus B_2) \cdot (A_1 \oplus B_1) \cdot (A_0 \oplus B_0).$$

If all $A_i \neq B_i$, X will generate the value 1. The number of quantum gates is calculated as follows: $4 \times 1 + 4 \times 3 = 16$.

6.3 The quantum 2-to-1 multiplier circuit

The quantum multiplexer (MUX) block carries out the function as $C_{\text{out}} = C_{\text{in}} \cdot X \cdot C_o \cdot X$ where X comes from the QCC and C_{out} comes from the last quantum full-adder in the four-bit quantum ripple carry adder. Its quantum

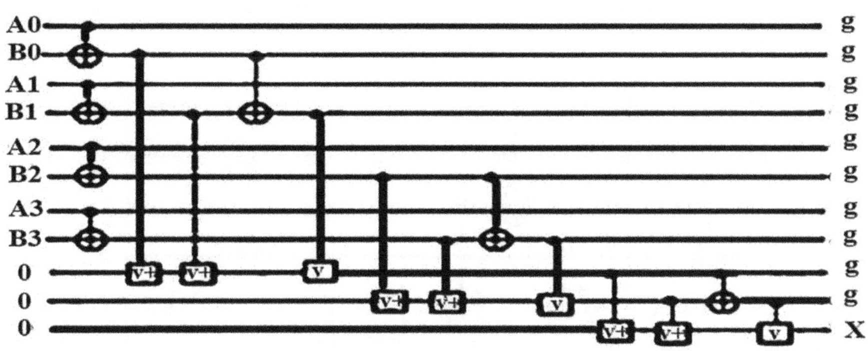

Figure 6.1. The quantum comparison circuit.

design is illustrated in figure 6.2, which has a CNOT gate and two quantum Peres gates. The number of quantum gates is $1 \times 1 + 4 \times 2 = 9$.

6.4 The design of a quantum carry skip adder

The CSA is constructed by cascading a four-bit CSA. Each four-bit CSA contains a four-bit ripple carry adder, a comparison circuit, and a 2-to-1 MUX. If the two inputs are completely reversed, then the carry entering the carry skip adder will simply be propagated to the next four-bit CSA. Thus, there is no need for the next CSA to wait for the carry-in created by this current four-bit CSA. Therefore, the delay generated by the carry bit can be shortened. Each comparison block is used for un-equivalence in input bits. A MUX is responsible for selecting a generated carry or a propagated (previous) carry with its selection line being the output of the comparison circuit just described. If each four-bit carry skip block $A_i \neq B_i$, then it can be said that a carry can skip over the block. Otherwise, if $A_i = B_i$ it can be said that the carry must be generated in the block.

In the following subsections a four-bit QCSA circuit is constructed and then an n-bit QCSA circuit is shown.

6.4.1 The four-bit quantum carry skip adder

A four-bit QCSA circuit is designed using a four-bit ripple carry adder along with a comparison circuit. The comparison circuit ensures that the comparison of each block is indeed unequal. In this regard, a MUX circuit is used that is responsible for selecting a generated carry or a propagated (previous) carry with its selection line, which is considered as the output of the comparison circuit.

The quantum four-bit ripple carry adder circuit is constructed by cascading a double Peres gate (DPG) quantum circuit where the DPG works as a full-adder, depicted in figure 6.3 where two four-bit binary additions are performed, such as A and B together with the input carry C_{in} being first added in the quantum ripple carry adder circuit to produce the binary sum and the carry out. The number of quantum gates of this circuit is $6 \times 4 = 24$ and the delay is 18Δ. Then the quantum comparison gate (QCG) constructs the quantum 2-to-1 MUX module for generating the carry, shown in figures 6.1 and 6.2, respectively. Finally, all circuits have been combined and the four-bit QCSA is obtained, shown in figure 6.4. The design procedure of a four-bit QCSA circuit is shown using algorithm 6.1.

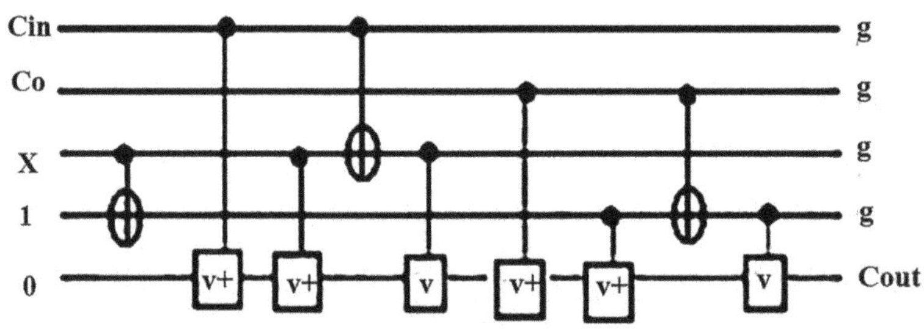

Figure 6.2. The quantum 2-to-1 MUX circuit. Reproduced with permission from [18]. Copyright 2014 IEEE.

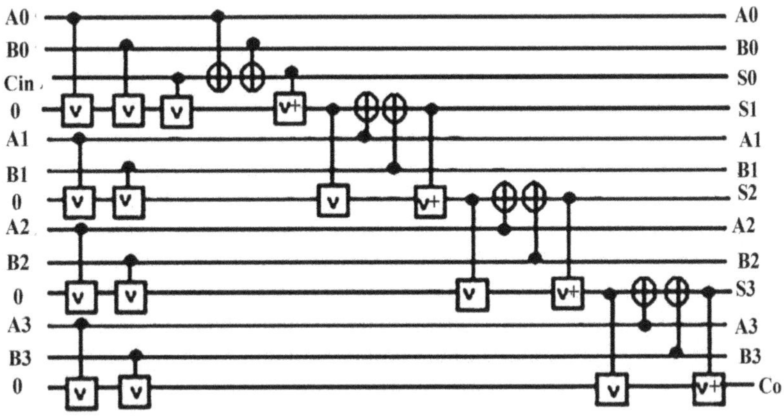

Figure 6.3. The four-bit quantum ripple carry adder. Reproduced with permission from [18]. Copyright 2014 IEEE.

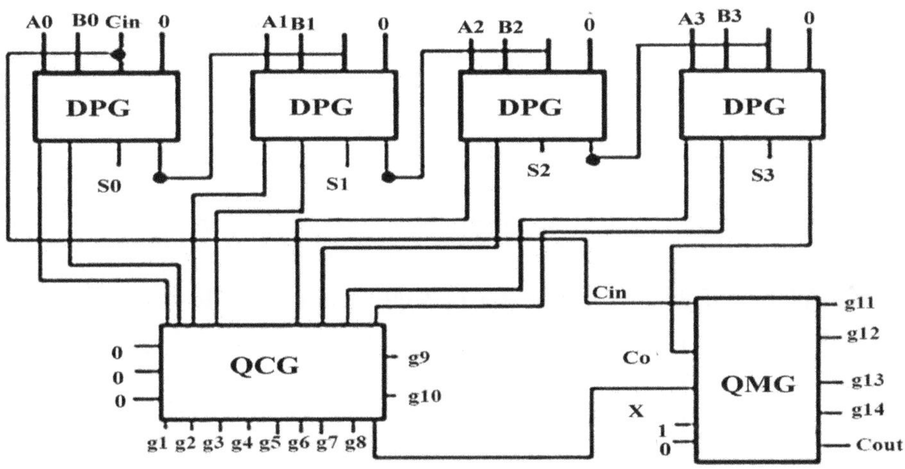

Figure 6.4. The four-bit QCSA. Reproduced with permission from [18]. Copyright 2014 IEEE.

Algorithm 6.1. Construction of a four-bit QCSA.

Input: $A(A_3A_2A_1A_0)$, $B(B_3B_2B_1B_0)$, C_{in}
Output: $S(S_3S_2S_1S_0)$, C_o, X, C_{out}
1: START
2: **for** $i = 0$ to 3 **do**
3: Apply a DPG quantum circuit with
4: Input := { A_i, B_i, C_{in}, 0 }
5: Output := { A_i, B_i, S_i, C_{oi}}
6: **end for**
7: Apply a QCG gate with

Input := A_0, B_0, A_1, B_1, A_2, B_2, A_3, B_3, 0, 0, 0
Output := g_1, g_2, g_3, g_4, g_5, g_6, g_7, g_8, g_9, g_{10}, X
Apply a QMG gate $(a, b, c, d, e)(p, q, r, s, t)$
Input := C_{in}, C_{03}, X, 1, 0
Output := g_{11}, g_{12}, g_{13}, g_{14}, C_{out}
END

6.4.2 The *n*-bit quantum carry skip adder

A four-bit QCSA circuit is constructed using a four-bit quantum ripple carry adder, a QCC, and a quantum 2-to-1 MUX gate (QMG), shown in figure 6.4. Therefore, an n-bit QCSA circuit can be constructed by cascading ($N = n/4$) four-bit QCSA circuits, shown in figure 6.5. The process is described by algorithm 6.2.

Algorithm 6.2. Construction of an n-bit QCSA.

START
for $i = 1$ to $n/4$ **do**
    Call Algorithm 6.1
end for
END

6.4.3 Calculation of the area and power of a quantum carry skip adder circuit

Area is an important issue in designing a circuit. If the area of a circuit is very large, the cost of that circuit will increase. The area of a gate is defined by the feature size. This size varies according to the quantum cost of the gate. As basic quantum gates are fabricated using quantum dots, with sizes ranging from several to tens of nanometers (10^{-9} m) in diameter, the size of the basic quantum gates ranges from 50 Å–300 Å. Quantum circuits can be implemented with the basic quantum gates and the quantum cost of a gate depends on the number of basic quantum gates needed to implement it. Thus the area of a gate can be defined as follows: $A = N_q \times S_q$, where

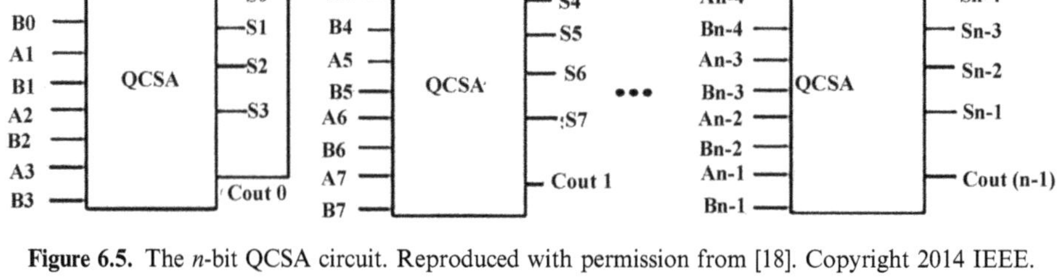

Figure 6.5. The n-bit QCSA circuit. Reproduced with permission from [18]. Copyright 2014 IEEE.

A = area, N_q = number of quantum gates, and S_q = size of basic quantum gates. According to the feature size of the DPG quantum circuit the area is ((50 Å × 6) – (300 Å × 6)) = (300 Å – 1800 Å), where the number of quantum gates of the DPG circuit is six. Algorithm 6.3 describes the area calculation of the n-bit QCSA circuit.

Algorithm 6.3. Calculation of area for a n-bit QCSA circuit.

1: START
2: Area = 0, Q_{QCSA} = 0, $N = n/4$
3: **for** $i = 1$ to N **do**
4: Obtain the number of quantum gates of a four-bit QCSA circuit
5: $Q_{QCSA} = Q_{QCSA} + Q$. QCSA
6: **end for**
7: Area = 50 × Q_{QCSA}
8: **return** Area
9: END

Power is also an important issue in designing a circuit. One should design a circuit which needs little power. The power of a gate is defined by the energy. The energy of the basic quantum gate is 142.3 meV. Quantum circuits can be implemented with basic quantum gates and the quantum cost of a gate depends on the number of basic quantum gates needed to implement it. Thus the power of a gate can be defined as follows: $P = N_q \times E_q$, where P = power, N_q = number of quantum gates, and E_q = energy of basic quantum gates. For example, the energy of the DPG quantum circuit is (6 × 142.3) meV = 853.8 meV, where the number of quantum gates of the DPG circuit is six. Algorithm 6.4 describes the power calculation of the n-bit QCSA circuit.

Algorithm 6.4. Calculation of power for an n-bit QCSA circuit.

1: START
2: Power = 0, Q_{QCSA} = 0, $N = n/4$
3: **for** $i = 1$ to N **do**
4: Obtain the number of quantum gates of a four-bit QCSA
5: $Q_{QCSA} = Q_{QCSA} + Q$. QCSA
6: **end for**
7: Power = 142.3 × Q_{QCSA}
8: **return** Power
9: END

6.4.4 Complexity of the *n*-bit quantum carry skip adder circuit

The complexities of a quantum circuit in terms of quantum gates, garbage output, and delay, etc, are very important as they have a great impact during physical implementation. In this subsection the different complexities of an *n*-bit QCSA circuit are presented.

Property 6.1. An *n*-bit QCSA circuit requires $49N$ quantum gates where $N = n/4$.

Proof. The above statement is proved by mathematical induction. As the number of quantum gates of a DPG quantum circuit is six, the number of quantum gate of a four-bit quantum ripple carry adder is 24.

A four-bit QCSA requires a four-bit quantum ripple carry adder, a QCC for which the number of quantum gates is 16, and a quantum 2-to-1 MUX for which the number of quantum gates is nine. Thus the total number of quantum gates (NOG) required to construct a four-bit QCSA circuit is $NOG_4 = NOG_{DPG} \times 4 + NOG_{QCG} + NOG_{QMG} = 6 \times 4 + 16 + 9 = 49$.

Thus the statement holds for base case $n = 4$.

Assume that the statement holds for $n = k$. Thus a *k*-bit QCSA can be realized with $49K$ quantum gates where $K = k/4$.

A $(k + 1)$-bit QCSA requires $(k + 1)$ QCSA blocks. As a result, the total number of quantum gates required to construct a $(k + 1)$-bit QCSA is $49 \times (K + 1)$, where $K = k/4$. Thus the statement holds for $n = k + 1$.

Therefore an *n*-bit QCSA circuit can be realized with $49N$ quantum gates where $N = n/4$.

Property 6.2. An *n*-bit QCSA circuit produces $43N\Delta$ quantum delay, where Δ is the unit delay and $N = n/4$.

Proof. A QCSA circuit consists of a quantum ripple carry adder circuit, a QCC, and a quantum 2-to-1 MUX circuit. A one-bit quantum ripple carry adder circuit requires 6Δ delay. So the total delay for an *n*-bit quantum ripple carry adder circuit is a $6n\Delta$ delay.

A QCC and a quantum 2-to-1 MUX require 16Δ and 9Δ delay, respectively. As there are $2(n - 1)$ controlled-V gates at the same level of an *n*-bit quantum ripple carry adder circuit, $2(n - 1)\Delta$ delays can be deducted from the total delay of the quantum ripple carry adder circuit. Thus the delay of a four-bit QCSA circuit is $6 \times 4\Delta + 16\Delta + 9\Delta - 2\,(4 - 1)\Delta = 24\Delta + 16\Delta + 9\Delta - 8\Delta + 2\Delta = 49\Delta - 6\Delta = 43\Delta$.

Therefore, the total delay of an *n*-bit QCSA circuit is $43N\Delta$, where $N = n/4$.

Example 6.1. When $n = 4$, the total delay of a four-bit QCSA circuit is $(43 \times 1)\Delta = 43\Delta$, where $N = n/4 = 4/4 = 1$.

Property 6.3. An n-bit QCSA circuit requires $18N\sigma + 31N\Omega$ quantum gate calculation complexity, where σ is the CNOT gate calculation complexity, Ω is the controlled-V (controlled-V^+) gate calculation complexity, and $N = n/4$.

Proof. From property 6.2 it is found that a QCSA circuit consists of a quantum ripple carry adder circuit, a QCC, and a quantum 2-to-1 MUX circuit. A one-bit quantum ripple carry adder circuit has 2σ CNOT quantum gate calculation complexity and 4Ω controlled-V (controlled-V^+) quantum gate calculation complexity. Thus the quantum gate calculation complexity of a four-bit quantum ripple carry adder circuit is $(2 \times 4)\sigma = 8\sigma$ for eight CNOT gates, $(4 \times 4) = 16\Omega$ for 16 controlled-V (controlled-V^+) gates, and the complexity of a QCC is 7σ for seven CNOT gates and 9Ω for nine controlled-V (controlled-V^+) gates, and the complexity of a quantum 2-to-1 MUX is 3σ for three CNOT gates and 6Ω for six controlled-V (controlled-V^+) gates, i.e. $8\sigma + 16\Omega + 7\sigma + 9\Omega + 3\sigma + 6\Omega = 18\sigma + 31\Omega$ quantum gate calculation complexity.

Therefore, the total quantum gate calculation complexity for an n-bit QCSA circuit is $18N\sigma + 31N\Omega$ where $N = n/4$.

Example 6.2. When $n = 4$, the total quantum gate calculation complexity of a four-bit QCSA circuit is $(18 \times 1)\sigma + (31 \times 1)\Omega = 18\sigma + 31\Omega$, where $N = n/4 = 4/4 = 1$.

Property 6.4. An n-bit QCSA circuit requires a $2450N$ Å area, where Å is the unit of area and $N = n/4$.

Proof. From property 6.2 it is found that a QCSA circuit consists of quantum ripple carry adder circuit, a QCC, and a quantum 2-to-1 MUX circuit. A one-bit quantum ripple carry adder circuit requires 6Δ quantum gates. Thus the total number of quantum gates for an n-bit quantum ripple carry adder circuit is $6n$. Moreover, a QCC and a quantum 2-to-1 MUX require 16 and 9Δ quantum gates, respectively. Thus the area for a four-bit QCSA circuit is $((6 \times 4 + 16 + 9) \times 50)$ Å = 2450 Å. Therefore, the total area for an n-bit QCSA circuit is $2450N$ Å, where $N = n/4$.

Property 6.5. An n-bit QCSA circuit produces $6972.7N$ meV power, where meV is the unit of power and $N = n/4$.

Proof. As an n-bit QCSA has $6n$ quantum gates for a quantum ripple carry adder, 16 quantum gates for a QCC, and nine quantum gates for a quantum MUX, as described in property 6.1, the power for a four-bit QCSA circuit is $((6 \times 4 + 16 + 9) \times 142.3)$ meV = 6972.7 meV.

Therefore, the total power for an n-bit QCSA circuit is $6972.7N$ meV, where $N = n/4$.

Property 6.6. An n-bit QCSA circuit can produce $14N$ garbage outputs, where $N = n/4$.

Proof. The above statement is proved by mathematical induction.

In a four-bit QCSA, a quantum ripple carry adder does not produce any garbage, but each QCC and quantum MUX produces ten and four garbage outputs, respectively, as shown in figure 6.3. Thus a four-bit QCSA produces $G_4 = G_{QCG} + G_{QMG} = 10 + 4 = 14$ garbage outputs. Thus the statement holds for base case $n = 4$.

Assume that the statement holds for $n = k$. Thus a k-bit QCSA circuit produces $14K$ garbage outputs, where $K = k/4$.

A $(k + 1)$-bit QCSA requires $(k + 1)$ QCSA blocks. As a result, the total number of garbage outputs produced for a $(k + 1)$-bit QCSA is $14 \times (K + 1)$, where $K = k/4$. Thus the statement holds for $n = k + 1$.

Therefore, an n-bit QCSA circuit can be realized with $14N$ garbage outputs, where $N = n/4$.

Example 6.3. When $n = 4$, see figure 6.5, the total garbage outputs of a four-bit QCSA are as follows:

$(14 \times 1) = 14$, where $N = n/4 = 4/4 = 1$.

6.5 The quantum BCD adder

A BCD adder comprises three units, a four-bit binary adder, an over-9-detection unit, and a correction unit. The first part is a binary adder which performs on two four-bit BCD digits and a one-bit carry input. The second part, the over-9-detector, recognizes if the result of the first part is more than 9 or not. Finally, in the third part, if the output of the detector overflow detection bit (ODB) is 1 the sum is added to by 6, otherwise do nothing. The four-bit binary adder is the cascade of a four-bit carry–propagate adder. The detection part is constructed using two AND gates and one OR gate. The correction unit adds 0 to the binary number if the binary result is less than 10 and adds 6 to the binary result if it is more than 9.

Let A and B be two one-digit BCD numbers. The BCD addition procedure of these numbers is shown in figure 6.6(a) and the detailed architecture of a conventional or irreversible BCD adder is shown in figure 6.6(b).

$A_3A_2A_1A_0$ and $B_3B_2B_1B_0$ are the two BCD numbers to be added and the resulting number is represented by $Z_3Z_2Z_1Z_0$. The carry out is represented by $C_4C_3C_2C_1$. C_4 is 1 when the resulting number is greater than binary 1111, i.e. decimal 15. As the single digit of the BCD number goes up to 9 (binary 1001), the maximum sum of the one-digit two BCD numbers is 19 (= 9 + 9 + 1), shown in table 6.1.

In this section the quantum BCD adder is presented. Four quantum full-adders can be used to design a four-bit quantum full-adder (the quantum full-adder is described in section 4.1.1). Two quantum Fredkin gates and one quantum Toffoli gate can be used as the overflow detection unit. Finally, one quantum Peres gate, one quantum full-adder, and one quantum Ex-OR gate can be used to construct the correction unit. The whole design of the one-digit quantum BCD adder is shown in figure 6.7.

A:	**0111**	**7**
B:	**0101**	**5**
A+B:	**0 1100**	**0 12 if the sum is greater than 9**
	0110	**6 then add 6 to the sum**
	1 0010	**1 2**
Sum:	**0010**	**BCD representation**
C_{out}:	**1**	**BCD overflow**

(a)

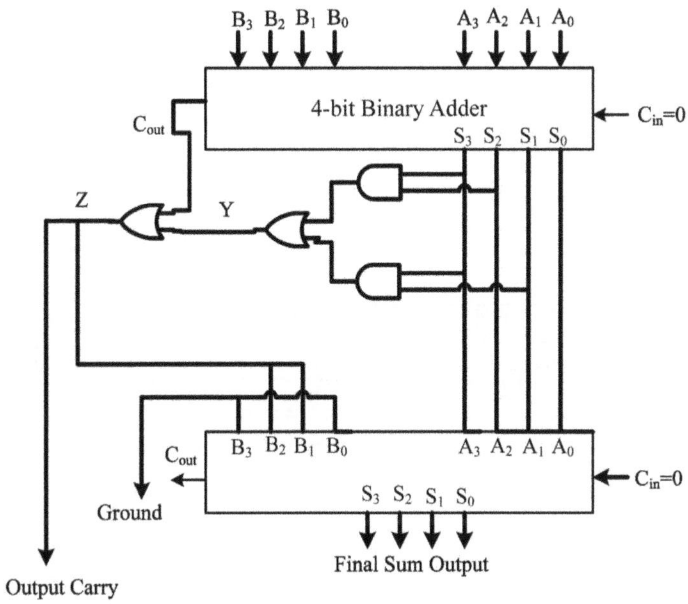

(b)

Figure 6.6. BCD addition.

Property 6.7. Let g_{pa} be the minimum number of garbage outputs for a quantum four-bit parallel adder, g_{od} be the minimum number of garbage outputs produced by the overflow detection circuit, and g_{ocl} be the minimum number of garbage outputs for overflow correction logic. Let g_{BCD} be the number of garbage outputs for a quantum BCD adder, then $g_{BCD} \geqslant g_{pa} + g_{od} + g_{ocl}$, where $g_{pa} \geqslant 8$, $g_{od} \geqslant 0$, and $g_{ocl} \geqslant 2$.

Proof. A quantum BCD adder consists of a four-bit quantum parallel adder, overflow detection logic, and overflow correction logic circuits. A four-bit parallel

Table 6.1. Binary sum of two BCD numbers.

BCD_1	BCD_2	Binary sum	
$A_3A_2A_1A_0$	$B_3B_2B_1B_0$	$Z_3Z_2Z_1Z_0$	
0 0 0 0	0 0 0 0	0 0 0 0	
.	.	.	
.	.	.	
1 0 0 1	1 0 0 1	1 0 0 1 0	(decimal 18, if $\text{carry}_{\text{in}} = 0$)
1 0 0 1	1 0 0 1	1 0 0 1 1	(decimal 19, if $\text{carry}_{\text{in}} = 1$)

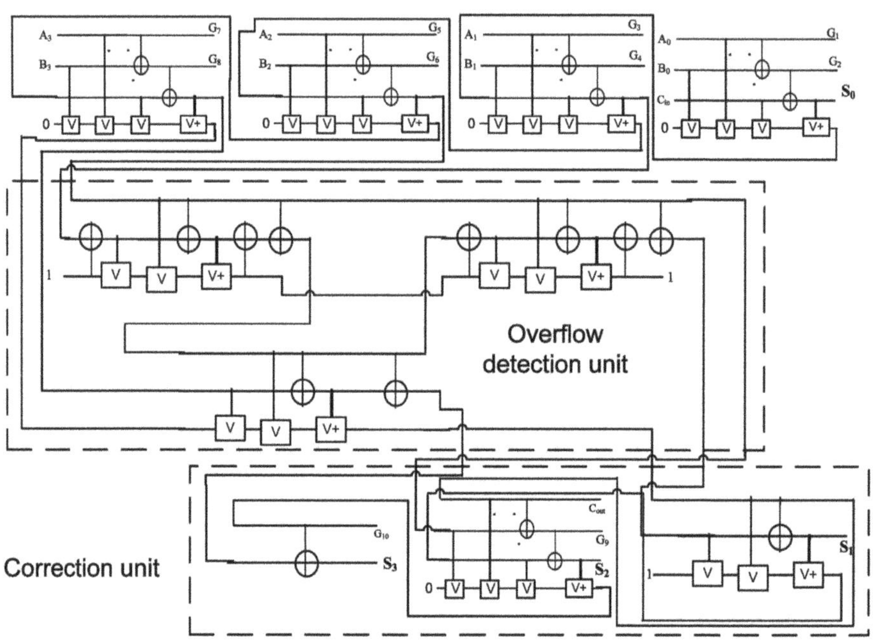

Figure 6.7. The one-bit quantum BCD adder.

adder consists of four full-adders and according to figure 6.7 a four-bit parallel adder will be realized with at least eight garbage outputs. Thus the minimum number of garbage outputs for a quantum four-bit parallel adder is $g_{\text{pa}} = 8$.

In the overflow detection logic circuit, the overflow expression, $F = (Z_1 + Z_2)Z_3 \oplus C_4$ is realized and, according to figure 6.7, a quantum BCD overflow detection unit can be realized using at least zero garbage outputs. Thus, the minimum number of garbage outputs for an overflow detection unit $g_{\text{od}} \geqslant 0$.

In the overflow correction logic, overflow F is propagated. According to the design presented in this section, the overflow correction logic generates only two garbage outputs. Thus the minimum number of garbage outputs for overflow correction logic is $g_{\text{ocl}} \geqslant 2$.

As a result, the total number of garbage outputs for a quantum BCD adder is $g_{BCD} \geqslant g_{pa} + g_{od} + g_{ocl}$, where $g_{pa} \geqslant 8$, $g_{od} \geqslant 0$, and $g_{ocl} \geqslant 2$.

Property 6.8. Let gt_{pa} be the minimum number of quantum gates for a quantum four-bit parallel adder, gt_{od} be the minimum number of quantum gates required by the overflow detection circuit and gt_{ocl} be the minimum number of quantum gates for the overflow correction logic circuit. Let gt_{BCD} be the total number of quantum gates for a quantum BCD adder, then

$gt_{BCD} = gt_{pa} + gt_{od} + gt_{ocl}$, where $g_{pa} = 24$, $g_{od} = 15$, and $g_{ocl} = 11$.

Proof. A quantum BCD adder consists of four-bit quantum parallel adder, overflow detection logic, and overflow correction logic circuits. A four-bit parallel adder consists of four quantum full-adders and each quantum full-adder is realized by six quantum gates. Thus the minimum number of gates for a quantum four-bit parallel adder is $gt_{pa} = 4 \times 6 = 24$.

In the overflow detection logic, three blocks are used, each of which has five quantum gates. Thus the minimum number of gates for the overflow detection logic is $gt_{od} = 15$.

In the overflow correction logic, one quantum Peres gate, one quantum full-adder, and one quantum Ex-OR gate are use to construct the correction unit. The quantum Peres gate consists of four quantum gates, and one quantum full-adder consists of six quantum gates. Thus the minimum number of quantum gates for the overflow correction logic is $gt_{ocl} = 4 + 6 + 1 = 11$.

As a result, the total number of quantum gates for a quantum BCD adder is $gt_{BCD} = gt_{pa} + gt_{od} + gt_{ocl}$, where $g_{pa} = 24$, $g_{od} = 15$, and $g_{ocl} = 11$.

6.6 Summary

This chapter presents the design methodology of an n-bit quantum carry skip adder circuit with optimal delay and a quantum BCD adder. An efficient algorithm is shown for designing a compact quantum carry skip adder. The BCD adder is designed using quantum full-adder circuits. The efficiency of the design is proved by presenting several properties. It is shown that the circuit has been constructed with the optimum number of quantum gates, delays, number of garbage outputs, quantum gate calculation complexity, area, and power. The adder is the basic arithmetic unit of a computer system and it can be used in diverse areas, such as complex quantum arithmetic circuits.

The design which is presented in this chapter will help the reader understand the next chapter better. In addition, the concept of designing a quantum carry skip adder will help in designing quantum multiplier–accumulator circuits.

Further reading

[1] Álvarez-Sánchez J J, Álvarez-Bravo J V and Nieto L M 2008 A quantum architecture for multiplying signed integers *J. Phys.: Conf. Series* **128** 012013

[2] Babu H M H and Chowdhury A R 2005 Design of a reversible binary coded decimal adder by using reversible 4-bit parallel adder *18th Int. Conf. on VLSI Design and 4th Int. Conf. on Embedded Systems Design* pp 255–60

[3] Babu H M H and Chowdhury A R 2006 Design of a compact reversible binary coded decimal adder circuit *J. Syst. Archit.* **52** 272–82

[4] Balandin A A and Wang K L 1999 Implementation of quantum controlled-not gates using asymmetric semiconductor quantum dots *Quantum Computing and Quantum Communications* (Berlin: Springer) pp 460–7

[5] Biswas A K, Hasan M M, Chowdhury A R and Babu H M H 2008 Efficient approaches for designing reversible binary coded decimal adders *Microelectron. J.* **39** 1693–703

[6] Biswas A K, Hasan M M, Hasan M, Chowdhury A R and Babu H M H 2008 A novel approach to design BCD adder and carry skip BCD adder *21st Int. Conf. on VLSI Design (VLSID 2008)* pp 566–71

[7] Bose A and Babu H M H 2015 Optimized designs of reversible fault tolerant BCD adder and fault tolerant reversible carry skip BCD adder *18th Int. Conf. on Computer and Information Technology (ICCIT)* pp 202–7

[8] Bose A, Babu H M H and Gupta S 2015 Design of compact reversible online testable ripple carry adder *IEEE Int. WIE Conf. on Electrical and Computer Engineering (WIECON-ECE)* pp 556–60

[9] Bruce J W, Thornton M A, Shivakumaraiah L, Kokate P S and Li X 2002 Efficient adder circuits based on a conservative reversible logic gate *Proc. IEEE Computer Society Annual Symp. on VLSI* (Piscataway, NJ: IEEE) pp 83–8

[10] Guyot A, Hochet B and Muller J-M 1987 A way to build efficient carry-skip adders *IEEE Trans. Comput.* **10** 1144–52

[11] Haghparast M 2011 Design and implementation of nanometric fault tolerant reversible BCD adder *Australian J. Basic Appl. Sci.* **5** 896–901

[12] Hamacher V C, Vranesic Z G and Zaky S G 2002 Computer Organization (New York: McGraw-Hill)

[13] Haque M U, Sworna Z T, Babu H M H and Biswas A K 2018 A fast FPGA-based BCD adder *Circ. Syst. Signal Process.* **37** 4384–408

[14] Hung W N, Hung X, Song G, Yang J and Perkowski M 2006 Optimal synthesis of multiple output Boolean functions using a set of quantum gates by symbolic reachability analysis *IEEE Trans. Comput.-Aided Des. Integr. Circuits Syst.* **25** 1652–63

[15] Islam M *et al* 2010 Fault tolerant variable block carry skip logic (VBCSL) using parity preserving reversible gates (arXiv:1009.3819)

[16] Islam M S, Rahman M M, Begum Z and Hafiz M Z 2009 Fault tolerant reversible logic synthesis: carry look-ahead and carry-skip adders *Int. Conf. on Advances in Computational Tools for Engineering Applications* (Piscataway, NJ: IEEE) pp 396–401

[17] Lin Y S and Radhakrishnan D 2008 Delay efficient 32-bit carry-skip adder *VLSI Des.* 9

[18] Lisa N J and Babu H M H 2014 A compact realization of an *n*-bit quantum carry skip adder circuit with optimal delay *2014 NASA/ESA Conf. on Adaptive Hardware and Systems* pp 270–7

[19] Lisa N J and Babu H M H 2015 Design of a compact ternary parallel adder/subtractor circuit in quantum computing *IEEE Int. Symp. on Multiple-Valued Logic* pp 36–41

[20] Mano M M and Kime C R 2008 *Logic and Computer Design Fundamentals* (Upper Saddle River, NJ: Pearson Prentice Hall)

[21] Mohammadi M M, Eshghi M, Haghparast M and Bahrololoom A 2008 Design and optimization of reversible BCD adder/subtractor circuit for quantum and nanotechnology based systems *World Appl. Sci. J.* **4** 787–92

[22] Mohammadi M, Haghparast M, Eshghi M and Navi K 2009 Minimization and optimization of reversible BCD-full adder/subtractor using genetic algorithm and don't care concept *Int. J. Quantum Inform.* **7** 969–89

[23] Nagendra C, Irwin M J and Owens R M 1996 Area-time-power tradeoffs in parallel adders *IEEE Trans. Circuits Syst.* **43** 689–702

[24] Pang Y, Wang J and Wang S 2012 A 16-bit carry skip adder designed by reversible logic *5th Int. Conf. on Biomedical Engineering and Informatics (BMEI)* (Piscataway, NJ: IEEE) pp 1332–5

[25] Parida D 2010 A novel high speed CMOS comparator with low power dissipation and low offset *PhD Thesis*

[26] Thapliyal H and Vinod A P 2006 Transistor realization of reversible TSG gate and reversible adder architectures *IEEE Asia Pacific Conf. on* (Piscataway, NJ: IEEE) pp 418–21

[27] Viamontes G F, Markov I L and Hayes J P 2005 High-performance quantum circuit simulation *QuIDDPro* http://vlsicad.eecs.umich.edu/Quantum/qp/ (Accessed: 19 March 2020)

[28] Yao A C-C 1993 Quantum circuit complexity *Proc. IEEE 34th Annual Foundations of Computer Science* pp 352–61

Chapter 7

The quantum multiplier–accumulator

All quantum signal processing (QSP) algorithms use the multiplier–accumulator (MAC) operation for high-performance quantum processing systems. This operation is needed in filters, Fourier transforms, etc, since it eases the computation of convolution. An MAC unit comprises a multiplier, an adder, and an accumulator. The multiplier multiplies the inputs and gives the result to the adder, which adds the multiplier result to the previously accumulated result. An MAC unit is used to perform the multiplication and accumulator operations together to avoid unnecessary overhead on the processor in terms of processing time and on-chip memory requirements.

High-speed multiplication has always been a fundamental requirement of high-performance processors and systems. In QSP applications, multiplication is one of the most utilized arithmetic operations. Improving multiplier design directly benefits the high-performance embedded processors and QSP applications used in consumer and industrial electronic products. Moreover, quantum information processing (QIP) is a high-impact research area in quantum information science for constructing a quantum computer. The main goal of QIP is to harness the fundamental laws of quantum mechanics to improve all aspects (e.g. acquisition, transmission, and processing) of information processing dramatically, as well as enhance the performance of quantum computers. There are several tasks for which a quantum computer will be useful. The one that is mentioned most frequently is that quantum computers will be able to read secret messages communicated over the Internet using current technologies such as RSA, Diffie–Hellman, and other cryptographic protocols that are based on the difficulty of number-theoretic problems such as factoring and discrete logarithms. In addition, quantum computers are also useful for scientists for conducting virtual experiments and searching in huge amounts of data.

doi:10.1088/978-0-7503-2747-3ch7

7.1 The importance of the quantum multiplier–accumulator

The scope of the optimization of time complexity as well as the optimization of garbage outputs and constant inputs has been addressed, which has motivated the introduction of a completely new and efficient technique of multiplication for quantum multipliers with optimum time complexity. Time complexity reduction also minimizes the number of operations and hardware complexity. It is necessary to introduce an efficient and faster multiplication algorithm to optimize time complexity. Thus a new, faster, more efficient, and optimal tree-based multiplication technique which works like the breadth fast search (BFS) algorithm is introduced in this chapter.

To implement the quantum multiplier using the quantum multiplication technique, a new quantum full-adder (QFA) circuit and a new quantum ANDing circuit (QAC) are introduced with the fewest number of gates and less area, power, and gate level delay. In this chapter, two algorithms for the construction of quantum partial product generation (PPG) and partial product addition (PPA) circuits have been shown using QFA circuits and QACs that work in the manner of a heuristic algorithm, and the divide and conquer algorithms, respectively, and a combination of them implements the quantum multiplier using the quantum multiplication technique and reduces the number of quantum gates, garbage outputs, and constant inputs, and the amount of area, power, and delay compared to the best known multipliers.

A large number of ancillary inputs and garbage outputs results in increased area, delay, power, and quantum cost. Thus it is important to design the quantum multiplier as well as the quantum MAC unit in such a way that they reduce the ancillary inputs and garbage outputs. For this purpose, in this chapter a new algorithm for designing the quantum MAC unit is described which is similar to that of the greedy algorithm. A quantum MAC unit is formed by combining the quantum multiplier, full-adder, and the quantum accumulator circuits. Thus the quantum MAC circuit has reduced ancillary inputs, garbage outputs, delay, area, power, quantum costs, and implementation complexities. Furthermore, this circuit is compact and efficient and requires a minimum number of quantum gates. In addition, this chapter presents a very novel and unparalleled design of a quantum multiplier as well as a quantum MAC unit that reduces the garbage outputs and ancillary inputs.

7.2 The multiplication technique

The multiplier is a very important device for arithmetic logic units. With advances in technology, researchers are trying to design multipliers with high speed and low power consumption to make them suitable for various high-speed, low power, and compact VLSI implementations. To achieve high speed and low power consumption in a multiplier, a tree-based multiplication technique is shown.

Property 7.1. A *tree* is a finite set T such that:

1. One node $r \in T$ is designated as the root.
2. The remaining nodes are partitioned into disjoint sets $T_1, T_2, \dots, T_k$, where each of the sets is a tree and $k \geqslant 0$.

Property 7.2. In a tree, each node can have any number of child nodes.

Property 7.3. The *partial product addition tree* is a tree consisting of nodes in levels where the nodes in each level contain the products and their additions.

Property 7.4. A *node* of a tree contains the information or data. A node may be a child or parent node.

Property 7.5. The *level* of a node is the distance of that node from the root where the distance between two vertically adjacent nodes is one.

Example 7.1. A tree, and its child and parent node with the root and level are depicted in figure 7.1. In this figure node A is at the top of the tree and is called the root. Nodes which are just the next level of the root (B, C, and D in figure 7.1) are the parent node and nodes at the bottom level are called the leaf or child nodes of the previous parent node (I and J are the child nodes of the parent node D, as shown in figure 7.1). To go to node E from root node A of the tree of figure 7.1, first we need to go to node B from root node A and then we go to node E from node B. The distance from root node A to node B is 1 and from node B to node E is 1. Thus the total distance of node E from root node A is 2. Therefore, node E is at level 2.

Property 7.6. The set of qubits that is to be multiplied with other qubits is called the *multiplicand.*

Property 7.7. The qubits which are multiplied with the qubits of the multiplicand are called the *multiplier.*

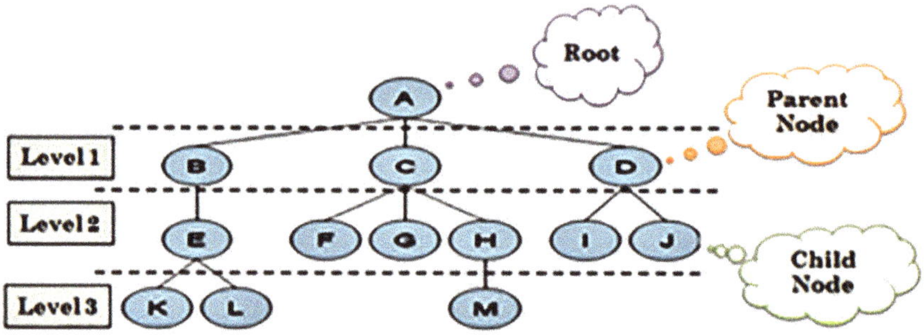

Figure 7.1. Examples of tree, root, node, and level. Reproduced with permission from [4]. Copyright 2017 Springer (GmbH).

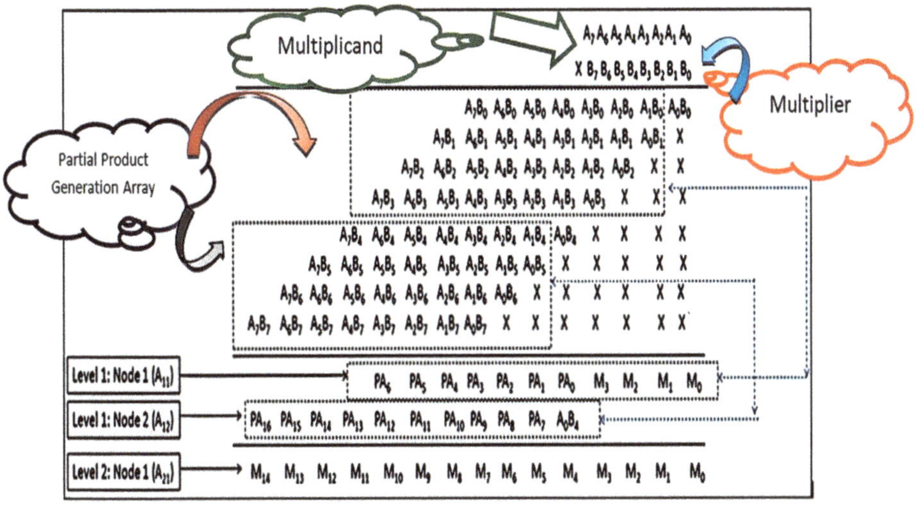

Figure 7.2. An example of the multiplication method. Reproduced with permission from [4]. Copyright 2017 Springer (GmbH).

Example 7.2. The multiplier and multiplicand qubits are depicted in figure 7.2. $|A7\rangle\,|A6\rangle\,|A5\rangle\,|A4\rangle\,|A3\rangle\,|A2\rangle\,|A1\rangle\,|A0\rangle$ is the set of multiplicand qubits and $|B7\rangle\,|B6\rangle\,|B5\rangle\,|B4\rangle\,|B3\rangle\,|B2\rangle\,|B1\rangle\,|B0\rangle$ is the set of multiplier qubits, as shown in figure 7.2.

Quantum logic preserves the properties of the linear combination of multiple states (superposition), whereas the binary logic preserves only the properties of two states. Thus a multiplication algorithm based on the superposition states of quantum computing is shown. However, the multiplication technique is a generalized algorithm which is applicable for both quantum and classical computations.

The technique has two basic steps. First, partial products are generated in such a way that the copying of multiplicand qubits, multiplier qubits, and ANDing operations are performed in parallel, which results in the faster generation of partial products. Second, the partial product additions are performed using a tree structure, where the tree has $((\log_2 n) - 1)$ levels. In the first layer of the tree, four successive rows of the partial product array are added by adders in parallel. In the following levels, the rate of increase of the number of adders will be $(n + 2 + i)$, where $i = 1, 2, \ldots, ((\log_2 n) - 2)$ and n is the number of qubits. This technique is described by algorithm 7.1. The working principle of algorithm 7.1 is presented using example 7.3.

Algorithm 7.1. Multiplication technique.

1: START

2: Create a partial product generation array where the ANDing and copying operations are performed in parallel to obtain a compact quantum partial product generation circuit.

3: Create a partial product addition tree with $((\log_2 n) - 1)$ levels.

4: Generate $n/4$ nodes at the first level of the partial product addition tree, where each node represents the addition of the successive four rows of the partial product array.
5: Define the next levels of the tree where the addition operations are performed between two predecessor nodes in which the rate of increase of the number of adders will be $(n + 2 + i)$ and $i = 1, 2, \ldots, ((\log_2 n) - 2)$.
6: Obtain the multiplication result in the $((\log_2 n) - 1)$th level of the tree.
7: END

Example 7.3. The method is described using the following example.

To multiply two eight-qubit numbers, such as $\{|A7\rangle\,|A6\rangle\,|A5\rangle\,|A4\rangle\,|A3\rangle\,|A2\rangle\,|A1\rangle\,|A0\rangle\}$ (multiplicand qubits) and $\{|B7\rangle\,|B6\rangle\,|B5\rangle\,|B4\rangle\,|B3\rangle\,|B2\rangle\,|B1\rangle\,|B0\rangle\}$ (multiplier qubits), the partial product generation and partial product addition processes using the designed method for the given numbers are shown in figure 7.2 and 7.3, respectively. In figure 7.2, the partial products are generated using the partial product generation array and the symbol 'x' indicates the absence of the partial product term. In this figure A_iB_i indicates the partial product terms, PA_i is the result of the addition of the partial products, and M_i is the final multiplication result. Since in this example the number of qubits is $n = 8$, the partial product addition tree requires $((\log_2 n) - 1) = 2$ levels as shown in figure 7.3. At the first level, the partial products are added in two different nodes A_{11} and A_{12}, and at the next level there is a node A_{21} in which partial addition results are added to form the final result of the multiplication, as shown in figure 7.3 where the symbol '+' inside a node indicates the partial product addition operation.

Property 7.8. The time complexity of the multiplication approach is at least $O((3\log_2 n) + 1)$, where n is the number of multiplicand and multiplier qubits.

Proof. Property 7.8 is proved by mathematical induction. For an n-qubit multiplier and multiplicand, a total of $\log_2 n^2$ ANDing and copying operations are required in

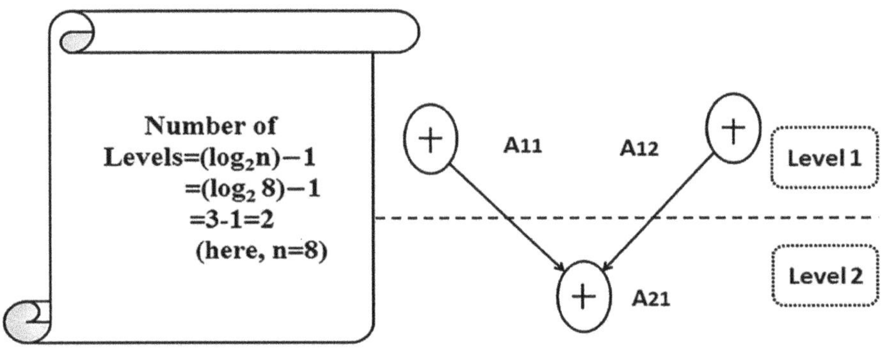

Figure 7.3. Partial product addition using the tree structure. Reproduced with permission from [4]. Copyright 2017 Springer (GmbH).

the partial product generation array. Thus the time for the ANDing and copying operations is $T(A) = O(\log_2 n^2)$.

To generate the partial product addition tree and to perform the addition of the partial products in the partial product addition tree, $\log_2 2n$ operations are required when the multiplier and multiplicand are n qubits. Thus the total time required to generate the partial product addition tree and perform the partial product addition is

$$T(P) = O(\log_2 2n).$$

When $n = 1$, we have the one-qubit multiplier and multiplicand. Thus the time required for this multiplication is

$$T(1) = 1 = O(\log_2 n^2 + \log_2 2n) = O((3\log_2 1) + 1).$$

Therefore, the statement holds for the base case $n = 1$.

Assume that the statement is true for $n = k$, we have the k-qubit multiplier and multiplicand. Thus the time for this multiplication is $T(k) = O(\log_2 k^2 + \log_2 2k) = O((3\log_2 k) + 1)$.

When $n = (k + 1)$, i.e. we have the $(k + 1)$-qubit multiplier and multiplicand and the total required time for this multiplication is

$$T(k + 1) = O(\log_2 (k + 1)^2 + \log_2 2(k + 1)) = O((3\log_2 (k + 1)) + 1).$$

Therefore, the statement is true for $n = (k + 1)$.

Thus, for all n, the time required for the multiplication process of an n-qubit multiplicand and multiplier is

$$\begin{aligned} T(n) &= T(A) + T(P) = O(\log_2 n^2) + O(\log_2 2n) \\ &= O(2\log_2 n + \log_2 n + \log_2 2) = O(3\log_2 n + 1). \end{aligned}$$

Thus the time complexity of the multiplication approach is at least $O((3\log_2 n) + 1)$.

It is important to note that this multiplication algorithm is also suitable for quantum computation. In quantum computation the smallest particles are known as qubits, which are always in motion and can be superpositioned at any time. Therefore, the execution time of any quantum computation becomes random and probabilistic due to the superposition of the particles. Whenever the quantum particles are in the 'cat state' (a special pure quantum state, where the quantum particles are in a fixed equal superposition of all being $|0\rangle$ and $|1\rangle$, i.e. $|00\ldots0\rangle + |11\ldots1\rangle$), one can compute the execution time when performing any quantum operation, which in turns measures the time complexity of the quantum operation. As a result, in the case of quantum computation the time complexity of the multiplication method will be similar when both the multiplicand qubits and multiplier qubits are in the 'cat state'.

Property 7.9. The multiplication algorithm holds multiple states of the superposition of quantum logic.

Proof. In the multiplication algorithm, the partial product addition is performed using a tree-based structure. Let T be the partial product addition tree of $|\phi\rangle$ where the qubit $|\phi\rangle$ is used to denote a multiplier or a multiplicand qubit. By applying an addition the operator A_i on the ith qubit simply changes the superposition $\alpha|0\rangle_i + \beta|1\rangle_i$ (where α and β are the probabilities of being the $|0\rangle$ and $|1\rangle$ states, respectively) of that qubit located at any arbitrary leaf (nodes at the bottom level of the tree) to a different superposition $\alpha'|0\rangle_i + \beta'|1\rangle_i$.

Thus the individual sum of the partial products can be represented by the partial product addition tree (T) using algorithm 7.1, which holds the multiple states of the superposition of quantum logic.

A quantum multiplication circuit is designed using the multiplication method, which is shown in section 7.4.

7.3 Reduction of the garbage outputs and ancillary inputs of quantum circuits

Quantum computers of many qubits are extremely difficult to realize. This sets the major objective of optimizing the number of ancillary (constant) input qubits and the number of garbage outputs in quantum circuits. In other words, in the case of choosing between increasing the garbage outputs and increasing the number of gates in a quantum circuit implementation, preference should be given to the design method that delivers the minimum amount of garbage outputs, as reducing garbage outputs and constant inputs in turn improves the logical complexity and quantum costs of the circuits. Sometimes in quantum circuits a huge number of ancillary (constant) inputs leads to an increase in the size (area) of the circuit as well as gate level delays and power consumption. This is, in particular, a problem in quantum computation, since in quantum circuits input lines are a highly limited resource (caused by the fact that the number of circuit lines corresponds to the number of qubits). Furthermore, if the number of constant inputs in quantum circuits increase, this may decrease the reliability of the circuits. Thus keeping the number of constant inputs as small as possible is an important issue. As the optimization of ancillary input bits and the garbage outputs may produce a compact circuit in terms of the quantum cost, power, and delay, it is always the primary focus to optimize the number of ancillary input bits and the garbage outputs when designing a circuit.

Different minimization algorithms and tree-based designs are given to decrease the number of ancillary inputs and garbage outputs. In addition to the algorithms, a new quantum adder circuit, quantum ANDing circuit, and quantum multiplier circuit are used in order to reduce the ancillary inputs and garbage outputs in the final circuit.

This solution enables us to overcome the overheads of increasing gate level delay, quantum cost, area, and power due to decreasing the number of garbage outputs and constant inputs in quantum circuits.

7.4 The design of a quantum multiplier circuit

In this section a quantum multiplier circuit is designed using a multiplication algorithm. The techniques of fast multiplication as well as the quantum ANDing circuit and quantum full-adder, and the design of a compact quantum multiplier and quantum $n \times n$ multiplier circuit are presented in this subsection.

7.4.1 The quantum ANDing circuit

In this subsection a new 3×3 QAC is shown. The input vector (I_v) and the output vector (O_v) of the circuit are as follows:

$$|I_v\rangle = \{A\rangle, |B\rangle, |C\rangle\}; \text{ and}$$
$$|O_v\rangle = \{A\rangle, |AC \oplus B\rangle, |C\rangle\}.$$

Figure 7.4 shows a diagram of the 3×3 QAC and the circuit can also be realized by the decomposition of the quantum Toffoli gate. In the QAC, five basic quantum gates are used. Thus the quantum cost is 5. The output $|Q\rangle$ is the result of the ANDing of qubits $|A\rangle$ and $|C\rangle$. The truth table corresponding to the QAC circuit is shown in table 7.1. It can be verified from the truth table that the input pattern corresponding to a particular output pattern can be uniquely determined. To perform the 'AND' operation using a QAC, the $|B\rangle$ input qubit in figure 7.4 has to remain $|0\rangle$. Thus the QAC can be used to design the partial product generation circuit.

7.4.2 The quantum full-adder circuit

A full-adder is a circuit that performs the arithmetic sum of three input bits. It consists of three inputs and two outputs. If we take three input variables A, B, C_{in} and two output variables S (sum) and C_{out} (carry), the following equations can be used to obtain the outputs:

$$S = A \oplus B \oplus C_{\text{in}}$$
$$C_{\text{out}} = AB \oplus BC_{\text{in}} \oplus C_{\text{in}}A.$$

Full-adders are versatile and widely used building blocks in quantum arithmetic circuits. In this subsection a new 4×4 QFA circuit is shown.

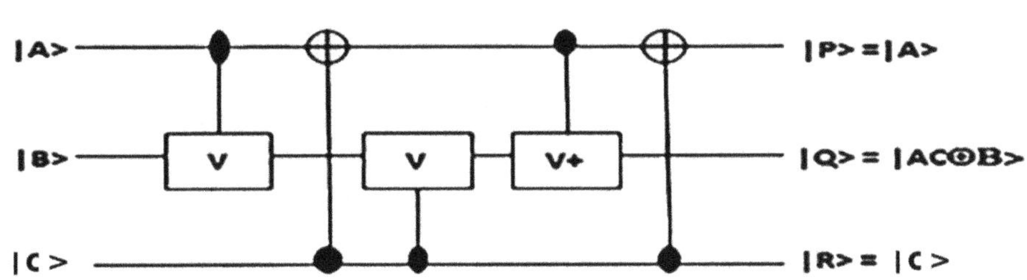

Figure 7.4. Quantum circuit for ANDing operation. Reproduced with permission from [4]. Copyright 2017 Springer (GmbH).

Table 7.1. Reversibility of a 3 × 3 QAC.

Input			Output		
A	B	C	P	Q	R
0	0	0	0	0	0
0	0	1	0	0	1
0	1	0	0	1	0
0	1	1	0	1	1
1	0	0	1	0	0
1	0	1	1	1	1
1	1	0	1	1	0
1	1	1	1	0	1

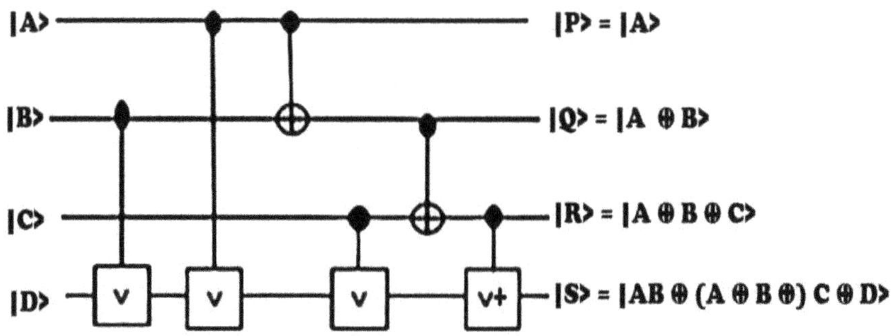

Figure 7.5. The quantum full-adder circuit, when $|D\rangle = |0\rangle$. Reproduced with permission from [4]. Copyright 2017 Springer (GmbH).

The input vector I_v and the output vector O_v of the circuit are as follows:

$$|I_v\rangle = \{|A\rangle, |B\rangle, |C\rangle, |D\rangle\}; \text{ and}$$
$$|O_v\rangle = \{A\rangle, |A \oplus B\rangle, |A \oplus B \oplus C\rangle, |AB \oplus (A \oplus B)C \oplus D\}.$$

Figure 7.5 shows the diagram of the 4 × 4 QFA circuit. In this figure, the output $|R\rangle$ is the S (sum) and $|S\rangle$ is the C_{out} (carry). The number of quantum gates of the QFA circuit is six and so the quantum cost is 6. The truth table corresponding to the QFA circuit is shown in table 7.2. It can be verified from the truth table that the input pattern corresponding to a particular output pattern can be uniquely determined. The QFA circuit is designed in such a way that it can be efficiently used as a full-adder by setting the fourth input qubit as a constant zero $|0\rangle$. Algorithm 7.2 describes the design of the QFA circuit as a full-adder.

Table 7.2. Reversibility of the 4 × 4 QFA circuit.

Input				Output			
A	*B*	*C*	*D*	*P*	*Q*	*R*	*S*
0	0	0	0	0	0	0	0
0	0	0	1	0	0	0	1
0	0	1	0	0	0	1	0
0	0	1	1	0	0	1	1
0	1	0	0	0	1	1	0
0	1	0	1	0	1	1	1
0	1	1	0	0	1	0	1
0	1	1	1	0	1	0	0
1	0	0	0	1	1	1	0
1	0	0	1	1	1	1	1
1	0	1	0	1	1	0	1
1	0	1	1	1	1	0	0
1	1	0	0	1	0	0	1
1	1	0	1	1	0	0	0
1	1	1	0	1	0	1	1
1	1	1	1	1	0	1	0

Algorithm 7.2 Construction of the quantum full-adder.

Input: $|A\rangle$, $|B\rangle$, $|C_{\text{in}}\rangle$ Output: $|\text{Sum}\rangle$, $|\text{Carry}\rangle$
1: START
2: Take an QFA circuit ($|A\rangle$, $|B\rangle$, $|C\rangle$, $|D\rangle$)($|P\rangle$, $|Q\rangle$, $|R\rangle$, $|S\rangle$)
3: Create a partial product addition tree with $((\log_2 n) - 1)$ levels.
4: Set QFA.a = $|A\rangle$; QFA.b = $|B\rangle$; QFA.c = $|C_{in}\rangle$; QFA.d = $|0\rangle$;
5: Output $|\text{Sum}\rangle$ = QFA.r
6: END

7.4.2.1 The quantum multiplier circuit

The design of the quantum multiplier is composed of two components: the quantum PPG circuit and quantum PPA circuit. The quantum PPG circuit and quantum PPA circuit are described in the following sections. The number of constant input and garbage output lines has been reduced significantly by using quantum memory elements.

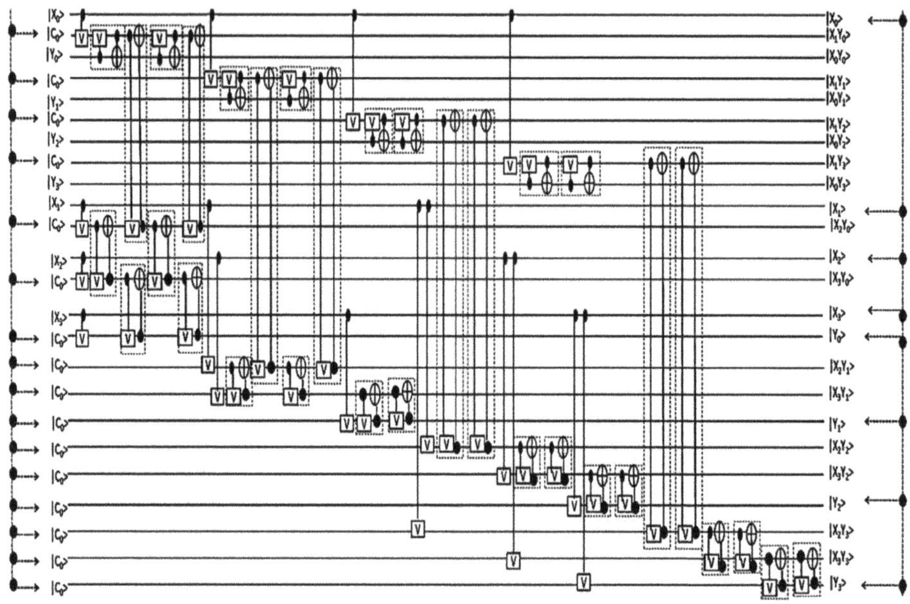

Figure 7.6. The quantum PPG circuit of a 4 × 4 multiplier. Reproduced with permission from [4]. Copyright 2017 Springer (GmbH).

7.4.2.2 The quantum partial product generation circuit

The quantum PPG which is constructed using the QACs, takes the multiplicand and multiplier qubits as inputs and produces the partial products as outputs. The multiplier qubits and the multiplicand qubits are fed to a component which produces partial products. The QAC has been used to construct this part.

Example 7.4. To construct a 4 × 4 quantum PPG circuit, sixteen QACs are needed to perform the ANDing operations for generating the partial products. In figure 7.6 the detailed quantum partial product generation circuit for a 4 × 4 multiplier is shown. In this figure the qubit $|C_0\rangle$ is used as the constant input lines, and qubits $|X_i\rangle$ and $|Y_i\rangle$ are the input qubits of the multiplicand and multiplier, respectively. The partial products denoted by the qubits $|X_i Y_i\rangle$ are the outputs of the PPG circuit, where the symbol '··· ' is used to denote the single constant input line and also the single garbage output line, which are reused in the circuit wherever necessary.

7.4.2.3 The quantum partial product addition circuit

The quantum PPA circuit adds column-wise the partial products that have been generated by the quantum PPG circuit. The PPA circuit is constructed using the QFA circuits. This construction procedure of the quantum PPA circuit is as follows.

To construct a quantum PPA circuit a tree structure is followed, consisting of levels. A PPA tree is created with $((\log_2 n) - 1)$ levels. At the first level of the PPA tree, there are $n/4$ nodes, where each node represents the addition by an adder of the successive four rows of a partial product array. In the following levels of the tree, the addition operations are performed between two predecessor nodes, where the rate of increase of the number of adders will be $(n + 2 + i)$, where $i = 1, 2, \ldots, ((\log_2 n) - 2)$ and n is the number of qubits. Finally, the multiplication result is obtained in the $((\log_2 n) - 1)$th level of the tree.

Example 7.5. To construct a 4 × 4 quantum PPA circuit, ten QFA circuits are needed to perform the addition operations for generating the PPA. In figure 7.7 the detailed quantum PPA circuit for a 4 × 4 multiplier is shown. In this figure the qubit $|C_0\rangle$ is used as the constant in the input lines. This PPA circuit takes the partial products of qubits ($|X_i Y_i\rangle$) as inputs and generates the result of addition denoted by the qubit $|S_i\rangle$ as output, where the symbol '…' is used to denote the single constant input line and also the single garbage output line, which are reused in the circuit wherever necessary.

7.4.3 The *n* × *n*-qubit quantum multiplier

A 4 × 4 quantum multiplier is constructed using QACs and QFA circuits. First, a quantum PPG circuit is constructed using a QAC to generate partial products for a 4 × 4 multiplier. The quantum PPA circuit is constructed for PPA using QFA circuits and obtains a 4 × 4 quantum multiplier. The generalized block diagrams for $n \times n$ quantum PPG and quantum PPA circuits are shown in figures 7.8 and 7.9, respectively, where the qubit $|C\rangle$ is the constant input and the qubit $|S\rangle$ stands for the

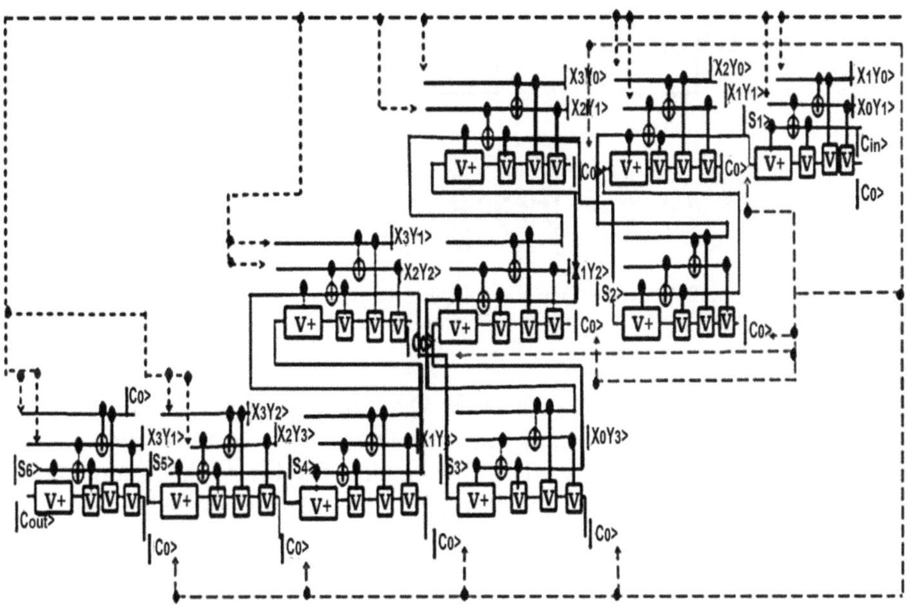

Figure 7.7. The quantum PPA circuit of a 4 × 4 multiplier. Reproduced with permission from [4]. Copyright 2017 Springer (GmbH).

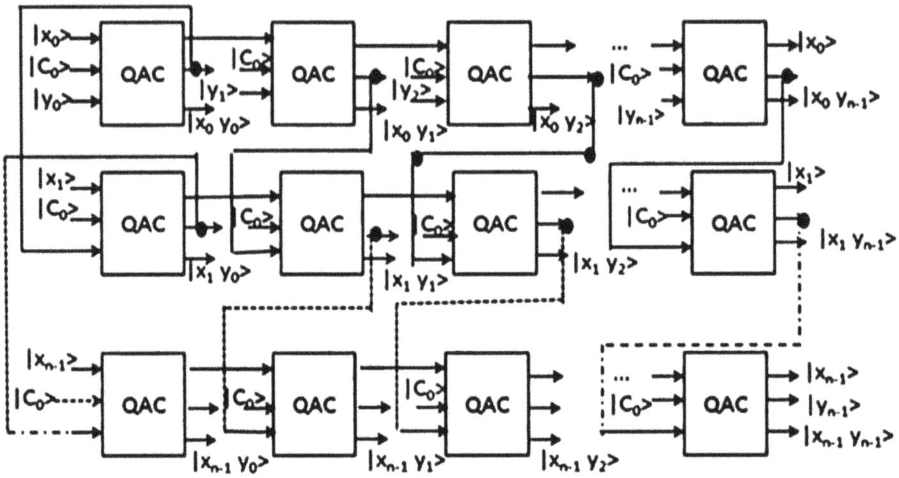

Figure 7.8. The quantum PPG circuit of an $n \times n$ multiplier. Reproduced with permission from [4]. Copyright 2017 Springer (GmbH).

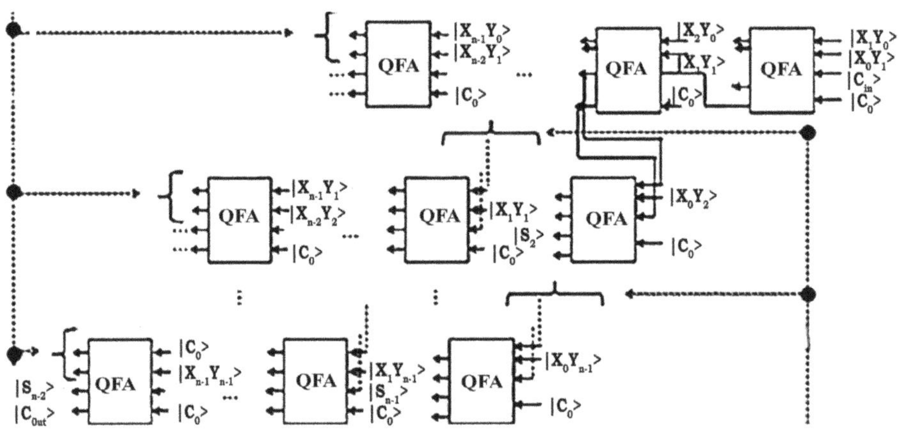

Figure 7.9. The quantum PPA circuit of an $n \times n$ multiplier. Reproduced with permission from [4]. Copyright 2017 Springer (GmbH).

sum of the qubits. In figure 7.8 each of the QAC blocks takes a multiplicand qubit $|X_i\rangle$ and a multiplier qubit $|Y_i\rangle$, and generates the product terms of qubits $|X_iY_i\rangle$ as outputs, which are carried out to the PPA circuits. The block diagram of the PPA circuit as shown in figure 7.9 is constructed by the QFA blocks where each QFA block takes two partial products of qubits $|X_iY_i\rangle$ and performs the addition of the products in which the sum ($|S\rangle$) is the output of the block. Thus the construction procedure of an $n \times n$ quantum multiplier circuit is as follows: first an $n \times n$ quantum PPG circuit is generated and then the quantum PPA circuit of an $n \times n$ multiplier is constructed using algorithm 7.3.

Algorithm 7.3. Quantum partial product addition circuit.

```
START
Divide the PPG array row-wise into (n/4) parts
for i := 0 to n/4 do
    for j := 0 to (n – 1) do
        Apply j QFA gates in j columns to obtain the addition of nodes of ith layers
    end for
    Apply QFA gates to obtain the addition between layers
end for
END
```

Property 7.10. The depth of the design of the quantum multiplier circuit is $O(9n + 14\log_2 n)$, where n is the number of qubits of the multiplicand and multiplier.

Proof. The above statement is proved by the method of mathematical induction. A 2×2 PPG and a 2×2 PPA circuit are required for constructing a 2×2 quantum multiplier circuit using the design of the tree-based multiplier, where the total number of stages (depth) in a 2×2 PPG circuit is 20 and the numbers of stages (depth) in a 2×2 PPA circuit is 12.

When $n = 2$ for a 2×2 quantum multiplier circuit, the total depth of the 2×2 quantum multiplier circuit is

$$\begin{aligned} D_{2\times 2} &= D_{\mathrm{PPG}} + D_{\mathrm{PPA}} \\ &= 20 + 12 = 18 + 14 = 9n + 14\log_2 n; \quad [\because n = 2 \text{ here}]. \end{aligned}$$

Thus the statement holds for the base case $n = 2$.

Assume that the statement also holds for $n = k$, which is a $k \times k$ quantum multiplier circuit.

Thus $D_{k\times k} = 9k + 14\log_2 k$.

Therefore, the depth of the $(k + 1) \times (k + 1)$ quantum multiplier circuit is

$$D_{(k+1)\times(k+1)} = 9(k + 1) + 14\log_2(k + 1).$$

Thus the statement is also true for $n = (k + 1)$ and thus the depth of the design of the quantum multiplier circuit is $O(9n + 14\log_2 n)$.

It is important to note that the depth of the quantum multiplier circuit is exponential or linear, whereas the depth of the quantum multiplier is logarithmic.

7.5 Summary

This chapter presents the design methodology of a novel quantum multiplier–accumulator unit using quantum multiplier and quantum accumulator circuits. As the multiplier is the major component of a multiplier–accumulator unit, in this chapter a new design methodology of the quantum multiplier device is presented using a multiplication technique. The design methodology generates the partial products using the quantum ANDing circuit and performs the addition of partial products in parallel using the quantum full-adders with reduced ancillary and garbage bits. The unique tree-based multiplication technique along with the use of quantum full-adders and quantum ANDing circuits reduces the number of ancillary and garbage bits in the quantum multiplier design. A quantum accumulator circuit is also shown which is more compact and efficient. The design methodology can be integrated as part of a quantum central processing unit (CPU), quantum signal processing, and quantum arithmetic and logic unit (ALU) optimized in terms of ancillary inputs and garbage outputs.

Further reading

[1] Ömer B QCL—a programming language for quantum computers http://tph.tuwien.ac.at/~oemer/qcl.html (Accessed: 5 December 2018)

[2] Akbar E P A, Haghparast M and Navi K 2011 Novel design of a fast reversible Wallace sign multiplier circuit in nanotechnology *Microelectron. J.* **42** 973–81

[3] Álvarez-Sánchez J J, Álvarez-Bravo J V and Nieto L M 2008 A quantum architecture for multiplying signed integers *J. Phys.: Conf. Ser.* **128** 012013

[4] Babu H M H 2017 Cost-efficient design of a quantum multiplier–accumulator unit *Quantum Inf. Process.* **16** 30

[5] Cai X-D, Wu D, Su Z-E, Chen M-C, Wang X-L, Li L, Liu N-L, Lu C-Y and Pan J-W 2015 Entanglement-based machine learning on a quantum computer *Phys. Rev. Lett.* **114** 110504

[6] Chakrabarti A and Sur-Kolay S 2008 Designing quantum adder circuits and evaluating their error performance *Int. Conf. on Electronic Design* (Piscataway, NJ: IEEE) pp 1–6

[7] Cormen T H 2001 *Introduction to Algorithms* ed T H Cormen, C E Leiserson, R L Rivest and C Stein (Cambridge, MA: MIT Press)

[8] Hung W N N, Song X, Yang G, Yang J and Perkowski M 2006 Optimal synthesis of multiple output Boolean functions using a set of quantum gates by symbolic reachability analysis *IEEE Trans. Comput.-Aided Des. Integr. Circuits Syst.* **25** 1652–63

[9] Kaye P *et al* 2007 *An Introduction to Quantum Computing* (Oxford: Oxford University Press)

[10] Kitaev A Y 1997 Quantum computations: algorithms and error correction *Russ. Math. Surv.* **52** 1191–249

[11] Kotiyal S, Thapliyal H and Ranganathan N 2014 Circuit for reversible quantum multiplier based on binary tree optimizing ancilla and garbage bits *27th Int. Conf. on VLSI Design and 8th Int. Conf. on Embedded Systems* (Piscataway, NJ: IEEE) pp 545–50

[12] Li D-F, Wang R-J, Zhang F-L, Deng F-H and Baagyere E 2015 Quantum information splitting of arbitrary two-qubit state by using four-qubit cluster state and bell-state *Quantum Inf. Process.* **14** 1103–16

[13] Maynard C M and Pius E 2014 A quantum multiply–accumulator *Quantum Inf. Process.* **13** 1127–38

[14] Mogensen T Æ 2013 Garbage-free reversible constant multipliers for arbitrary integers *Int. Conf. on Reversible Computation* (Berlin: Springer) pp 70–83

[15] Nielsen M A and Chuang I 2002 *Quantum Computation and Quantum Information* 10th edn (Leiden: Cambridge University Press)

[16] Ramya P R and Vani Y S 2013 Optimization and implementation of reversible BCD adder in terms of number of lines *Int. J. Reconfig. Embedded Syst.* **2** 21–6

[17] Schubert M and Rana F 2006 Analysis of terahertz surface emitting quantum-cascade lasers *IEEE J. Quantum Electron.* **42** 257–65

[18] Viamontes G F, Markov I L and Hayes J P 2005 High-performance quantum circuit simulation *QuIDDPro* http://vlsicad.eecs.umich.edu/Quantum/qp/ (Accessed: 19 March 2020)

[19] Wille R, Soeken M and Drechsler R 2010 Reducing the number of lines in reversible circuits *Proc. of the XLVII Design Automation Conf.* (New York: ACM) pp 647–52

[20] Yao A C-C 1993 Quantum circuit complexity *Proc. IEEE 34th Annual Foundations of Computer Science* pp 352–61

IOP Publishing

Quantum Computing
A pathway to quantum logic design
Hafiz Md Hasan Babu

Chapter 8

The quantum divider

Among all the fundamental operations in a quantum computer, division is the most complex operation. Division invokes operations such as addition, multiplication, and subtraction. The quantum divider is an indispensable part of the quantum processor that comprises arithmetic sub-modules such as the quantum comparator, quantum adder, quantum multiplier, and quantum subtractor.

A computer processor is composed of a large number of units to execute instructions. Arithmetic units are important units in computer hardware because they have many uses in computer systems. Therefore, designing quantum circuits for arithmetic units is very important. One of these arithmetic units is the division unit. The division circuit is a somewhat complicated component in computer hardware.

8.1 Division algorithms

This section presents classical integer division algorithms as well as quantum algorithms for integer division.

8.1.1 Classical integer division algorithms

Integer division is a basic part of the elementary theory of computation. A number of classical integer division algorithms have some advantages and disadvantages. The school book division algorithm for integer division described by Knuth is similar to the conventional mathematical division where both the dividend and the divisor must be two-word integers and it has a time complexity of $O(n^3)$. This division algorithm is more complex and slower than other division algorithms. Moller showed an improved integer division algorithm which solved the problem of dividing a two-word integer by a single-word integer using single-word approximate reciprocals. This method requires a lower multiplication time and omits the most significant half of the products. However, this technique requires a time complexity of $O(n^2)$ and still needs more operations to perform the division. The restoring division algorithm first operates on fixed-point fractional numbers and uses an extra

doi:10.1088/978-0-7503-2747-3ch8

register to save the partial numerators obtained by the multiplication, and then reuses the result when the result of the subtraction is negative. This method has a time complexity of $O(n^2)$.

8.1.2 Quantum integer division algorithms

There is no explicit quantum integer division algorithm. All the algorithms incorporate quantum division in the classical integer division algorithms. One of the most well-known algorithms in the field of quantum computation is Shor's factorization algorithm. This is a quantum algorithm that finds the factors of a composite integer in polynomial time with great probability of success. Division is not incorporated in this factorization algorithm as well as in other implementations of the modular exponentiation part of the Shor's algorithm. The quantum modular exponentiation architecture is described for Shor's factoring algorithm, where division has been incorporated. It is currently one of the most renowned design architectures of integer division using Shor's quantum factoring algorithm. This quantum modular exponentiation circuit requires $(9n + 2)$ qubits and has a depth of $O(2000n^2)$ where n is the number of qubits to be factored. The time complexity of this method for quantum integer division is $O(72(\log_2 n)3)$.

Alireza devised a design to implement the quantum divider for division of two n-qubit fixed-point numbers in binary format using the restoring division algorithm. The design methodology reduces the number of ancillary qubits and eliminates the use of quantum arithmetic sub-modules, such as the quantum shift register, quantum adder, subtractor, etc. However, this architecture requires a time complexity of $(5n/6)^2 + O(n^2)$ and a total of $3n^3 + 6n^2 + n$ operations are required, where n is the number of qubits of the dividend and divisor.

All of the above mentioned quantum algorithms for integer division have high time complexity and a high total number of operations to perform the division operation of two arrays of qubits representing integers. Therefore it is expedient to incorporate a new quantum division algorithm for integer division with minimal time complexity and number of operations.

8.2 The importance of the quantum divider

In this chapter the optimization of time complexity as well as the reduction of the number of operations are addressed, which has motivated the introduction of a completely new and efficient technique of integer division with optimum time complexity. Time complexity is the amount of instructions that will be executed during the run time of that particular algorithm. Obviously, the greater the time complexity of the algorithm, the more operations there are and the greater is the delay of simulation. A large time complexity makes the algorithm slower. Increasing the time complexity of an algorithm is likely to increase the number of operations. This increase in turn increases the cost and delay of the execution of operations. Algorithmic complexity is the property of an algorithm which is mapped to the amount of computational resources used by the algorithm. This factor is analogous to engineering productivity for a repeating or continuous process. Thus if the complexity changes, the resources such as memory space, access time, cost, delay, area, and power of execution of the algorithm will also fluctuate proportionally. This will make the simulation of the algorithm more vulnerable, in particular in the case of large sets of data. The reduction of this time complexity minimizes the number of operations and hardware complexities as well. This optimization in turn makes the division operation faster. Thus it is expedient to reduce the time complexity of the division algorithm to make it faster and more efficient. Researchers are looking for an efficient and fast division algorithm to optimize time complexity. Therefore a new, faster, efficient, and optimal tree-based division technique is introduced in this chapter which works like the breadth fast search (BFS) algorithm.

8.3 The tree-based quantum division technique

In this section a tree-based quantum division algorithm is described to perform the division operation of integers, which is the first such ever presented in the literature. The quantum algorithms for integer division based on the classical division algorithms require a large time complexity and a greater number of operations to execute the division of the dividend by the divisor.

8.3.1 Definitions and properties of the division technique

In this subsection the terms divisor, dividend, quotient, remainder, partial numerator, node, leaf, and root, and the properties of the tree, that are used frequently throughout this chapter are elaborated to establish the main purpose and theme of this chapter.

Definition 8.1. The qubits representing the integer that is to be divided are called the *dividend.*

Definition 8.2. The qubits of the integer that divides the dividend are known as the *divisor.*

Definition 8.3. The *quotient* is the resulting qubits of the integer obtained by dividing the dividend using the divisor.

Definition 8.4. In any step of the division process when the quotient qubits are multiplied with the divisor qubits, it generates *partial numerator* qubits that are subtracted from the original dividend qubits.

Definition 8.5. *Remainder* qubits representing integers are the result obtained after subtracting the current partial numerator qubits from the original divisor qubits.

Example 8.1. Suppose a qubit string of integer $\{|1\rangle|1\rangle|0\rangle|1\rangle|0\rangle\}$ is divided by $\{|1\rangle|0\rangle|1\rangle\}$, then the first one is the dividend and the second qubit string is the divisor. The

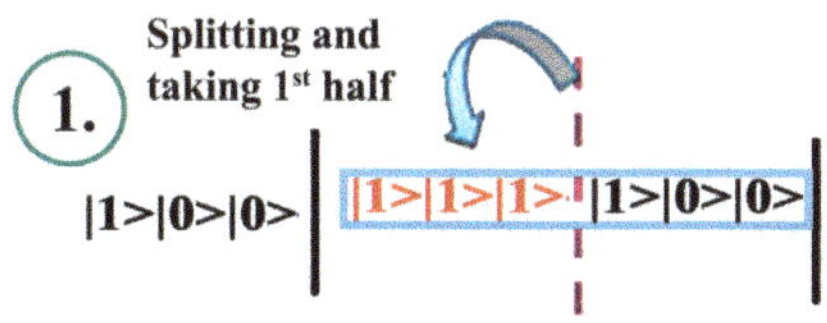

Figure 8.1. The divisor, dividend, remainder, quotient, and partial numerator.

quotient is {|1⟩|0⟩|1⟩}, the partial numerator is {|1⟩|0⟩|1⟩}, and the remainder is {|1⟩}. Figure 8.1 illustrates this scenario.

Property 8.1. A connected graph G is called a *tree* if:

The removal of any of its edges makes G disconnected.

Top most node $r \in G$ is designated as the root.

The remaining nodes are partitioned into disjoint sets $G_1, G_2, \ldots, G_k$, where each of the sets is a tree and $k \geqslant 0$.

G is connected and contains no closed path or cycles.

Definition 8.6. A *node* in a tree contains the information or data. The node may be a child or parent node.

Definition 8.7. The *level* of a node is the unique path of that node from the root of the tree.

Example 8.2. A tree, its child and parent nodes, root, and level are depicted in figure 8.2.

8.3.2 The algorithm of the division technique

In the algorithm the tree structure approach is used for simplicity and generality as well as optimizing the time complexity through parallelism. A tree-based division algorithm makes the operation faster and it also simplifies the circuits.

The technique of division has five basic steps:

Step 1. First the mid-point qubit of the dividend is determined and the dividend is split using the quantum dividend splitter (QDS) circuit, as shown in figure 8.3. The first half will be taken after splitting, as shown in figure 8.1.

Step 2. Then the partial numerator qubits are generated in such a way that the copying and ANDing operations of the divisor and quotient qubits are performed in parallel in the split dividend for faster operation. In this step, first an additional |0⟩ is added at the beginning of the first half of step 1 to form the divisor qubits and then the quotient qubits are multiplied with the divisor qubits to obtain the results which are stored in the partial numerator qubits generation (PQG) array. Figure 8.3 illustrates the procedure.

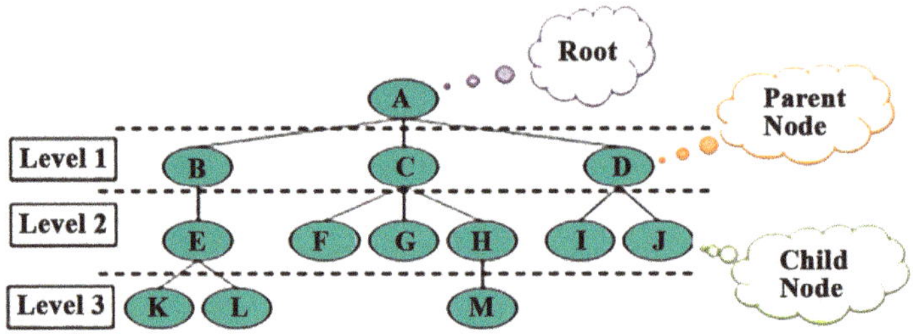

Figure 8.2. Examples of tree, root, node, and level.

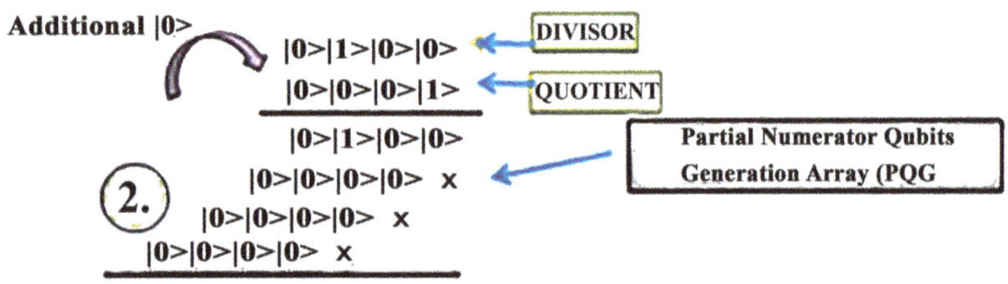

Figure 8.3. Step 2, multiplication of the divisor and quotient, and storing the result in PQG.

Step 3. In this step the partial numerator additions are performed in a tree structure known as the partial numerator addition (PNA) tree where the tree has $((\log_2 n) - 1)$ levels for numerator generation and addition. In the first level, four concurrent rows of the PQG array are combined by adders in parallel, and in the next levels (up to $((\log_2 n) - 1)$th) the results of the additions between the individual qubits of the divisor and quotient are added together to form the partial numerator. According to example 8.3, we have $n = 4$. Thus we need only $((\log_2 n) - 1) = 1$ levels to complete the additions of the results of multiplication stored in the PQG, as shown in figure 8.4.

Step 4. Now a partial numerator subtractor (PNS) tree with only one level is necessary for subtracting the partial numerator from the original dividend. Figure 8.5 illustrates the procedure.

Step 5. If there is no remainder after subtraction, then the division is complete, where the quotient qubits form the result. Otherwise, the remainders are added at the beginning positions of the rest of the qubits of the dividend before splitting the newly formed dividend. Repeat the same operation until a remainder exists. In example 8.3 there are remainders after step 4. Thus the same process is repeated as shown in figure 8.6.

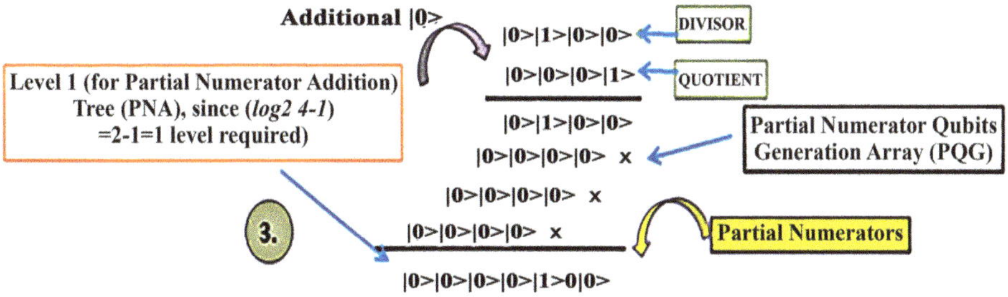

Figure 8.4. Step 3, the partial numerator obtained by addition in the PNA tree.

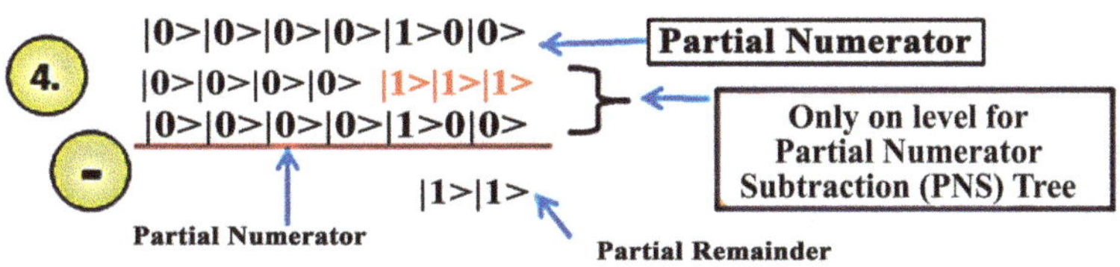

Figure 8.5. Step 4, subtraction of the partial numerator from the original dividend in the PNS tree.

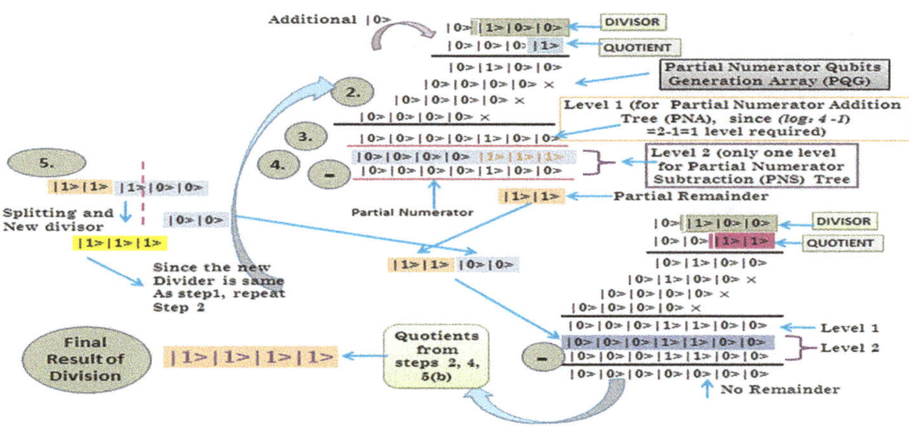

Figure 8.6. Step 5, repetition of the same process until there is a remainder.

Algorithm 8.1 describes the division technique and example 8.3 illustrates it.

Algorithm 8.1. Division technique.

Input: Two qubit arrays $|Di[i]\rangle$ and $|Dd[j]\rangle$ for the divisor and dividend, respectively.
Output: $(|Di[j]\rangle)/(|Dd[i]\rangle)$, i.e. the resulting qubits of the division algorithm.
1: START
2: $(|Dd[i]\rangle)$ = array containing the qubits of the dividend
3: $(|Di[j]\rangle)$ = array containing the qubits of the divisor
4: $(|Qa[i]\rangle)$ = array that stores the PNA qubits
5: $(|Qg[i]\rangle)$ = array that generates the partial numerator qubits
6: $(|Q[i]\rangle)$ = array that contains the quotients
7: $(|R[i]\rangle)$ = array that contains the remainders
8: n = the size of the dividend qubit array
9: m = the size of the divisor qubits array
10: mid = **ceil**$(n/2)$
11: **if** mid >= m **then**
12: Divide $(Dd[i]\rangle)$ by $(Di[j]\rangle)$, where $0 \leqslant i \leqslant$ mid and $0 \leqslant j \leqslant m$
13: **end if**
14: Generate the PQG using $(|Qg[i]\rangle)$, for copying and ANDing operation of $(|Dd[i]\rangle)$ and $(|Q[i]\rangle)$, which is to be done in parallel
15: Create the PNA tree with $(\log_2 n - 1)$ levels.
16: Generate $(n/4)$ nodes at the first level of the tree where each node represents the addition of four successive rows of the PQG array.
17: Store the result of the PNA tree in $(|Qa[i]\rangle)$
18: Save the quotient in the $(|Q[i]\rangle)$
19: Create the PNS tree with 1 level
20: Subtract the $(|Qa[i]\rangle)$ from $(|Dd[i]\rangle)$
21: Store the result in $(|R[i]\rangle)$
22: **if** $|R[i]\rangle == |0\rangle$ for all qubit positions and all qubits are executed
23: The result of the division would be $(|Q[i]\rangle)$
24: **else**
25: Add the values of $(|R[i]\rangle)$ in the beginning positions of the remaining qubits of $(|Dd[i]\rangle)$
26: **end if**
27: **loop**
28: Steps 10 to 26 for new qubits of the $(|Dd[i]\rangle)$ with the same divisor
29: **end loop**
30: **print** (Division is not possible since the length of the divisor is larger than the dividend);
31: END

Example 8.3. Let the divisor be a qubit string $\{|1\rangle|0\rangle|0\rangle\}$ and the dividend be $\{|1\rangle|1\rangle|1\rangle|1\rangle|0\rangle|0\rangle\}$. First, it is necessary to split the dividend according to algorithm 8.1. After splitting, a dividend consisting of the first three qubits, i.e. $\{|1\rangle|1\rangle|1\rangle\}$ is obtained. In the next step the partial numerator qubits $(|1\rangle, |0\rangle, |0\rangle...)$ are generated by copying and ANDing operations of the divisor $\{|1\rangle|0\rangle|0\rangle\}$ and quotient qubit $\{|1\rangle\}$.

Then the PNA tree is generated and the partial numerator $\{|1\rangle|0\rangle|0\rangle\}$ is produced by the addition operations of the partial numerator qubits generated in the previous step. Then a PNS tree with a single level is generated and the partial numerator $\{|1\rangle|0\rangle|0\rangle\}$ is subtracted from the split dividend $\{|1\rangle|1\rangle|1\rangle\}$. In the next step the remainder $\{|1\rangle|1\rangle\}$ is added at the beginning of the rest of the dividend qubits $\{|1\rangle|0\rangle|0\rangle\}$ and the newly formed dividend becomes $\{|1\rangle|1\rangle|1\rangle|0\rangle|0\rangle\}$. Now again, after splitting, the dividend becomes three qubits, i.e. $\{|1\rangle|1\rangle|1\rangle\}$. Again the partial numerator produced by the PNA tree using quotient $\{|1\rangle\}$ is $\{|1\rangle|0\rangle|0\rangle\}$, which is subtracted from the splitted dividend in the PNS tree and the remainder $\{|1\rangle|1\rangle\}$ is added at the beginning of the rest of the dividend qubits and that becomes $\{|1\rangle|1\rangle|0\rangle|0\rangle\}$. No more splitting is needed and using the divisor $\{|1\rangle|0\rangle|0\rangle\}$ and quotient $\{|1\rangle|1\rangle\}$, the partial numerator $\{|1\rangle|1\rangle|0\rangle|0\rangle\}$ is produced, which is then subtracted from the dividend $\{|1\rangle|1\rangle|0\rangle|0\rangle\}$; there is no remainder. Thus the quotient qubits form the result of the division, i.e. $\{|1\rangle|1\rangle|1\rangle|1\rangle\}$. This whole procedure is clarified in figures 8.1 and 8.3–8.6.

Property 8.2. The time complexity of the division algorithm is at least $O(3n((\log_2 n) + 1) - n)$, where n is the number of qubits of the dividend.

Proof. Let n be the size of the dividend and m be the size of the divisor arrays of qubits, $n \geqslant m$. For a one-qubit division, the best known method requires a time complexity of $(5/6)^2 + O(1)$, whereas the best case time complexity of the division algorithm for one qubit is $O(2)$.

According to the algorithm, in the first step the dividend is split and it requires the best case time complexity of $T(P) = O(\log_2 n)$, where n is the number of qubits of the dividend.

In the second step a PNA tree is produced to perform the addition of the partial numerators. Assume that T is the PNA tree and $N(n)$ is the total number of levels of the tree. Thus at first the required time for the numerator generation and addition of the partial numerators is calculated using a tree structure by the method of mathematical induction. It is assumed that the time of numerator generation and addition operations using the tree is $O((\log_2 n) - 1)$.

When $n = 1$, T is a single level tree. So $N(1) = 1 = |(\log_2 1) - 1|$. Thus the statement holds for the base case which is $n = 1$.

Assume $N(i) = |(\log_2 i) - 1|$ for $0 \leqslant i < n$. A partial product addition tree consists of $(n - 1)$ levels and the root node. Thus

$N(n) = 2 \times N(n - 1) + 1 = \log_2(2n - 2) - 2 + 1$

$= (\log_2 n) + (\log_2 2) - (\log_2 2) - 1 = (\log_2 n) - 1.$

Therefore in the best case, the time for the generation and the addition operation of the numerator using a PNA is $T(G) = O((\log_2 n) - 1)$.

In the third step it is needed to access arrays. The array access time is linear and it is $O(1)$. According to the algorithm, three arrays have to be accessed: one for storing partial numerators, one for storing quotients, and the last one for storing partial remainders. Thus, in the best case, the total access time for three arrays is $T(A) = O(3)$.

In the fourth step the subtraction of the partial numerator from the partitioned dividend is performed which requires $T(S) = O(\log_2 n)$ time in the best case. Thus in the best case, the total time complexity of the division algorithm is $T(N) = T(P) + T(G) + T(A) + T(S) = O(\log_2 n) + O((\log_2 n) - 1) + O(3) + O(\log_2 n) = O(3((\log_2 n) + 1) - 1)$.

Therefore the division algorithm requires a time complexity of $O(3((\log_2 n + 1) - 1)$ in the best case, where n is the number of dividend qubits. Thus the division algorithm requires a time complexity of $O(3n((\log_2 n) + 1) - n)$ for the n times iteration of the division process.

8.4 The design of a quantum divider circuit

Quantum divider design techniques are a very important topic at present. All the quantum divider circuits are designed using naive approaches. The designs of these quantum divider circuits require high area, delay, power, and quantum costs. As these designs produce quantum divider circuits with higher depths, the design techniques are not efficient enough. Thus a tree-based design algorithm is introduced. The splitting operation of the dividend is first performed in the root node of a tree. The partial numerator generator and addition circuits are considered as the parent nodes of the tree. In the last level of the tree there will be leaf nodes for the subtractors, which are the child nodes of the numerator generator and addition nodes.

8.4.1 A technique to minimize the number of ancillary inputs in the quantum circuit realization

In the quantum circuit an important parameter to measure the performance is the number of qubits in the circuit required for it to operate. It is expected that the number of qubits is to be optimized during the realization of quantum circuits. The number of input lines of the circuit is the indicator of the required qubits as the input lines are limited resources in quantum circuits. Extra ancillary input lines help to increase the number of input lines. Therefore, in this subsection, a merging technique is shown to optimize the number of ancillary inputs in large quantum circuits. This procedure works like the merge join algorithm. First it is necessary to divide the initial quantum circuit into sub-modules. Then find the number of ancillary input lines in those sub-modules. Now combine the sub-modules and form the original structure up to the $m(2^n)$th ancillary input lines where n is the number of qubits of each operand and m is the number of ancillary inputs in each sub-module.

Finally, calculate whether the number of quantum gates in the ancillary lines of the sub-modules are even or odd. If the value is odd, combine the quantum gates of the ancillary input lines of the sub-modules with the quantum gates in each of the $m(2^n + 1)$th and $m(2^n + 2)$th ancillary input lines of the initial circuit and then combine the first half of the quantum gates in the ancillary input lines of the odd sub-modules with quantum gates in the $m(2^n + 1)$th ancillary input line of the initial circuit.

Then combine the other half of the quantum gates in the ancillary input lines of the other half of the odd sub-modules with the gates in the ancillary input lines of the even sub-modules and the gates of the $m(2^n + 2)$th ancillary input line of the initial circuit. Algorithm 8.2 gives the procedure of the minimization of the number of ancillary input lines and example 8.4 describes the whole scenario.

Algorithm 8.2. Reduction of the number of ancillary input lines in the design of the quantum divider circuit.

Input: (2^{n+1}) sub-modules of quantum circuits
Output: n-qubit circuit with $m(2^n + 2)$ ancillary input lines

```
1: START
2: m = the number of ancillary inputs in each sub-module
3: p = the number of quantum gates in the ancillary line of the sub-module
4: c = the number of ancillary inputs in large circuit (initially 0)
5: for each sub-module do
6:    combine them together without explicit change
7:    c = c + m
8:    if (c==(m(2^n + 1))) then
9:       Step 3
10:   else
11:      Step 5
12:   end if
13:   if (p%2==0) then
14:      combine the first p/2 quantum gates with quantum gates in the m(2^n + 1)th ancillary input line
15:      combine the other p/2 along with the p gates of another sub-module with the p gates of the
         m(2^n + 2)th ancillary input line
16:   else
17:      combine the p gates with gates in the m(2^n + 1)th ancillary input line
18:      combine p gates of another sub-module with the p gates of the m(2^n + 2)th ancillary input
         line
19:   end if
20: end for
21: END
```

Example 8.4. Consider the sub-modules of the quantum ANDing circuit (QAC) presented in figure 8.7. Suppose we need to combine eight ($2^{n+1} = 2^{2+1} = 8$) of these sub-modules (the QACs) of the quantum divider circuit to form a two-qubit quantum numerator generator (QNG) circuit, as shown in figure 8.8, where the number of ancillary inputs in each sub-module (the QAC circuit) is one ($m = 1$) and the number of quantum gates in the ancillary line of the sub-module is three. Thus according to algorithm 8.2, we combine the first four ($2^n = 2^2 = 4$) sub-modules (the QAC circuits) without joining the ancillary input lines. Thus the QNG will have four ancillary input lines at its input from the four QAC circuits.

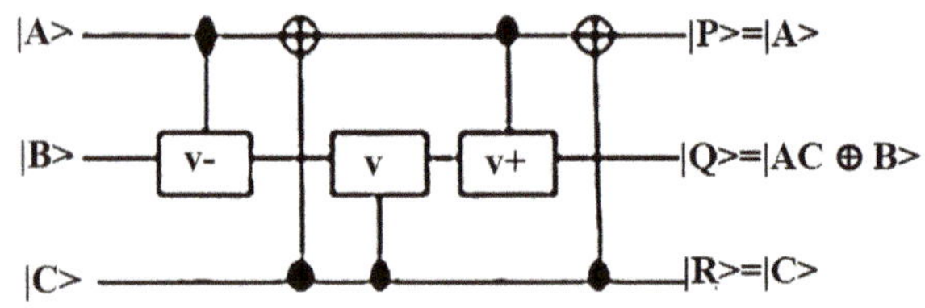

Figure 8.7. The quantum circuit for the ANDing operation.

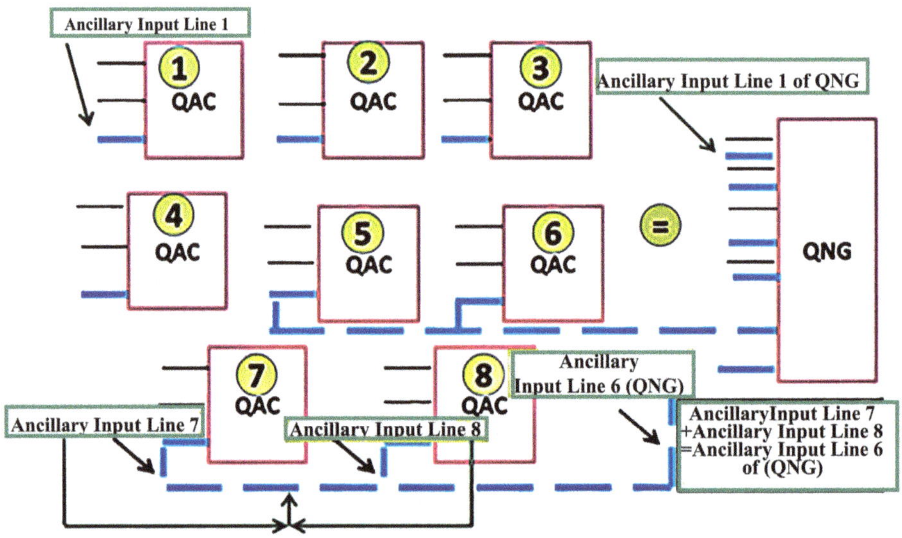

Figure 8.8. Block diagram to illustrate algorithm 8.2 to reduce ancillary inputs in a two-qubit QNG.

Combine three quantum gates on the ancillary input line of sub-module 7 (ancillary input line 7) with the three quantum gates on the ancillary input line of sub-module 8 (ancillary input line 8) and place them on the sixth ($m(2^n + 2)$th = 6th) ancillary input line of the initial QNG circuit. Figure 8.8 provides the block diagram to show how the sub-modules are combined to reduce the number of ancillary inputs according to algorithm 8.2 to form a two-qubit QNG circuit. In this figure the ancillary input line 7 and ancillary input line 8 are joined together and form the ancillary input line 6 of the QNG. Similarly, ancillary input line 5 and ancillary input line 6 are joined together and finally form the QNG circuit with reduced ancillary inputs. The algorithm requires only six ancillary inputs instead of eight, as shown in figure 8.8. A detail of this implementation in the gate level is illustrated in figure 8.9.

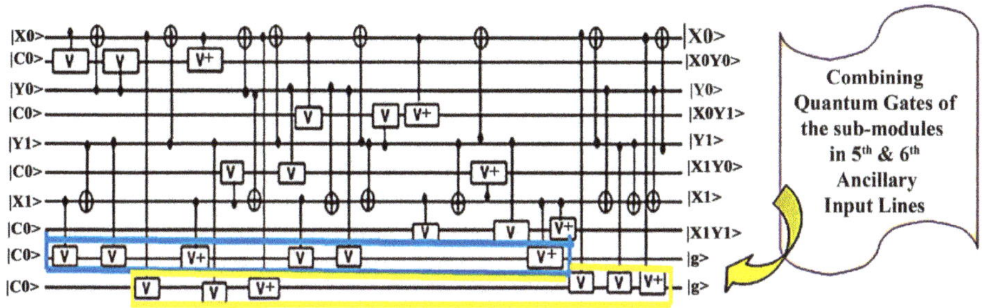

Figure 8.9. A two-qubit QNG circuit.

8.4.2 The components of the quantum divider circuit

The main components of the quantum divider circuit are the QAC, quantum full-subtractor (QFS) circuit, QNG circuit, quantum numerator addition (QNA) circuit, and QDS circuit. These circuit components of the quantum divider are elaborated as follows.

8.4.2.1 The quantum ANDing circuit

In this subsection a new 3 × 3 QAC is presented, which performs the ANDing operation. The input vector, I_v, and the output vector, O_v, of the circuit are as follows:

$|I_v\rangle = \{|A\rangle, |B\rangle, |C\rangle\}$; and

$|O_v\rangle = \{|A\rangle, |AC \oplus B\rangle, |C\rangle\}$.

Here the input $|B\rangle$ is the target qubit and $|A\rangle$ and $|C\rangle$ are two control qubits. The QAC is constructed by two controlled-V, one controlled-V^+, and two CNOT gates, as shown in figure 8.7. Hence the quantum cost of the QAC is 5. The corresponding truth table of the circuit is shown in table 8.1. To perform the 'AND' operation using the QAC the $|B\rangle$ input qubit in figure 8.7 has to remain $|0\rangle$.

8.4.2.2 The quantum full-subtractor circuit

A QFS is a quantum circuit that performs the arithmetic subtraction of three input qubits. It consists of three inputs and two outputs. If we take three input variables as $|A\rangle$, $|B\rangle$, and $|C\rangle$, and two output variables as $|D\rangle$ (difference) and $|B_{out}\rangle$ (borrow), then the equations as mentioned below can be used to obtain the outputs:

$|D\rangle = |A\rangle \oplus |B\rangle \oplus |C\rangle$

$|B_{out}\rangle = (|C\rangle.(\overline{|A \oplus B\rangle})) \oplus (|\overline{A}. B\rangle)$.

In this subsection, a new 4 × 4 QFS circuit is shown. The input vector (I_v) and the output vector (O_v) of the circuit are as follows:

$|I_v\rangle = \{|B\rangle, |D\rangle, |C\rangle, |A\rangle\}$; and

$|O_v\rangle = \{|B\rangle, (|C\rangle.(\overline{|A \oplus B\rangle})) \oplus (|\overline{A}.B\rangle), |C\rangle, |A\rangle \oplus |B\rangle \oplus |C\rangle\}$.

Table 8.1. Reversibility of a 3 × 3 QAC.

Input			Output		
A	*B*	*C*	*P*	*Q*	*R*
0	0	0	0	0	0
0	0	1	0	0	1
0	1	0	0	1	0
0	1	1	0	1	1
1	0	0	1	0	0
1	0	1	1	1	1
1	1	0	1	1	0
1	1	1	1	0	1

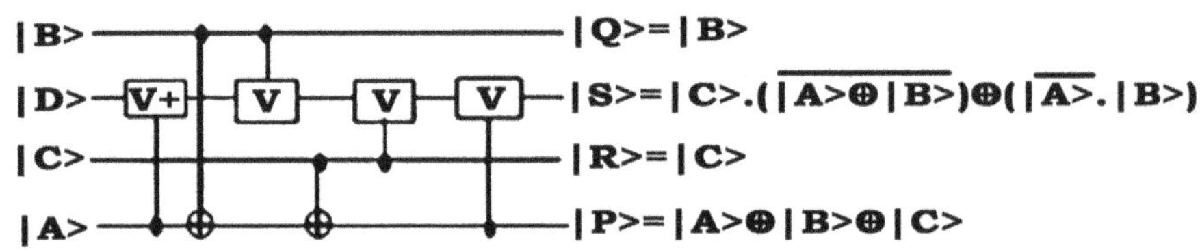

Figure 8.10. The QFS, when $|D\rangle = |0\rangle$.

Here the input $|D\rangle$ is the target qubit and $|A\rangle$, $|B\rangle$, and $|C\rangle$ are control qubits. The QFS circuit is constructed by three controlled-V, one controlled-V^+, and two CNOT gates, as shown in figure 8.10. Hence the quantum cost of the QFS circuit is 6. The corresponding truth table of the circuit is shown in table 8.2. Input $|A\rangle$ is used to take the role of the borrow in and output $|S\rangle$ is the borrow out, as shown in figure 8.10. Output $|P\rangle$ is the difference of the inputs $|B\rangle$, $|C\rangle$, and borrow in $|A\rangle$. The QFS circuit is designed in such a way that it can act as a full-subtractor by setting $|D\rangle = |0\rangle$.

8.4.2.3 The quantum dividend splitter circuit

A 4 × 4 QDS circuit is shown that takes the string of dividend qubits and divides it into two parts which are carried out to the QFS circuit. The input vector I_v and the output vector O_v of the circuit are as follows:

$$|I_v\rangle = \{|Dd[i/2]\rangle,\ Di[j/2]\rangle,\ |0\rangle,\ |0\rangle\};\ \text{and}$$
$$|O_v\rangle = \{|Dd[i/2]\rangle,\ \overline{(|Dd[i/2]\rangle \oplus |Di[j/2]\rangle)},\ \overline{(|Di[j/2]\rangle} \cdot |Dd[i/2]\rangle),\ \overline{(|Dd[i/2]\rangle} \cdot |Di[j/2]\rangle)\}.$$

In figure 8.11 the last input is the target qubit ($|0\rangle$) and other inputs are control qubits. The QDS circuit is constructed by two controlled-V, one controlled-V^+, one NOT, and four CNOT gates, as shown in figure 8.11. Hence the quantum cost of the QDS circuit is 8. The main purpose of the QDS circuit is to separate the set of qubits

Table 8.2. The reversibility of the 4 × 4 QFS circuit.

Input				Ouput			
A	*B*	*C*	*D*	*P*	*Q*	*R*	*S*
0	0	0	0	0	0	0	0
0	0	0	1	0	0	0	1
0	0	1	0	1	0	1	1
0	0	1	1	1	0	1	0
0	1	0	0	1	1	0	1
0	1	0	1	1	1	0	0
0	1	1	0	0	1	1	1
0	1	1	1	0	1	1	0
1	0	0	0	1	0	0	0
1	0	0	1	1	0	0	1
1	0	1	0	0	0	1	0
1	1	0	0	0	1	0	0
1	1	0	1	0	1	0	1
1	1	1	0	1	1	1	1
1	1	1	1	1	1	1	0

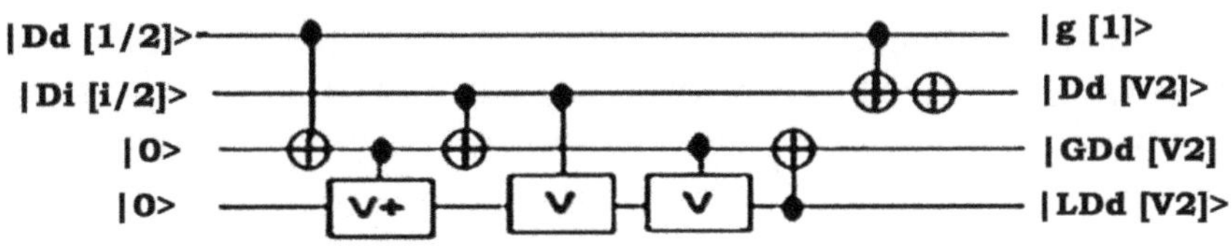

Figure 8.11. The QDS splitter circuit.

of the dividend. If the dividend consists of four qubits, the mid-point qubit (second qubit) passes as (|*Dd*[*i*/2]⟩) output. Then the inputs, the first and third input qubits, are Ex-ORed, which goes out as outputs |*GDd*[*i*/2]⟩ and |*LDd*[*i*/2]⟩, respectively, as shown in figure 8.11.

8.4.2.4 The quantum numerator generator circuit

The QNG circuit takes the divisor and quotient as inputs and produces the partial numerators as outputs by performing all possible ANDing operations using the divisor and quotient qubits. The divisor qubits and the quotient qubits are fed to this component which produces the product terms of the qubits, known as partial numerators. The QAC has been used to construct this part. If the n-qubit divisor and n-qubit quotients are used, the QNG circuit requires (2^{n+1}) QACs and $m(2^n + n)$ ancillary input lines where m is the number of ancillary qubits in each QAC. Thus, including the divisor and quotient qubits, the total $2n + m(2^n + n)$ qubits are required to construct the n-qubit QNG circuit.

Example 8.5. To construct a two-qubit QNG circuit, $(2^{n+1}) = 8$ QACs and $m(2^n + n) = 6$ ancillary input lines are required, where m is the number of ancillary qubits in each QAC (here, $m = 1$). In figure 8.9 the detailed two-qubit QNG circuit is shown. From figure 8.9, it is clear that the two-qubit QNG circuit needs $2n + m\,(2^n + n) = 4 + 6 = 10$ qubits according to the ancillary input reduction as shown in algorithm 8.2.

8.4.2.5 The quantum numerator addition circuit

The QNA circuit adds the partial numerators column-wise that have been generated by the QNG circuit. The quantum full-adder (QFA) is used to construct the QNA circuit. For an n-qubit QNA circuit $(5n/2)$ QFAs are needed. Thus for that purpose $m(2^n + n)$ ancillary input lines are needed, where m is the number of ancillary qubits in each QFA. If the number of QFAs requires constructing QNA greater than five (when $n \geqslant 4$), then it is necessary to reduce the ancillary input lines by combining these lines of the quantum gates according to reduction algorithm.

Example 8.6. To construct a two-qubit QNA circuit, $(5n/2) = 5$ QFA circuits are required to perform the addition operations of the partial numerators generated by the QNG. Since the number of QFAs is five, merging of the quantum gates is not required to reduce the number of ancillary input lines. Thus the five ancillary input lines are required. In figure 8.12 the detailed QNA circuit for a two-qubit divider is shown.

8.4.2.6 The n*-qubit quantum divider*

The tree-based design structure of the quantum divider circuit reduces the depth of the circuit. The design also minimizes the number of ancillary inputs. In this subsection, the details of the design of a two-qubit quantum divider circuit are presented using algorithm 8.3, as shown in figure 8.13. In addition, a block diagram

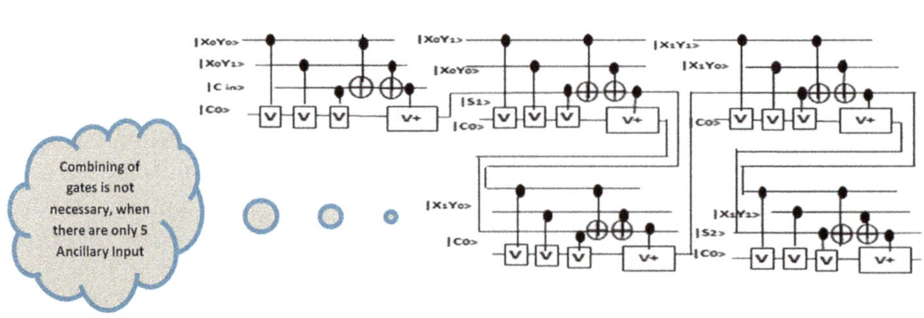

Figure 8.12. The two-qubit QNA circuit.

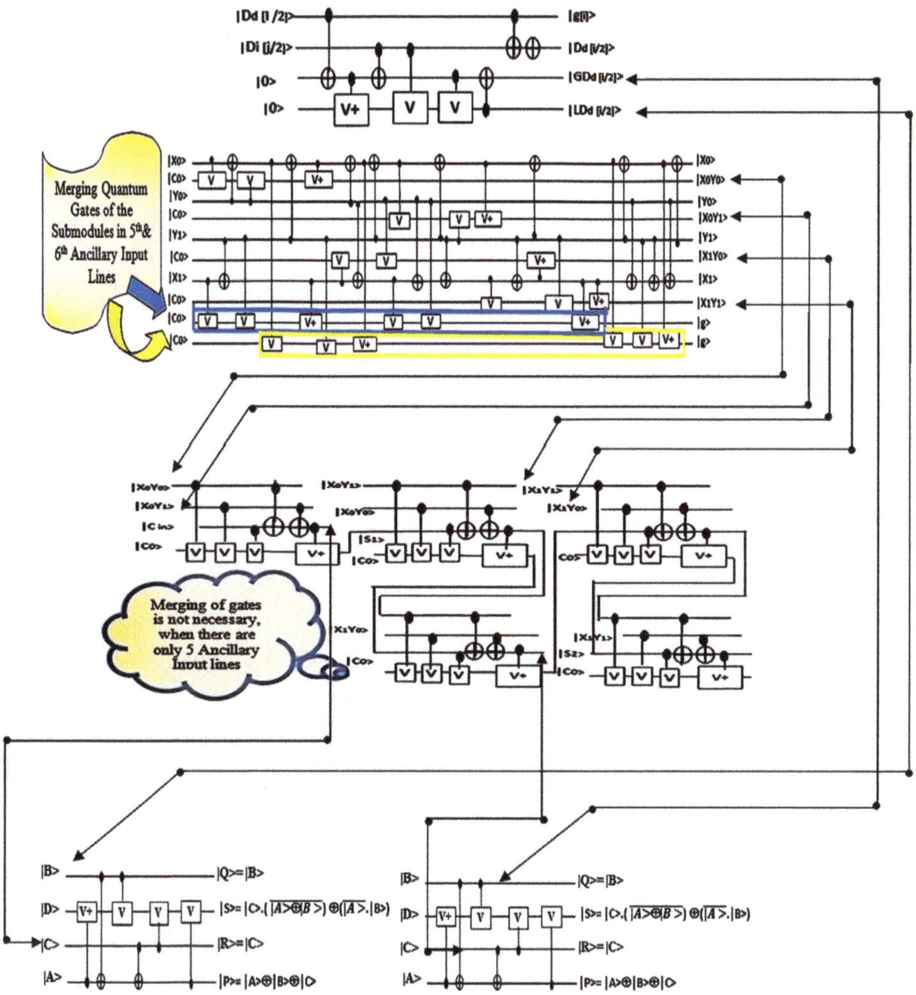

Figure 8.13. The two-qubit quantum divider circuit.

of an n-qubit quantum divider circuit using the two-qubit quantum divider circuit is presented using algorithms 8.2 and 8.3, as shown in figure 8.14.

The algorithm for the tree-based architecture of an n-qubit quantum divider circuit works recursively. This algorithm at first creates the design for the two-qubit divider circuit using a procedure of micro-division and then it is called recursively to generate an n-qubit quantum divider circuit, where in the procedure of micro-division of algorithm 8.3, at first, the QDS splits the dividend and sends it to the subtractor for subtraction operation. Then, the QNG generates the partial numerators that are added using the QNA circuit and the outputs of the QNA circuits are carried to the QDS circuit for subtraction operation. In figure 8.14, a block diagram of an n-qubit quantum divider is presented by applying algorithm 8.3. The n-qubit

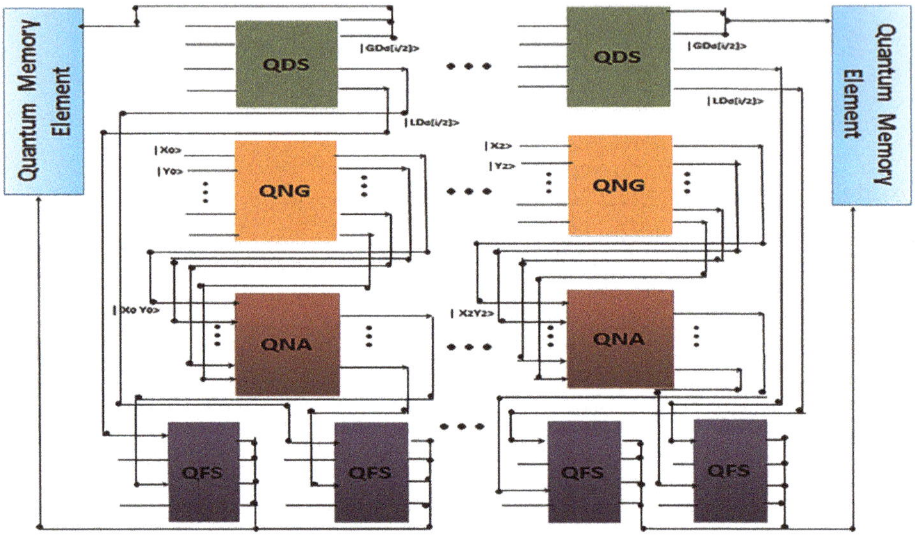

Figure 8.14. Block diagram of the n-qubit quantum divider circuit.

quantum divider circuit consists of (n/2) two-qubit QNG circuits, (n/2) two-qubit QNA circuits, n QFS circuits, and ($\log_2 n$) QDS circuits.

The number of quantum wires can be minimized using the quantum memory elements in a quantum circuit. As a consequence the unwanted outputs (garbage outputs) can be reduced using quantum memory elements.

Here, this concept of the quantum memory element is used to reduce the number of garbage outputs. There is a single garbage output line that collects garbage from different locations and stores it in the memory locations for further reuse and thus shrinks the number of garbage output lines in the circuit design, as shown in figure 8.14.

Algorithm 8.3. Design of an n-qubit quantum divider circuit.

Input: n-qubit dividend $\{|Z_1\rangle, |Z_2\rangle, ..., |Z_n\rangle\}$ and n-qubit divisor $\{|X_1\rangle, |X_2\rangle, ..., |X_n\rangle\}$
Output: Result (quotient qubits) of n-qubit division: $\{|Z_1\rangle, |Z_2\rangle, ..., |Z_n\rangle\}/\{|X_1\rangle, |X_2\rangle, ..., |X_n\rangle\}$

```
START
for(log2n times) do
    Split the n-qubit dividend into (log2n) numbers of a two-qubit divisor by the QDS circuit
    Each time put the qubits into a separate QFS circuit
    Call the micro-division procedure
end for
Micro-division procedure
Put the two-qubit divisor and two-qubit quotient into the circuit with (n/2) two-qubit QNG
  circuits
Send the output of the QNG circuit to (n/2) two-qubit QNA circuits
```

(*Continued*)

```
Subtract the output of the QNA from the dividend qubits in the QFS
Apply two QFS circuits to subtract the qubit of each position of the dividend
if the output of each QFS is (|0⟩) in each QFS then
    Take the quotient as result of the division
else
    Add the remainders as the beginning of the other split and continue the same process again
  until no remainder (|0⟩) state is obtained
end if
END
```

Example 8.7. A four-qubit quantum divider circuit is designed using a two-qubit quantum divider, as shown in figure 8.13, by applying algorithm 8.3. According to the design procedure of algorithm 8.3, a four-qubit quantum divider circuit requires two ($\log_2 n = \log_2 4 = 2$) QDS circuits, two ($n/2 = 4/2 = 2$) two-qubit QNGs, two ($n/2 = 4/2 = 2$) two-qubit QNA circuits, and four QFS circuits. First QDS splits the dividend in half until it is a two-qubit dividend. Then this two-qubit goes to the QFS where the partial numerator qubit is subtracted from the dividend qubit which is obtained by the QDS and the recursive procedure of micro-division of algorithm 8.3 (two-qubit divider design). If the result of the subtraction is $|0\rangle$, then the quotient is the result of the division of the four-qubit quantum divider circuit as constructed by algorithm 8.3, otherwise the remainder qubits are added at the beginning positions of the initial dividend to form a new dividend which is split further. To obtain a four-qubit quantum divider circuit, the procedure of micro-division of algorithm 8.3 continues until the remainder exists. The four-qubit quantum divider circuit is shown in figure 8.15.

Property 8.3. The depth of the design of the quantum divider circuit is at least $O(32n + 8\log_2 n)$, where n is the number of qubits of the dividend.

Proof. The above mentioned statement is proved by the method of mathematical induction. A QDS circuit, a two-qubit QNG, a two-qubit QNA, and two QFS circuits are required for constructing a two-qubit quantum divider circuit using the design method of the tree-based divider in the best case where there is no remainder after the first iteration. The total number of stages in a QDS circuit is eight and the number of stages in a two-qubit QNG, a two-qubit QNA, and a QFS are 34, 18, and 6, respectively.

When $n = 2$, i.e. for a two-qubit quantum divider circuit, the total depth of the two-qubit quantum divider circuit in the best case is

$$D_{\text{two-qubit}} = D_{\text{QDS}} + D_{\text{QNG}} + D_{\text{QNA}} + 2 \times D_{\text{QFS}}$$
$$= 8 + 34 + 18 + 2 \times 6 = 64 + 8 = 32n + 8\log_2 n; \; [\because n = 2 \text{ here}].$$

Thus the statement holds for the base case $n = 2$.

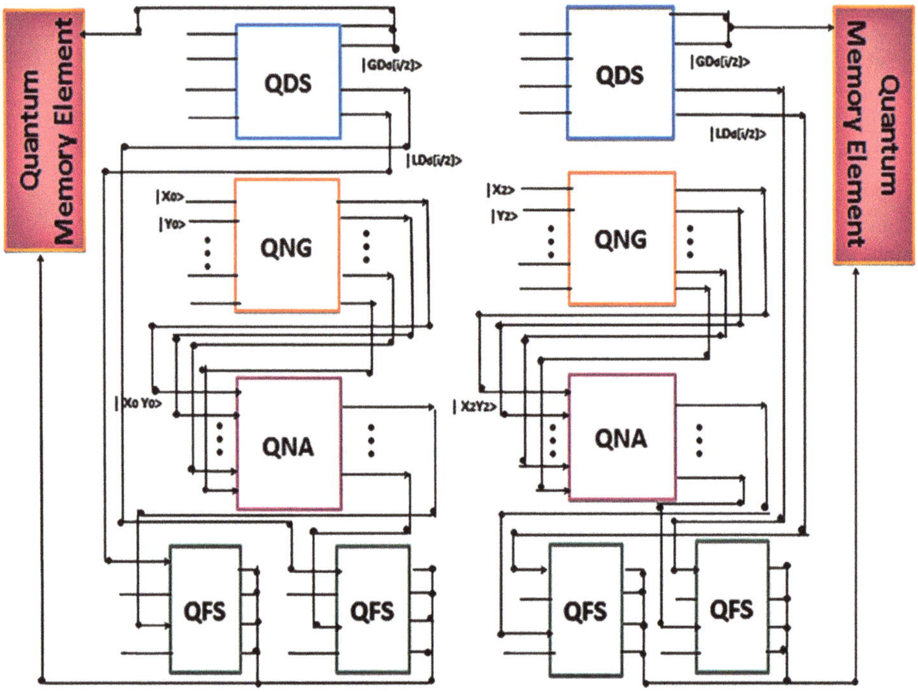

Figure 8.15. Block diagram of a four-qubit quantum divider circuit.

Assume that the statement also holds for $n = k$, which is a k-qubit quantum divider circuit.

Thus $D_{k\text{-qubit}} = 32k + 8\log_2 k$.

Now in the best case, $(k+1)/2$ two-qubit QNGs, $(k+1)/2$ two-qubit QNAs, $(k+1)$ QFS circuits, and $\log_2(k + 1)$ QDS circuits are needed for the $n = (k + 1)$-qubit quantum divider circuit.

Thus the depth of the $(k + 1)$-qubit quantum divider circuit is:

$D_{(k+1)\text{-qubit}} = \log_2(k + 1) \times D_{QDS} + ((k + 1)/2) \times D_{QNG} + ((k + 1)/2) \times D_{QNA} + (k + 1) \times D_{QFS} = 8\log_2(k + 1) + 17(k + 1) + 9(k + 1) + 6(k + 1) = 32 (k + 1) + 8\log_2(k + 1)$.

Therefore, the statement is also true for $n = (k + 1)$.

Thus, the depth of the design of the quantum divider circuit is at least $O(32n + 8\log_2 n)$. □

8.5 Summary

In this chapter the tree-based quantum integer division algorithm is developed. This division technique is a paragon in the emerging technologies of computer aided design, since it has reduced the time complexity to a large extent. The technique reduces the number of ancillary input lines in the quantum circuit, which in turn enables the reduction of the required qubits for the circuit to operate. In addition to this, an algorithm is introduced using a tree structure that is applicable to designing compact quantum divider circuits with optimum depth. The quantum divider design algorithm with optimized depth is an epoch-making achievement in the field of quantum computation.

Further reading

[1] Abdessaied N, Wille R, Soeken M and Drechsler R 2013 Reducing the depth of quantum circuits using additional circuit lines *Int. Conf. Reversible Computation RC 2013 (Lecture Notes in Computer Science* vol 7948*)* ed G W Dueck and D M Miller (Berlin: Springer) pp 221–33

[2] Aggarwal N, Asooja K, Verma S S and Negi S 2009 An improvement in the restoring division algorithm (needy restoring division algorithm) *2009 2nd IEEE Int. Conf. on Computer Science and Information Technology* (Piscataway, NJ: IEEE) pp 246–9

[3] Barenco A, Bennett C H, Cleve R, DiVincenzo D P, Margolus N, Shor P, Sleator J A and Weinfurter H 1995 Elementary gates for quantum computation *Phys. Rev.* A **52** 3457

[4] Bera D, Green F and Homer S 2007 SIGACT News Complexity Theory Column 55 https://mathcs.clarku.edu/~fgreen/papers/constQDepthSurvey.pdf

[5] Cheng K-W and Tseng C-C 2002 Quantum full adder and subtractor *Electron. Lett.* **38** 1343–4

[6] Cormen T H 2001 *Introduction to Algorithms* ed T H Cormen, C E Leiserson, R L Rivest and C Stein (Cambridge, MA: MIT Press)

[7] Cuccaro S A, Draper T G, Kutin S A and Moulton D P 2004 A new quantum ripple-carry addition circuit (arXiv: quant-ph/0410184)

[8] Deutsch D E 1989 Quantum computational networks *Proc. R. Soc. Lond.* A **425** 73–90

[9] Haghparast M and Navi K 2007 A novel reversible full adder circuit for nanotechnology based systems *J. Appl. Sci.* **7** 3995–4000

[10] Kaye P *et al* 2007 *An Introduction to Quantum Computing* (Oxford: Oxford University Press)

[11] Khosropour A, Aghababa H and Forouzandeh B 2011 Quantum division circuit based on restoring division algorithm *2011 8th Int. Conf. on Information Technology: New Generations* (Piscataway, NJ: IEEE) pp 1037–40

[12] Knuth D E 2007 Seminumerical algorithms *The Art of Computer Programming* 3rd edn, vol 2 (Reading, MA: Addison-Wesley)

[13] Niels M and Granlund T 2011 Improved division by invariant integers *IEEE Trans. Comput.* **60** 165–75

[14] Nielsen M A and Chuang I L 2000 Quantum computation *Quantum Information* (Cambridge: Cambridge University Press)

[15] Pavlidis A and Gizopoulos D 2012 Fast quantum modular exponentiation architecture for Shor's factorization algorithm (arXiv:1207.0511)

[16] Radhika Ramya P and Sudha Vani Y 2013 Optimization and implementation of reversible BCD adder in terms of number of lines *Int. J. Reconfig. Embedded Syst.* **2** 21–6

[17] Sultana S and Radecka K 2014 Reversible architecture of computer arithmetic *Int. J. Comput. Appl.* **93** 6

[18] Chi-Chih Yao A 1993 Quantum circuit complexity *Proc. 1993 IEEE 34th Annual Foundations of Computer Science* (Piscataway, NI: IEEE) pp 352–61

Chapter 9

The quantum BCD priority encoder

In quantum circuit design, energy dissipation has become a crucial factor that needs to be taken into consideration. Since irreversible computing is one of the most significant causes of energy dissipation, designing quantum circuits using quantum gates is the efficient way to reduce the energy dissipation of the circuit. Although binary coded decimal (BCD) encoders are one of the most common types of unit circuits, they are all designed in an irreversible way.

The encoder is an essential element in many quantum devices. The BCD encoder is one of the most notable and common encoders in quantum systems. However, the conventional BCD encoder has certain faults in addition to energy dissipation, such as confused output codes when two or more inputs are valid. Considering these drawbacks, it is obviously necessary to focus on identifying a useful quantum BCD priority encoder circuit which not only resolves the problem of confused codes, but also dissipates little energy.

In this chapter the design of a quantum BCD priority encoder circuit that consists of quantum gates is analyzed.

9.1 The properties of an encoder

Unlike a multiplexer that selects one individual data input line and then sends these data to a single output line or switch, a quantum encoder takes all its data inputs one at a time and then converts them into a single encoded output. Thus a quantum encoder is a multi-input quantum logic circuit that converts the logic level 1 data at its inputs into an equivalent code at its output.

Generally, quantum encoders produce outputs of two-bit, three-bit or four-bit codes, depending on the number of data input lines. An n-bit quantum encoder has 2^n input lines and n-bit output lines. A 4-to-2 quantum encoder circuit is shown in figure 9.1.

doi:10.1088/978-0-7503-2747-3ch9

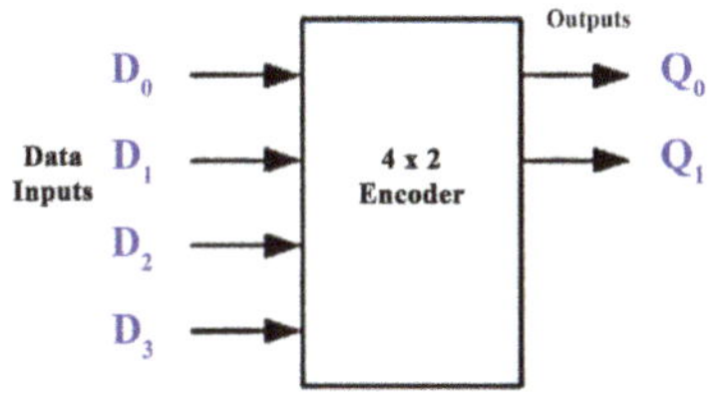

Truth Table

Inputs				Outputs	
D_3	D_2	D_1	D_0	Q_1	Q_0
0	0	0	1	0	0
0	0	1	0	0	1
0	1	0	0	1	0
1	0	0	0	1	1
0	0	0	0	x	x

Figure 9.1. Block diagram of the 4-to-2 quantum encoder circuit and its truth table.

The output lines of a quantum encoder generate the input line whose value is equal to 1 and are available to encode either a decimal or hexa-decimal input pattern to, typically, a binary or BCD output code.

One of the main disadvantages of standard quantum encoders is that they can generate the wrong output code when there is more than one input present at logic level 1. For example, if we make inputs D_1 and D_2 HIGH at logic 1 at the same time, the resulting output is neither at 01 or at 10, but will be at 11 which is an output that is different to the actual input present. Also, an output code of all logic 0s can be generated when all of its inputs are at 0 OR when the input D_0 is equal to one.

One simple way to overcome this problem is to 'prioritize' the level of each input pin and thus, if there was more than one input at logic level 1, the actual output code would only correspond to the input with the highest designated priority. This type of quantum encoder is commonly known as a priority encoder or P-encoder for short.

The priority encoder solves the problems mentioned above by allocating a priority level to each input. The priority encoder's output corresponds to the currently active input which has the highest priority. Thus when an input with a higher priority is present, all other inputs with a lower priority will be ignored.

The priority encoder comes in many different forms, with an example of an eight-input priority encoder along with its truth table shown in figure 9.2.

9.2 The design of a quantum BCD priority encoder circuit

In this section a quantum 2^n-to-n BCD priority encoder circuit is presented. An optimized 2^n-to-n BCD priority encoder circuit is designed using quantum gates. The working principle of an 8-to-3 quantum priority encoder is shown in figure 9.2.

9.2.1 The quantum BCD priority encoder circuit

A traditional 8-to-3 BCD encoder has eight inputs which are I_0 to I_7, and three outputs which are A, B, and C. All the inputs and outputs are required to be binary. The function of a traditional BCD encoder can be represented as

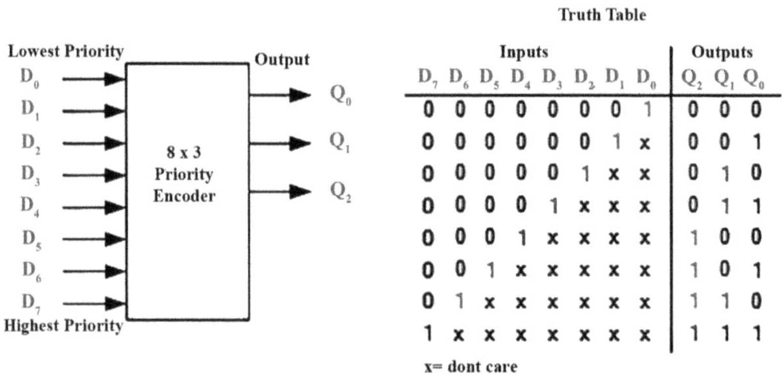

Figure 9.2. Block diagram of the 8-to-3 quantum priority encoder and its truth table.

Table 9.1. The truth table of the quantum BCD priority encoder.

Input								Output		
I_7	I_6	I_5	I_4	I_3	I_2	I_1	I_0	A	B	C
1	X	X	X	X	X	X	X	1	1	1
0	1	X	X	X	X	X	X	1	1	0
0	0	1	X	X	X	X	X	1	0	1
0	0	0	1	X	X	X	X	1	0	0
0	0	0	0	1	X	X	X	0	1	1
0	0	0	0	0	1	X	X	0	1	0
0	0	0	0	0	0	1	X	0	0	1
0	0	0	0	0	0	0	1	0	0	0

$$A = I_7 + I_6 + I_5 + I_4$$
$$B = I_7 + I_6 + I_3 + I_2$$
$$C = I_7 + I_5 + I_3 + I_1.$$

Different to traditional BCD encoders, the function of a BCD priority encoder has priority levels. To be more precise, I_7 has the highest priority level and the other inputs' priority levels decrease in turn, so that I_0 has the lowest. For example, if I_5 and I_3 are both 1 and the rest of the inputs are 0, then since I_5 is in a higher priority level, the encoder will ignore I_3 and regard it as a situation where only I_5 is 1 and the rest of the inputs are all 0. The truth table of a BCD priority encoder is shown in table 9.1. 'X' represents uncertain values of inputs.

In order to design a quantum BCD priority encoder circuit, irreversible logic gates such as AND and OR gates are not allowed in the circuit. Thus the functions of a traditional priority BCD encoder are transformed to a form which can be conveniently converted into a circuit diagram. An 8-to-3 quantum BCD priority encoder is shown in figure 9.3, where quantum NOT gates in all inputs are utilized to obtain the inverse. In the second step the OR function is completed using quantum gates. Similarly, figure 9.4 presents a 4-to-2 quantum BCD priority encoder circuit.

For example, consider the function of A. We obtain the inputs I_7', I_6', I_5', and I_4' at the first step, then connect the inputs I_5', I_4' and the additional input 1 with the quantum gates. Hence it produces $I_5 + I_4$. In the same way we can also obtain $I_7 + I_6$. Finally, by cascading $I_5 + I_4$, $I_7 + I_6$, and the additional input line 1, we obtain $I_7 + I_6 + I_5 + I_4$, which is illustrated in figure 9.3. The generalized algorithm to construct a quantum 2^n-to-n BCD priority encoder circuit is presented in algorithm 9.1.

Algorithm 9.1. Constructing a 2^n-to-n quantum BCD priority encoder circuit.

INPUT: 2^n-bit operand I $(I_{2n}, I_{2n-1}, \ldots, I_1)$
OUTPUT: n-bit operand A $(A_n, A_{n-1}, \ldots, A_1)$
START
for $i = 1$ **to** $2^n - 1$ **do**
    Apply $(2^n - 1)$ quantum NOT gates at each input line to obtain the inverse input
end for
for $i = 1$ **to** $[2(\frac{2^n}{2} - 1) + 2^{n-2}(n-2)]$ **do**
    if $(i = 1$ or $i = [2(\frac{2^n}{2} - 1) + 2^{n-2}(n-2)])$ **then**
        Apply $(\frac{2^n}{2} - 1)$ quantum Toffoli gates to obtain the output $(A_n$ or $A_1)$
    else

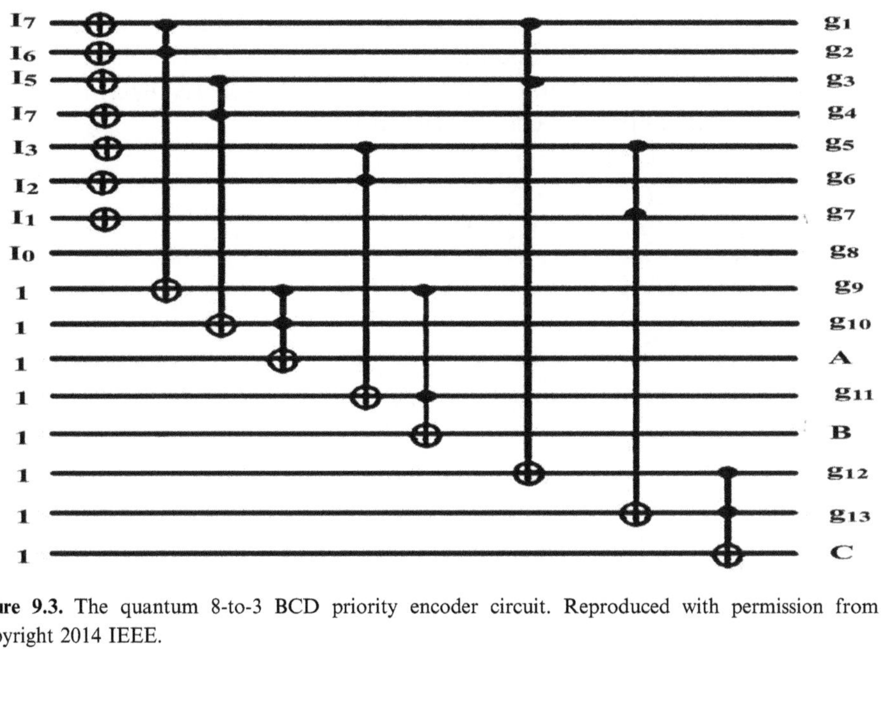

Figure 9.3. The quantum 8-to-3 BCD priority encoder circuit. Reproduced with permission from [6]. Copyright 2014 IEEE.

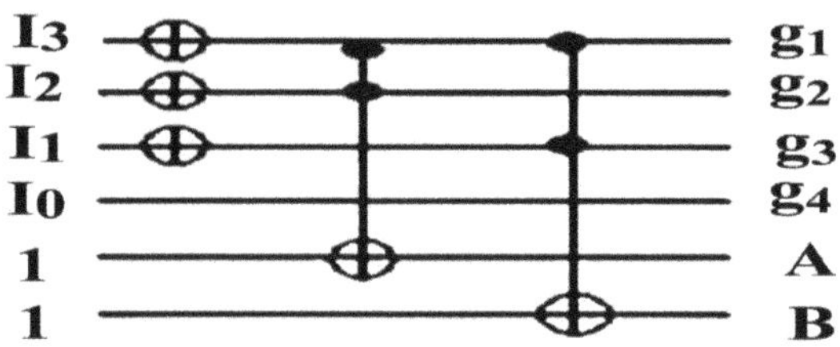

Figure 9.4. The quantum 4-to-2 BCD priority encoder circuit. Reproduced with permission from [6]. Copyright 2014 IEEE.

Apply 2^{n-2} $(n - 2)$ quantum Toffoli gates to obtain the rest of the outputs (A_{n-1}, A_{n-2}, A_{n-3}, …, A_2)
end if
end for
END

9.2.2 Analysis of the properties of the encoder circuit

In this subsection the properties of a quantum 2^n-to-n BCD priority encoder circuit are presented.

Property 9.1. A quantum 2^n-to-n BCD priority encoder circuit ($n \geqslant 2$) can be realized with $[5\,[2(\frac{2^n}{2} - 1) + 2^{n-2}(n - 2)] + (2^n - 1)]$ quantum gates, where n is the number of output bits.

Proof. The property 9.1 is proved by mathematical induction. A 4-to-2 BCD encoder circuit consists of NOT quantum gates and quantum Toffoli gates, and is shown in figure 9.4. Thus the total number of quantum gates (NOQG) required to construct a 4-to-2 BCD priority encoder circuit is

$$\begin{aligned}
&NOQG_{4\text{-to-}2} = NOQG_{NOT} + 5 \times NOQG_{Toffoli}\\
&= (2^2 - 1) + 5 \times [2(\tfrac{2^2}{2} - 1) + 2^{2-2}\,(2 - 2)]\\
&= (4 - 1) + 5 \times [2(2 - 1) + 2 \times 0)]\\
&= 3 + 5 \times 2\\
&= 13\\
&= (4 - 1) + 5 \times [2(2 - 1) + 2 \times 0)].
\end{aligned}$$

Thus, the statement holds for base case $n = 2$.

Assume that the statement holds for $n = k$. Thus a quantum 2^k-to-k BCD priority encoder circuit can be realized with $[5\,[2(\frac{2^k}{2} - 1) + 2^{k-2}(k - 2)] + (2^k - 1)]$ quantum gates.

A 2^{k+1}-to-$(k + 1)$ BCD priority encoder circuit requires $(2^{k+1} - 1)$ quantum NOT gates and $5[2(\frac{2^{k+1}}{2} - 1) + 2^{k+1-2}(k + 1 - 2)]$ quantum Toffoli gates. As a result, the total number of quantum gates required to construct a 2^{k+1}-to-$(k+1)$ BCD priority encoder circuit is $(2^{k+1}-1)+5[2(\frac{2}{2}^{k+1}-1)+2^{k+1-2}(k+1-2)]$.

Thus, the statement holds for $n=k+1$.

Therefore, a quantum 2^n-to-n BCD priority encoder circuit can be realized with $[5\,[2(\frac{2}{2}^{n}-1)+2^{n-2}(n-2)]+(2^n-1)]$ quantum gates.

Property 9.2. A quantum 2^n-to-n BCD priority encoder circuit produces $[5\,[2(\frac{2}{2}^{n}-1)+2^{n-2}(n-2)]+1]\Delta$ quantum delay where Δ is the unit delay and n is the number of output bits.

Proof. A quantum BCD priority encoder circuit consists of quantum Toffoli gates and quantum NOT gates. A quantum Toffoli gate require 5Δ delay and a quantum NOT gate require 1Δ delay. Thus, 2^n quantum Toffoli gates require $5[2(\frac{2}{2}^{n}-1)+2^{n-2}(n-2)]\Delta$ delays. Therefore, the total delay of a quantum 2^n-to-n BCD priority encoder circuit is $[5\,[2(\frac{2}{2}^{n}-1)+2^{n-2}(n-2)]+1]\Delta$, where 2^n-1 is the number of NOT gates at the same level.

Example 9.1. When $n=3$ in figure 9.3, the total delay of an 8-to-3 BCD priority encoder is $[5\,[2(\frac{2}{2}^{3}-1)+2^{3-2}(3-2)]+1]=41\Delta$.

Property 9.3. A quantum 2^n-to-n BCD priority encoder circuit requires $[(3\Omega+2\sigma)[2(\frac{2}{2}^{n}-1)+2^{n-2}(n-2)]+(2^n-1)\,]$ quantum gate calculation complexity, where n is the number of output bits, σ is the CNOT gate calculation complexity, and Ω is the controlled-V or controlled-V^+ gate calculation complexity.

Proof. From property 9.1 it is found that a quantum BCD priority encoder circuit consists of quantum Toffoli gates and quantum NOT gates. A quantum Toffoli gate has 2σ CNOT quantum gate calculation complexity and 3Ω controlled-V (controlled-V^+) quantum gate calculation complexity. Thus the total quantum gate calculation complexity for a 2^n-to-n quantum BCD priority encoder circuit is $2[2(\frac{2}{2}^{n}-1)+2^{n-2}(n-2)]\sigma$ for $2[2(\frac{2}{2}^{n}-1)+2^{n-2}(n-2)]$ CNOT gates, $3[2(\frac{2}{2}^{n}-1)+2^{n-2}(n-2)]\Omega$ for $3[2(\frac{2}{2}^{n}-1)+2^{n-2}(n-2)]$ controlled-V (controlled-V+) gates, and $(2^n-1)\rho$ for (2^n-1) NOT gates, i.e. $[(3\Omega+2\sigma)[2(\frac{2}{2}^{n}-1)+2^{n-2}(n-2)]+\rho(2^n-1)]$ quantum gate calculation complexity.

Example 9.2. When $n=3$ in figure 9.3 the total quantum gate calculation complexity of an 8-to-3 quantum BCD priority encoder circuit is $(3+2)[2(\frac{2}{2}^{3}-1)+2^{3-2}(3-2)]+(2^3-1)\,]=7+16\sigma+24\Omega$.

Property 9.4. A quantum 2^n-to-n BCD priority encoder circuit requires a $[250[2(\frac{2^{n}}{2} - 1) + 2^{n-2}(n - 2)] + 50(2^n - 1)]$ Å area, where n is the number of output bits and Å is the unit of measuring area.

Proof. From property 9.1 it is found that a quantum BCD priority encoder circuit consists of quantum Toffoli gates and quantum NOT gates. A quantum Toffoli gate has five quantum gates. Thus a quantum 2^n-to-n BCD priority encoder circuit requires $5[2(\frac{2^{n}}{2} - 1) + 2^{n-2}(n - 2)]$ quantum Toffoli gates. Moreover, a quantum 2^n-to-n BCD priority encoder needs an extra $(2^n - 1)$ quantum NOT gates to obtain inverted input. Therefore, the total area for a quantum 2^n-to-n BCD priority encoder circuit is $((5[2(\frac{2^{n}}{2} - 1) + 2^{n-2}(n - 2)] \times 50) + (2^n - 1) \times 50)$ Å $= (250[2(\frac{2^{n}}{2} - 1)+ 2^{n-2}(n - 2)] + (2^n - 1) \times 50)$ Å.

Property 9.5. A quantum 2^n-to-n BCD priority encoder circuit requires $142.3 \times [\ 5[2(\frac{2^{n}}{2} - 1) + 2^{n-2}(n - 2)] + (2^n - 1)]$ meV power, where n is the number of output bits and meV is the unit of measuring power.

Proof. From property 9.1 it is found that a quantum BCD priority encoder circuit consists of quantum Toffoli gates and quantum NOT gates. A quantum Toffoli gate has five quantum gates. Thus a quantum 2^n-to-n BCD priority encoder circuit requires $5[2(\frac{2^{n}}{2} - 1) + 2^{n-2}(n - 2)]$ quantum Toffoli gates. Moreover, a quantum 2^n-to-n BCD priority encoder requires an extra $(2^n - 1)$ quantum NOT gates to obtain inverted input. Therefore, the total power for a quantum 2^n-to-n BCD priority encoder circuit is $142.3 \times [5\ [2(\frac{2^{n}}{2} - 1) + 2^{n-2}(n - 2)] + (2^n - 1)]$ meV, where meV is the unit of measuring power.

Property 9.6. A quantum 2^n-to-n BCD priority encoder circuit produces $[2^n + [[2(\frac{2^{n}}{2} - 1) + 2^{n-2}(n - 2)] - n]]$ garbage outputs, where n is the number of output bits.

Proof. The above statement is proved by mathematical induction.

In a quantum realization of a 4-to-2 BCD encoder circuit, each input line produces $2^2 = 4$ garbage outputs, whereas a quantum Toffoli gate produces $[2(\frac{2^{2}}{2} - 1) + 2^{2-2}(2 - 2)] - 2 = [2 \times 1 + 1 \times 0] - 2 = 0$ garbage outputs, as shown in figure 9.4. Thus a 4-to-2 BCD produces $G_{4\text{-to-}2} = G_{\text{input\_lines}} + G_{\text{Toffoli}} = 2^2 + [2(\frac{2^{2}}{2} - 1) + 2^{2-2}(2 - 2) - 2] = 4 + [2 + 0 - 2] = 4$ garbage outputs.

Thus the statement holds for base case $n = 2$.

Assume that the statement holds for $n = k$. Thus a quantum 2^k-to-k BCD priority encoder circuit produces $[2^k + [[2(\frac{2^{k}}{2} - 1) + 2^{k-2}(k - 2)] - k]]$ garbage outputs.

A 2^{k+1}-to-$(k + 1)$ BCD priority encoder circuit produces 2^{k+1} garbage outputs from input lines and $[[2(\frac{2^{k+1}}{2} - 1) + 2^{k+1-2}(k + 1 - 2)] - (k + 1)]$ garbage outputs from quantum Toffoli gates. As a result, the total number of garbage outputs produced for an 2^{k+1}-to-$(k + 1)$ BCD priority encoder circuit is $[2^{k+1} + [2(\frac{2}{2}^{k+1} - 1) + 2^{k+1-2}(k + 1 - 2) - (k + 1)]]$.

Thus the statement holds for $n = k + 1$.

Therefore a quantum 2^n-to-n BCD priority encoder circuit can be realized with $[2^n + [[2(\frac{2}{2}^n - 1) + 2^{n-2}(n - 2)] - n]]$ garbage outputs.

Example 9.3. When $n = 3$ in figure 9.3 is the total garbage outputs of an 8-to-3 BCD priority encoder are $2^3 + [2(\frac{2}{2}^3 - 1) + 2^{3-2}(3 - 2)] - 3 = 8 + 8 - 3 = 13$.

Property 9.7. A quantum 2^n-to-n BCD priority encoder circuit requires $[2(\frac{2}{2}^n - 1) + 2^{n-2}(n - 2)]$ constant inputs.

Proof. From property 9.1 it is found that a quantum BCD priority encoder circuit consists of quantum Toffoli gates and quantum NOT gates. Quantum Toffoli gates are used to perform the OR operation with one constant input. As a quantum 2^n-to-n quantum BCD priority encoder circuit requires $[2(\frac{2}{2}^n - 1) + 2^{n-2}(n - 2)]$ quantum Toffoli gates, the total number of constant inputs is $[2(\frac{2}{2}^n - 1) + 2^{n-2}(n - 2)]$.

Example 9.4. When $n = 3$ in figure 9.3, the total constant inputs of an 8-to-3 BCD priority encoder are $[2(\frac{2}{2}^3 - 1) + 2^{3-2}(3 - 2)] = 6 + 2 = 8$.

9.3 Summary

This chapter presents a design methodology of a 2^n-to-n quantum BCD priority encoder circuit, where n is the number of output bits. An algorithm has been presented for designing a compact 2^n-to-n quantum BCD priority encoder circuit. The circuit has been constructed with the optimum number of quantum gates, garbage outputs, and constant inputs, and optimal delays, quantum gate calculation complexity, area, and power. The efficiency of the design has been proved by presenting several properties. The BCD encoder designed using quantum logic has the obvious advantages of low power dissipation, so it can be applied to wireless sensors, quantum signal processing, parallel circuits, etc.

Further reading

[1] Aspencore Priority encoder *Electronics Tutorials* https://www.electronics-tutorials.ws/combination/comb_4.html (Accessed: 12 May 2018)

[2] Abdel-Hafeez S and Harb S 2006 A VLSI high-performance priority encoder using standard CMOS library *IEEE Trans. Circuits Syst. II: Exp. Briefs.* **53** 597–601

[3] Richard C and Watrous J 2000 Fast parallel circuits for the quantum Fourier transform (arXiv: quant-ph/0006004)

[4] Delgado-Frias J G and Nyathi J 1998 A VLSI high-performance encoder with priority lookahead *Proc. of the 8th Great Lakes Symp. on VLSI* (Piscataway, NJ: IEEE) pp 59–64

[5] Fsaifes I, Lepers C, Obaton A-F and Gallion P 2006 DS-OCDMA encoder/decoder performance analysis using optical low-coherence reflectometry *J. Lightwave Technol.* **24** 3121

[6] Lisa N J and Babu H M H 2014 Minimization of a reversible quantum 2^n-to-n BCD priority encoder *2014 IEEE/ACM Int. Symp. on Nanoscale Architectures* (Piscataway, NJ: IEEE) pp 77–82

[7] Voyiatzis I, Gizopoulos D and Paschalis A 2005 Accumulator-based test generation for robust sequential fault testing in DSP cores in near-optimal time *IEEE Trans. Very Large Scale Integr. Syst.* **13** 1079–86

[8] Wang J-C, Pang Y and Xia Y 2012 A BCD priority encoder designed by reversible logic *Processing 2012 Int. Conf. on Wavelet Active Media Technology and Information* (Piscataway, NJ: IEEE) pp 318–21

[9] Wang W, Swamy M N S and Ahmad M O 2002 A new architecture of RRNS error-correcting QC encoder/decoder and its FPGA implementation *IEEE Int. Symp. on Circuits and Systems* vol 5 (Piscataway, NJ: IEEE) p V

[10] Yunru L, Yeqing Y, Shaojie C and Lijun C 2011 Reversible watermarking algorithm for wireless sensor network using method of multi-node negotiation *2011 Int. Conf. on E-Business and E-Government (ICEE)* (Piscataway, NJ: IEEE) pp 1–4

IOP Publishing

Quantum Computing
A pathway to quantum logic design
Hafiz Md Hasan Babu

Chapter 10

The quantum decoder

Multiple valued logic is a promising approach to reduce the width of quantum circuits. Moreover, quaternary logic is considered a good choice for future quantum computing technology and hence it is very suitable for the encoded realization of binary logic functions through its grouping of two bits together into quaternary values. The quaternary decoder is an essential unit of quaternary quantum systems.

A decoder is a fundamental building block in many computing systems, which can take the form of a multiple-input, multiple-output logic circuit that converts coded inputs into coded outputs, where the input and output codes are different. An n-to-2^n binary decoder converts binary information from n coded inputs to a maximum of 2^n unique outputs. Decoders have been used in memory systems and the I/O of microprocessors.

In this chapter a useful quantum decoder circuit is designed. The quantum decoder circuit that consists of a quantum 2-to-4 decoder circuit and quantum Fredkin gates is analyzed.

10.1 The characteristics of a decoder

The term 'decoder' means to translate or decode coded information from one format into another. Thus a quantum decoder transforms a set of quantum input signals into an equivalent output.

A quantum decoder is a quantum logic device that has inputs of two-bit, three-bit, or four-bit codes depending on the number of data input lines. Thus a decoder that has a set of two or more bits will be defined as having an n-bit code, and therefore it has 2^n possible values. A quantum decoder generally decodes a quantum value.

If a quantum decoder receives n inputs, it activates only one of its 2^n outputs based on that input, where all other outputs are deactivated.

For example, an inverter (NOT gate) can be classified as a 1-to-2 binary decoder as one input and two outputs (2^1) are possible, because with an input A it can produce two outputs A and $\bar{A}$ (not-A) as shown in figure 10.1.

doi:10.1088/978-0-7503-2747-3ch10

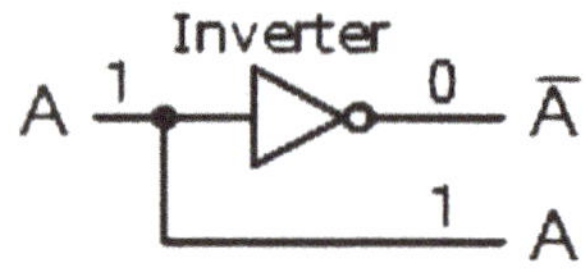

Figure 10.1. The digital inverter circuit.

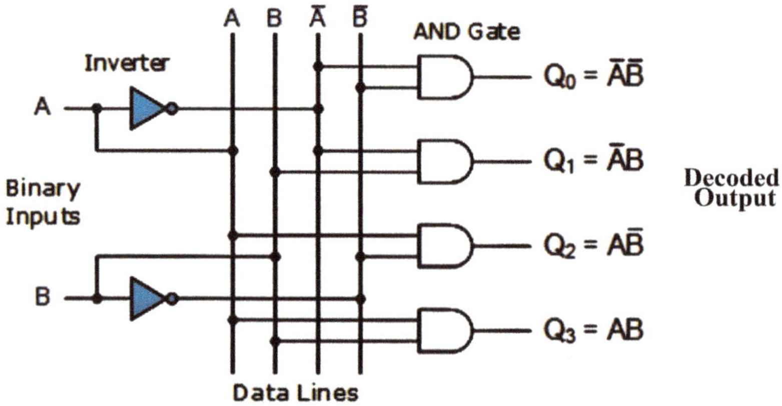

Figure 10.2. The decoder circuit.

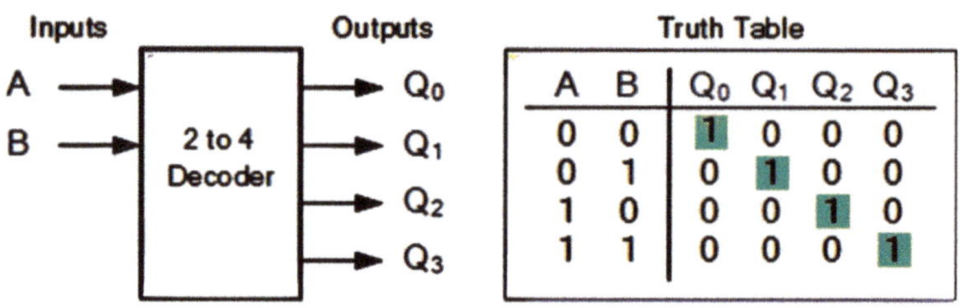

A	B	Q_0	Q_1	Q_2	Q_3
0	0	1	0	0	0
0	1	0	1	0	0
1	0	0	0	1	0
1	1	0	0	0	1

Figure 10.3. Block diagram of a 2-to-4 decoder circuit and truth table.

A standard combinational logic decoder is an n-to-m decoder, where $m \leqslant 2^n$, and whose output Q is dependent only on its present input states. In other words, a binary decoder looks at its current inputs, determines which binary code or binary number is present at its inputs, and selects the appropriate output which corresponds to that binary input.

Figure 10.2 presents a simple example of a 2-to-4 binary decoder circuit which consists of an array of four AND gates.

The block diagram of a 2-to-4 decoder circuit with its truth table is shown in figure 10.3. The two binary inputs labeled A and B are decoded into one of four outputs, where each output represents one of the miniterms of the two input variables (each output = a miniterm).

The binary inputs A and B determine which output line from Q_0 to Q_3 is HIGH with logic level 1 while the remaining outputs are LOW with logic 0. Thus only one output can be active (HIGH) at a time. Therefore, the output line with HIGH identifies its correspondence binary code located at the input. In other words it decodes the binary input.

Some binary decoders have an additional input pin labeled 'enable' that controls the outputs from the device. This extra input allows the decoder outputs to be turned ON or OFF as required. These types of binary decoders are commonly used as 'memory address decoders' in microprocessor memory applications.

It can be said that a binary decoder is a demultiplexer with an additional data line that is used to enable the decoder. An alternative way of looking at the decoder circuit is to regard inputs *A*, *B*, and *C* as address signals. Each combination of *A*, *B*, or *C* defines a unique memory address.

10.2 The design of a quantum decoder

In this subsection a quantum *n*-to-2*n* decoder circuit is shown. The optimized quantum *n*-to-2*n* decoder circuit is described using a 2-to-4 quantum decoder and quantum Fredkin gates.

10.2.1 The quantum decoder circuit

A 2-to-4 decoder generates four logical AND functions $\acute{A}\acute{B}$, $\acute{A}B$, $A\acute{B}$, AB. The quantum 2-to-4 decoder design generates all four necessary AND functions using one quantum Peres gate and three CNOT quantum gates, shown in figure 10.4. A 3-to-8 decoder is designed using one 2-*to*-4 quantum decoder circuit and four quantum Fredkin gates, shown in figure 10.5. Using the same approach, an n-to-2^n decoder can be designed. The design of a 4-to-16 decoder is presented in figure 10.6. The generalized algorithm for constructing a quantum n-to-2^n decoder circuit is presented in algorithm 10.1.

Algorithm 10.1 Constructing an n-to-2^n quantum decoder circuit.

INPUT: n-bit operand $A(A_n, A_{n-1},\ldots, A_1)$ (here, n starts from 3)
OUTPUT: 2^n-bit operand $D(D_{2n}, D_{2n-1},\ldots, D_1)$
START
if $n/2 = 0$ **then**
    The design needs $3(\lceil 2^{n-5} \rceil)$ quantum 2-to-4 gates
    decoder circuits and $(2^n/2)$ quantum Fredkin gates

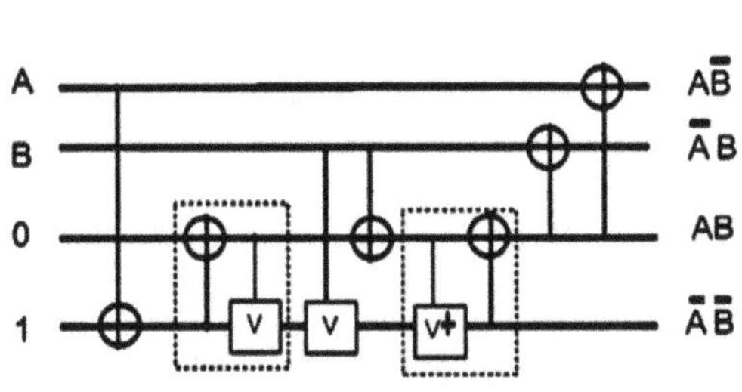

Figure 10.4. The quantum 2-to-4 decoder circuit. Reproduced with permission from [5]. Copyright 2013 IEEE.

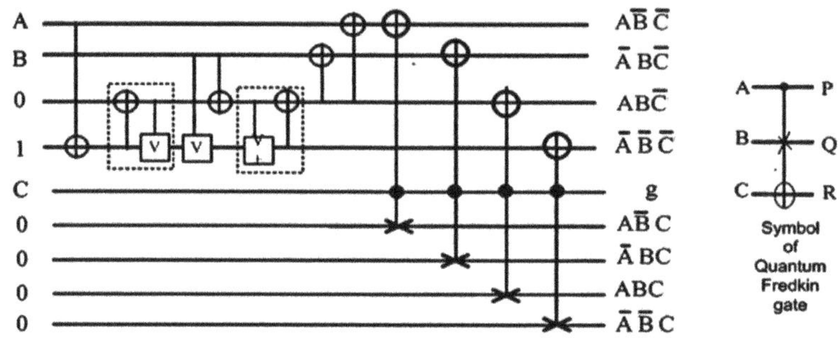

Figure 10.5. The quantum 3-to-8 decoder circuit. Reproduced with permission from [5]. Copyright 2013 IEEE.

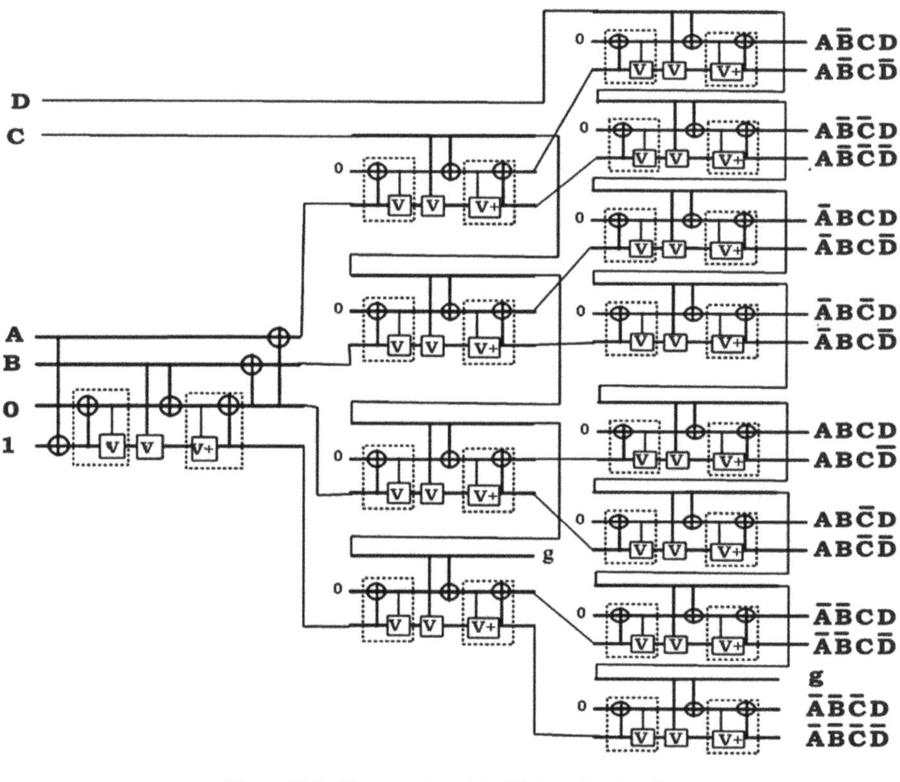

Figure 10.6. The quantum 4-to-16 decoder circuit.

(*Continued*)

---

else
  The design needs $3(\lceil 2^{n-6} \rceil)$ quantum 2-to-4 gates
  decoder circuits and $(2^n/2)$ quantum Fredkin gates
end if
END

10.2.2 Analysis of the properties of the circuits

In this section the properties of a quantum n-to-2^n decoder circuit are presented.

Property 10.1. A quantum n-to-2^n decoder circuit can be realized with $[7 \times (3 \times \lceil 2^{n-5} \rceil) + 5 \times (2^n/2)]$ quantum gates, when $n \geqslant 4$ and $n/2 = 2$, where n is the number of input bits.

Proof. The above statement is proved by mathematical induction. A 2-to-4 decoder circuit requires one quantum Peres gate and three quantum CNOT gates. Thus the number of quantum gates of a 2-to-4 decoder circuit is seven, where the numbers of quantum Peres gates and quantum CNOT gates are 4 and 1, respectively. A 4-to-16 decoder circuit consists of three 2-to-4 decoder circuits and eight quantum Fredkin gates, where the number of quantum Fredkin gates is 5. Thus the total number of quantum gates required to construct a 4-to-16 decoder circuit is $NOQG_{4\text{-to-}16} = 5 \times NOQG_{Fredkin} + 7 \times NOQG_{2\text{-to-}4} = 5 \times (2^4/2) + 7 \times (3 \times \lceil 2^{4-5} \rceil) = 61 = 5 \times (2^4/2) + 7 \times (3 \times \lceil 2^{4-5} \rceil)$. Thus the statement holds for base case $n = 4$. Assume that the statement holds for $n = k$. Thus a quantum k-to-2^k decoder circuit can be realized with quantum gates when $k \geqslant 4$ and $k/2 = 2$.

A $(k + 1)$-to-2^{k+1} decoder circuit requires $7 \times (3 \times (\lceil 2^{k+1-5} \rceil))$ quantum 2-to-4 decoders and quantum Fredkin gates. As a result, the total number of quantum gates required to construct a $(k + 1)$-to-2^{k+1} decoder circuit is $(7 \times (3 \times \lceil 2^{k+1-5} \rceil) + 5 \times (2^{k+1} + 2^n/2))$ when $k \geqslant 4$ and $k/2 = 2$. Thus the statement holds for $n = k + 1$. Therefore a quantum n-to-2^n decoder circuit can be realized with $[7 \times (3 \times \lceil 2^{n-5} \rceil) + 5 \times (2^n/2)]$ quantum gates when $n \geqslant 4$ and $n/2 = 2$.

Similarly, a quantum n-to-2^n decoder circuit can be realized with $[7 \times (3 \times \lceil 2^{n-6} \rceil) + 5 \times (2^n/2)]$ quantum gates when $n \geqslant 4$ and $n/2 \neq 2$.

Property 10.2. A quantum n-to-2^n decoder circuit produces $[7 \times (3 \times \lceil 2^{n-5} \rceil) + 5 \times (2^n/2)]\Delta$ quantum delay when $n \geqslant 4$ and $n/2 = 2$, where Δ is the unit delay.

Proof. From property 10.1 it is found that the number of quantum gates of an n-to-2^n decoder circuit are $[7 \times (3 \times \lceil 2^{n-5} \rceil) + 5 \times (2^n/2)]$, when $n \geqslant 4$ and $n/2 = 2$. Therefore the total delay of a quantum n-to-2^n decoder circuit is $[7 \times (3 \times \lceil 2^{n-5} \rceil) + 5 \times (2^n/2)]\Delta$ quantum delay when $n \geqslant 4$ and $n/2 = 2$.

Similarly, a quantum n-to-2^n decoder circuit produces $[7 \times (3 \times \lceil 2^{n-6} \rceil) + 5 \times (2^n/2)]\Delta$ quantum delay when $n \geqslant 4$ and $n/2 \neq 2$.

Property 10.3. A quantum n-to-2^n decoder circuit requires $350 \times (3 \times \lceil 2^{n-5} \rceil) + 250 \times (2^n/2)$ Å area when $n \geqslant 4$ and $n/2 = 2$, where Å is the unit of measuring area.

Proof. From property 10.1 a quantum 2-to-4 decoder has seven quantum gates and a quantum Fredkin gate has five quantum gates. So a quantum n-to-2^n decoder circuit requires $[7 \times (3 \times \lceil 2^{n-5} \rceil) + 5 \times (2^n/2)]$ quantum gates when $n \geqslant 4$ and $n/2 = 2$. Therefore the total area for a quantum n-to-2^n decoder circuit is $[50 \times 7 \times (3 \times \lceil 2^{n-5} \rceil) + 5 \times (2^n/2)] = 350 \times (3 \times \lceil 2^{n-5} \rceil) + 250 \times (2^n/2)$ Å when $n \geqslant 4$ and $n/2 = 2$.

Similarly, a quantum n-to-2^n decoder circuit requires $350 \times (3 \times \lceil 2^{n-6} \rceil) + 250 \times (2^n/2)$ Å area when $n \geqslant 4$ and $n/2 \neq 2$, where Å is the unit of measuring area.

Property 10.4. A quantum n-to-2^n decoder circuit requires $[996.1 \times (3 \times \lceil 2^{n-5} \rceil) + 711.5 \times (2^n/2)]$ meV power when $n \geqslant 4$ and $n/2 = 2$, where meV is the unit of measuring power.

Proof. From property 10.1 a 2-to-4 quantum decoder has seven quantum gates and a quantum Fredkin gate has five quantum gates. Thus a quantum n-to-2^n decoder circuit requires $[7 \times (3 \times \lceil 2^{n-5} \rceil) + 5 \times (2^n/2)]$ quantum gates when $n \geqslant 4$ and $n/2 = 2$. Therefore the total power for a quantum n-to-2^n decoder circuit is $(142.3 \times [7(3 \times \lceil 2^{n-5} \rceil) + 5 \times (2^n/2)])$ meV power when $n \geqslant 4$ and $n/2 = 2$, where meV is the unit of measuring power.

Similarly, a quantum n-to-2^n decoder circuit requires $[996.1 \times (3 \times \lceil 2^{n-6} \rceil) + 711.5 \times (2^n/2)]$ meV power when $n \geqslant 4$ and $n/2 \neq 2$, where meV is the unit of measuring power. □

Property 10.5. A quantum n-to-2^n decoder circuit can be realized with $(2(3 \times \lceil 2^{n-5} \rceil) + (2^n/2))$ constant inputs when $n \geqslant 4$ and $n/2 = 2$, where n is the number of input bits.

Proof. The above statement is proved by mathematical induction. The quantum 2-to-4 decoder circuit requires two constant inputs and a quantum Fredkin gate requires one constant input. A 4-to-16 decoder circuit requires six constant inputs from three 2-to-4 decoder circuits and eight constant inputs from eight quantum Fredkin gates. Thus the total number of constant inputs required to construct a 4-to-16 decoder circuit ($NOQG_{4\text{-to-}16}$) is $NOCI_{4\text{-to-}16} = NOCI_{Fredkin} + NOCI_{2\text{-to-}4} = (2^4/2) + 2 \times (3 \times \lceil 2^{4-5} \rceil) = 14 = (2^4/2) + 2 \times (3 \times \lceil 2^{4-5} \rceil)$. Thus the statement holds for base case $n = 4$. Assume that the statement holds for $n = k$. Thus a quantum k-to-2^k decoder circuit can be realized with $(2(3 \times \lceil 2^{k-5} \rceil) + (2^k/2))$ constant inputs when $k \geqslant 4$ and $k/2 = 2$.

A $(k + 1)$-to-2^{k+1} decoder circuit requires $(2^{k+1}/2)$ constant inputs for quantum Fredkin gates and $(2(3 \times \lceil 2^{k+1-5} \rceil))$ constant inputs for quantum 2-to-4 decoders. As a result, the total number of constant inputs required to construct a $(k + 1)$-to-2^{k+1} decoder circuit is $(2(3 \times \lceil 2^{k+1-5} \rceil) + (2^{k+1}/2))$ when $n \geqslant 4$ and $n/2 = 2$. Thus the statement holds for $n = k + 1$. Therefore a quantum n-to-2^n decoder circuit can be realized with $(2(3 \times \lceil 2^{n-5} \rceil) + (2^n/2))$ constant inputs when $n \geqslant 4$ and $n/2 = 2$, where n is the number of input bits.

Similarly, a quantum n-to-2^n decoder circuit can be realized with $(2(3 \times \lceil 2^{n-6} \rceil) + (2^n/2))$ constant inputs when $n \geqslant 4$ and $n/2 \neq 2$, where n is the number of input bits. □

Property 10.6. A quantum n-to-2^n decoder circuit requires $(3\Omega + 4\sigma)[3 \times \lceil 2^{n-5} \rceil) + (2^n/2)]$ quantum gate calculation complexity when $n \geqslant 4$ and $n/2 = 2$, where σ is a CNOT gate calculation complexity and Ω is a controlled-V or controlled-V^+ gate calculation complexity.

Proof. From property 10.1 a quantum 2-to-4 decoder and a quantum Fredkin gate both have 4σ CNOT quantum gate calculation complexity and 3Ω controlled-V (controlled-V^+) quantum gate calculation complexity. Thus the total quantum gate calculation complexity for an n-to-2^n decoder circuit is $(3\Omega + 4\sigma)[3 \times \lceil 2^{n-5} \rceil) + (2^n/2)]$ quantum gate calculation complexity when $n \geqslant 4$ and $n/2 = 2$, where σ is a CNOT gate calculation complexity and Ω is a controlled-V or controlled-V^+ gate calculation complexity.

Similarly, a quantum n-to-2^n decoder circuit requires $(3\Omega + 4\sigma)[3 \times \lceil 2^{n-6} \rceil) + (2^n/2)]$ quantum gate calculation complexity, when $n \geqslant 4$ and $n/2 \neq 2$, where σ is a CNOT gate calculation complexity and Ω is a controlled-V or controlled-V^+ gate calculation complexity.

10.3 Summary

This chapter presents a design methodology of an n-to-2^n quantum decoder circuit, where n is the number of input bits. A technique to calculate the quantum gate complexity of quantum circuits has been shown. The efficiency of the design is proved by several properties. The decoder designed using quantum logic has obvious advantages of low power dissipation. Thus it can be applied very well in wireless sensors, network components, quantum signal processing, etc.

Further reading

[1] Aspencore Binary decoder *Electronics Tutorials* https://www.electronics-tutorials.ws/combination/comb_5.html (Accessed: 5 December 2018)

[2] Weisstein E W Square root *Wolfram MathWorld* http://mathworld.wolfram.com/SquareRoot.html (Accessed: 22 December 2018)

[3] Chrzanowska-Jeske M 1993 Architecture and synthesis issues in FPGAS *Proc. of NORTHCON'93 Electrical and Electronics Convention* pp 102–5

[4] Haghparast M and Monfared A T 2017 Novel quaternary quantum decoder, multiplexer and demultiplexer circuits *Int. J. Theor. Phys.* **56** 1694–707

[5] Lisa N J and Babu H M H 2013 A compact realization of a reversible quantum n-to-2^n decoder *2013 IEEE 4th Int. Conf. on Electronics Information and Emergency Communication* (Piscataway, NJ: IEEE) pp 90–3

[6] Sharmin F, Polash M M A, Shamsujjoha M, Jamal L and Babu H M H 2011 Design of a compact reversible random access memory *4th IEEE Int. Conf. on Computer Science and Information Technology* vol 10 pp 103–7

[7] Voyiatzis I, Gizopoulos D and Paschalis A M 2005 Accumulator-based test generation for robust sequential fault testing in DSP cores in near-optimal time *IEEE Trans. Very Large Scale Integr. Syst.* **13** 1079–86

[8] Yunru L, Yeqing Y, Shaojie C and Lijun C 2011 Reversible watermarking algorithm for wireless sensor network using method of multi-node negotiation *Int. Conf. on E-Business and E-Government* (Piscataway, NJ: IEEE) pp 1–4

IOP Publishing

Quantum Computing
A pathway to quantum logic design
Hafiz Md Hasan Babu

Chapter 11

The quantum square root circuit

A square root of x is a number r such that $r^2 = x$. It is also written in the form $x^{1/2}$ or, in particular, $\sqrt{x}$. This chapter is focused on identifying a useful quantum square root circuit. The quantum conditions and design quantum square root circuit are analyzed. These circuits consist of quantum adder/subtractor (QAS) circuits and CNOT gates.

11.1 Properties of a square root function

The principal square root function $f(x) = \sqrt{x}$ (usually just referred to as the 'square root function') is a function that maps the set of nonnegative real numbers onto itself. In geometrical terms, the square root function maps the area of a square to its side length.

The square root of x is rational if and only if x is a rational number that can be represented as a ratio of two perfect squares. The square root function maps rational numbers into algebraic numbers (a superset of the rational numbers).

For all real numbers

$$\sqrt{x^2} = |x| = x = \begin{cases} x & \text{if } x \geqslant 0 \\ -x & \text{if } x < 0. \end{cases}$$

For all nonnegative real numbers x and y, $\sqrt{xy} = \sqrt{x}\sqrt{y}$ and $\sqrt{x} = x^{\frac{1}{2}}$.

The square root function is continuous for all nonnegative x and differentiable for all positive x. If f denotes the square root function, its derivative is given by $f'(x) = \frac{1}{2\sqrt{x}}$

The Taylor series of $\sqrt{1} + x$ about $x = 0$ converges for $|x| \leqslant 1$:

$$\sqrt{1+x} = \sum_{n=0}^{\infty} \frac{(-1)^n (2n)!}{(1-2n)(n!)(4^n)} x^n$$
$$= 1 + \frac{1}{2}x - \frac{1}{8}x^2 + \frac{1}{16}x^3 - \frac{5}{128}x^4 + \cdots.$$

The square root of a nonnegative number is used in the definition of the Euclidean norm (and distance), as well as in generalizations such as Hilbert spaces. It defines an important concept of standard deviation using the theory of probability and statistics. It has a major use in the formula for roots of a quadratic equation, such as quadratic fields and rings of quadratic integers which are based on square roots. These are important in algebra and have uses in geometry. Square roots frequently appear elsewhere in mathematical formulas, as well as in many physical laws.

11.2 The design of a quantum square root circuit

In this section an n-bit quantum square root circuit is shown. The optimized n-bit quantum square root circuit is designed using QAS circuits and CNOT gates.

11.2.1 The quantum adder/subtractor circuit

A QAS circuit is constructed using a Haghparast–Navi gate (HNG) quantum circuit and a CNOT quantum gates. It is constructed directly at the quantum level. A one-bit QAS circuit is a combinational quantum circuit which adds three bits, X, Y, Z (where Z is the carry-in for the circuit), and generates a sum $S(X \oplus Y \oplus Z)$ and a carry-out C_{out}, $((X \oplus Y)Z \oplus XY)$, or subtracts $X - Y - Z$ resulting in a difference D, $(X \oplus Y \oplus Z \oplus 1)$, and a borrow B_{out}, $((X \oplus Y)Z \oplus XY \oplus 1)$. Thus the inputs to this circuit are data signals, X, Y, and Z, a control signal A/S, and a constant input bit 0. The outputs are sum, carry, A/Sg, which is an A/S control signal propagating to the next block, and two garbage outputs g_1 and g_2. A CNOT quantum gate with a control signal A/S and a target Y, as well as a full-adder implemented by a HNG quantum circuit, contribute to the QAS block. When A/S is set to 0, then CNOT passes the true copy of Y. When A/S is 1, then Y' is produced at the output of CNOT, and $X + Y' + Z$ is performed, where $'$ denotes the complement sign. The A/Sg output of each block is used as a control signal A/S to the consecutive block. To obtain a true copy of input Y for further operation, another CNOT gate is added between lines A/S and $A/S \oplus Y$ with control at A/Sg, which performs the operation $A/S \oplus Y \oplus A/S = Y$. The structure of a QAS circuit is shown in figure 11.1.

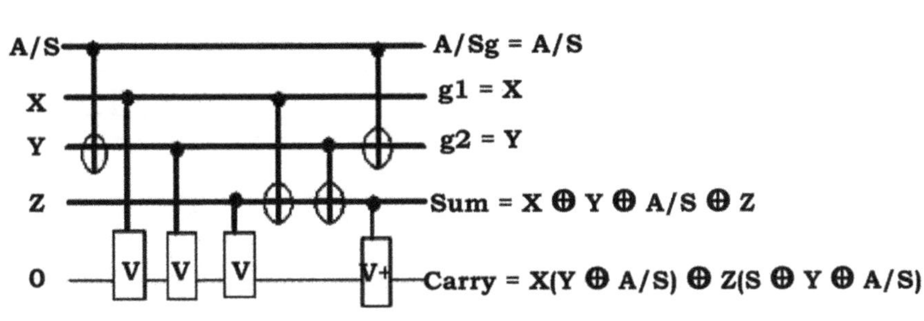

Figure 11.1. The QAS circuit.

11.2.2 The quantum square root circuit

An array structure of a non-restoring square root circuit uses classical controlled adder/subtractor blocks. The $(2 \times 4) = 8$-bit non-restoring circuit realizes a digit-by-digit scheme, where at each iteration computed in each row, only one digit of the square root is performed. Based on this structure the quantum square root circuit is created using the QAS circuit presented in figure 11.1. A few additional CNOT gates are added to provide fan-out signals. In the quantum square root circuit, the QAS circuit is incorporated from figure 11.1 and reuses the copy of input signal Y of the previous stage for the next stage square root operation. A $(2 \times 4) = 8$-bit quantum square root circuit with the input $x_1x_2x_3x_4x_5x_6x_7x_8$, the four-bit square root output $q_1q_2q_3q_4$, and the eight-bit remainder output $r_1r_2r_3r_4r_5r_6r_7r_8$ are presented in figure 11.2. The order of the QAS signal propagation is the direction from right to left. Hence the right-most QAS2 circuit in the top row of figure 11.2 takes the control input A/S. In the classical design, first the A/S signal is set to 1 in order to calculate the first digit of the square root. A/S input to 1 is also set. The outputs A/Sg and C_{out} of the QAS2 circuit are connected to the left QAS1 circuit. To provide the inverted A/S as well as A/S itself, each inverter in a classical implementation is replaced by a CNOT. In addition, since each square root bit from each row (q_i) is the control signal for the next row, as shown in figure 11.2, it is needed to fan-out q_i. In quantum implementation a CNOT gate is used to generate a copy of a required signal. For example, to obtain a copy of q_1, a CNOT gate is also used with the I/O mapping q_1, $0 \rightarrow q_1, q_1 \oplus 0$.

A two-bit quantum square root circuit with four-bit input $x_1x_2x_3x_4$, two-bit square root output q_1q_2, and four-bit remainder output $r_1r_2r_3r_4$ is constructed using six QAS circuits and six CNOT gates arranged in two rows. Therefore, an n-bit

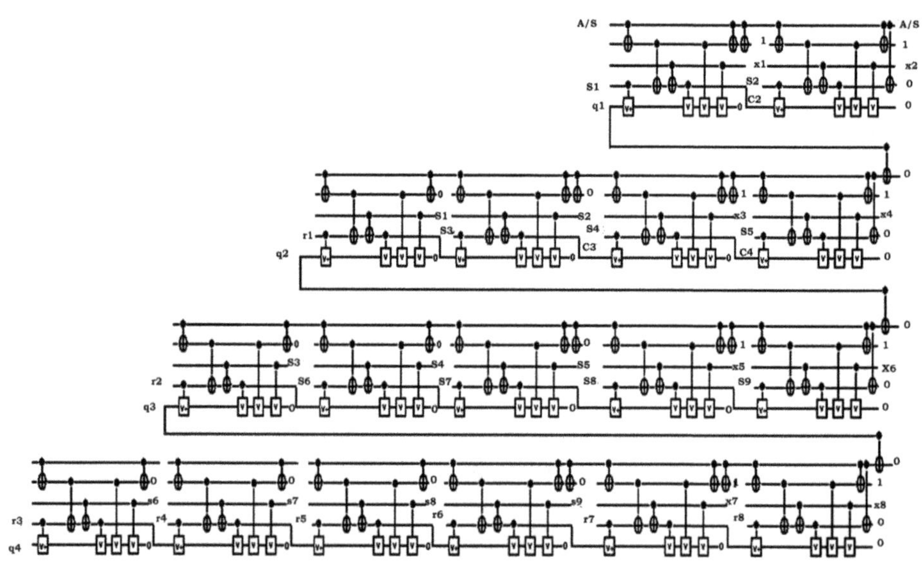

Figure 11.2. The $(2 \times 4) = 8$-bit quantum square root circuit.

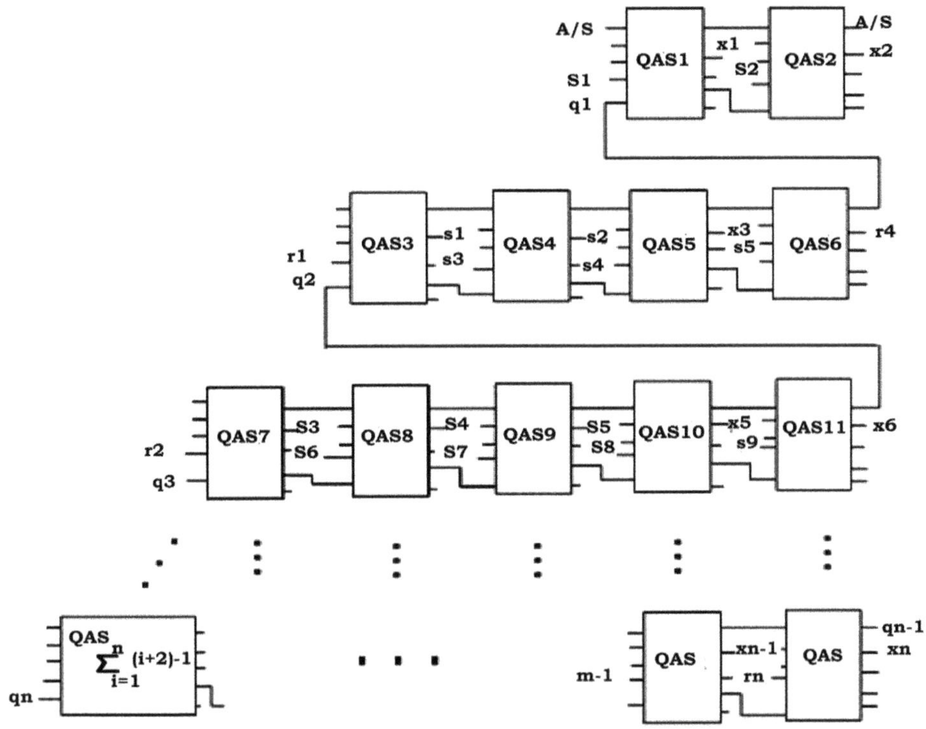

Figure 11.3. Block diagram of an n-bit quantum square root circuit.

quantum square root circuit with the $2n$-bit input $x_1x_2,\ldots, x_{2n}$, the n-bit square root output $q_1q_2,\ldots, q_n$, and the $2n$-bit remainder output $r_1r_2,\ldots, r_{2n}$ can be constructed using $[\sum_{i=1}^{n}(i+2)-1]$ QAS circuits and $4n-2$ CNOT gates arranged in n rows, presented in figure 11.3, where n is the number of output bits. In addition, $4n-2$ CNOT gates are used to control the fan-out signals. The generalized algorithm for constructing a compact n-bit quantum square root circuit is presented in algorithm 11.1.

Algorithm 11.1. Constructing an n-bit quantum square root circuit.

INPUT: $2n$-bit operand X
OUTPUT: n-bit quotient q
1: START
2: **for** $i = 1$ **to** $\sum_{k=1}^{n}(k+2)-1$ **do**
3: **if** $i = 1$ **then**
4: **for** $j = 2i$ **to** 1 **do**
5: **if** $j > 1$ **then**
6: Input := $\{A/S, 1, X_j, 0, 0\}$ and Output := $\{A/S, g, g, S_j, C_j\}$

else
Input := $\{A/S, 1, X_j, C_{j+1}, 0\}$ and Output := $\{A/S, g, g, r_j, q_j\}$
end if
end for
else if $i > 1$ **then**
for $j = 2i$ **to** 1 **do**
if $j = 2i$ **then**
Input := $\{A/S, 1, X_j, 0, 0\}$ and Output := $\{A/S, g, g, S_j, C_j\}$
else if $j > 2i - 2$ **then**
Input := $\{A/S, 1, X_j, C_{j+1}, 0\}$ and Output := $\{A/S, g, g, S_j, C_j\}$
else if $j > 1$ **then**
Input := $\{A/S, 0, S_{j-2i}, C_{j+1}, 0\}$ and Output := $\{A/S, g, g, S_j, C_j\}$
else
Input := $\{A/S, 0, S_{j-2i}, C_{j+1}, 0\}$ and Output := $\{A/S, g, g, r_j, q_j\}$
end if
end for
end if
end for
END

11.2.3 Analysis of the properties of the circuit

In this subsection the properties of an n-bit quantum square root circuit are presented.

Property 11.1. An n-bit quantum square root circuit can be realized with $[8[\sum_{i=1}^{n}(i+2) - 1] + 4n - 2]$ quantum gates, where n is the number of output bits.

Proof. The property 11.1 is proved by mathematical induction.

A two-bit square root requires six QAS circuits where the number of quantum gates of a QAS circuit is eight. Moreover, this circuit needs an extra six CNOT gates. Thus the total number of quantum gates required to construct a two-bit quantum square root (NOG_2) circuit is

$$\begin{aligned} NOG_2 &= \sum_{i=1}^{2}[(i+2) - 1] + (4 \times 2 - 2) \\ &= 8 \times [(1+2) + (2+2) - 1] + 6 \\ &= 8 \times [3 + 4 - 1] + 6 \\ &= 54. \end{aligned}$$

Thus, the statement holds for base case $n = 2$.

Assume that the statement holds for $n = k$. Thus a quantum k-bit square root circuit can be realized with $[8[\sum_{i=1}^{k}(i+2) - 1] + 4k - 2]$ quantum gates.

A $(k + 1)$-bit quantum square root circuit requires $[8[\sum_{i=1}^{k+1}(i + 2) - 1]]$ QAS circuits and $[4(k + 1) - 2]$ CNOT gates. As a result, the total number of quantum gates required to construct a $(k + 1)$-bit quantum square root circuit is $[8[\sum_{i=1}^{k+1}(i + 2) - 1] + 4(k + 1) - 2]$.

Thus, the statement holds for $n = k + 1$.

An n-bit quantum square root circuit can be realized with $8[\sum_{i=1}^{n}(i + 2) - 1) + 4n - 2$ quantum gates, where n is the number of output bits.

Property 11.2. An n-bit quantum square root circuit produces the following quantum delay:

$(8[\sum_{i=1}^{n}(i + 2) - 1] + 4n - 2 - [\sum_{i=1}^{n}(i + 2) - 1])\Delta$, where n is the number of output bits and Δ is the unit of delay.

Proof. A quantum square root circuit consists of QAS circuits and CNOT gates. In a one-bit QAS circuit, 8Δ delay is required. Thus the total delay for an n-bit QAS circuit is $(8[\sum_{i=1}^{n}(i + 2) - 1])\Delta$ delays. A quantum square root circuit requires $(4n - 2)\Delta$ additional delays. As there are $[\sum_{i=1}^{n}(i + 2) - 1]$ controlled-V^+ gates at the same level with the CNOT gate of an n-bit quantum square root circuit, the $[\sum_{i=1}^{n}(i + 2) - 1]\Delta$ delays can be deducted from the total delay of the quantum square root circuit. Therefore, the total delay of a quantum square root circuit is $(8[\sum_{i=1}^{n}(i + 2) - 1] + 4n - 2 - [\sum_{i=1}^{n}(i + 2) - 1])\Delta$.

Example 11.1. When $n = 4$ in figure 11.1, the total delay of a $(2 \times 4) = 8$-bit quantum square root circuit is $(8[\sum_{i=1}^{4}(i + 2) - 1] + 4 \times 4 - 2 - [\sum_{i=1}^{4}(i + 2) - 1])\Delta = 133\Delta$.

Property 11.3. An n-bit quantum square root circuit requires $[(4 + 4)[\sum_{i=1}^{n}(i + 2) - 1] + 4n - 2]$ quantum gate calculation complexity, where n is the number of output bits, σ is the CNOT gate calculation complexity, and Ω is a controlled-V or controlled-V^+ gate calculation complexity.

Proof. From property 11.1, it is found that a quantum square root circuit consists of a QAS circuit and CNOT gates. A one-bit QAS circuit has eight quantum gates. Thus an n-bit quantum square root circuit requires $8[\sum_{i=1}^{n}(i + 2) - 1]$ QAS circuits. Moreover, a quantum square root circuit needs an extra $4n - 2$ CNOT gates. Therefore, the total area for an n-bit quantum square root circuit is $((8[\sum_{i=1}^{n}(i + 2) - 1] \times 50) + (4n - 2) \times 50))$ Å $= 400[\sum_{i=1}^{n}(i + 2) - 1] + 200n - 100$ Å.

Property 11.4. An n-bit quantum square root circuit produces $1138.4[\sum_{i=1}^{n}(i + 2) - 1] + 569.2n - 284.6$ meV, where n is the number of output bits and meV is the unit of measuring power.

Proof. From property 11.1, it is found that a quantum square root circuit consists of a QAS circuit and CNOT gates. A one-bit QAS circuit has eight quantum gates. Thus an n-bit quantum square root circuit has $8[\sum_{i=1}^{n}(i + 2) - 1]$ QAS circuits.

Moreover, a quantum square root circuit needs an extra $4n - 2$ CNOT gates. Therefore, the total power for an n-bit quantum square root circuit is $((8\,[\sum_{i=1}^{n}(i + 2) - 1] \times 142.3) + (4n - 2) \times 142.3)$ meV $= 1138.4[\sum_{i=1}^{n}(i + 2) - 1] + 569.2n - 284.6$ meV.

11.3 Summary

This chapter presents the design methodology of an n-bit quantum square root circuit, where n is the number of output bits. An algorithm is shown for designing the compact n-bit quantum square root circuit using a quantum adder/subtractor circuit. A new technique to calculate the quantum gate complexity of quantum circuits is shown in this chapter. It is also shown that the circuit is constructed with the optimum number of quantum gates and garbage outputs, and optimal delays, quantum gate calculation complexity, area, and power. The design is more scalable and performs much better than its counterparts, e.g. the n-bit quantum square root circuit requires $(8[\sum_{i=1}^{n}(i + 2) - 1] + 4n - 2)$ quantum gates, $(8[\sum_{i=1}^{n}(i + 2) - 1] + 4n - 2 - [\sum_{i=1}^{n}(i + 2) - 1])\Delta$ delay, $((4\Omega + 4\sigma)[\sum_{i=1}^{n}(i + 2) - 1] + 4n - 2)$ quantum gate calculation complexity, $(400[\sum_{i=1}^{n}(i + 2) - 1] + 200n - 100)A^{\circ}$ area, and $(1138.4[\sum_{i=1}^{n}(i + 2) - 1] + 569.2n - 284.6)$ meV power, where n is the number of output bits, σ is the CNOT gate calculation complexity, Ω is the controlled-V (controlled-V^{+}) gate calculation complexity, ρ is the NOT gate calculation complexity, Δ is the unit delay, Å is the unit of measuring area, and meV is the unit of measuring power. The efficiency of the design is also proved by several properties. The square root is the basic arithmetic unit of a computer system. For example, numerical analysis, complex number computations, statistical analysis, computer graphics, and radar signal processing are among the fields where square root circuits can be used.

Further reading

[1] Square root *Wikipedia* https://en.wikipedia.org/wiki/Square_root (Accessed: 6 December 2018)

[2] Semiconductor Industry Association 2006 International technology roadmap for semiconductors *ITRS* http://www.itrs.net/

[3] Balandin A A and Wang K L 1999 Implementation of quantum controlled-NOT gates using asymmetric semiconductor quantum dots *Quantum Computing and Quantum Communications* (Berlin: Springer) pp 460–7

[4] Black P E, Kuhn D R and Williams C J 2002 Quantum computing and communication *Advances in Computers* vol 56 (Amsterdam: Elsevier) pp 189–244

[5] Gandhi S M, Devishree J and Mohan S S 2014 A new reversible SMG gate and its application for designing two's complement adder/subtractor with overflow detection logic for quantum computer-based systems *Computational Intelligence, Cyber Security and Computational Models* (Berlin: Springer) pp 259–66

[6] Hamacher V C, Vranesic Z G and Zaky S G 1987 *Computer Organization* (New York: McGraw-Hill)

[7] Jain V K and Lin L 1998 Nonlinear DSP coprocessor cells-one and two cycle chips *Proc. of the 1998 IEEE Int. Symp. on Circuits and Systems* vol 2 (Piscataway, NJ: IEEE) pp 264–7

[8] Samavi S, Sadrabadi A and Fanian A 2008 Modular array structure for non-restoring square root circuit *J. Syst. Archit.* **54** 957–66

[9] Shor P W 1999 Polynomial-time algorithms for prime factorization and discrete logarithms on a quantum computer *SIAM Rev.* **41** 303–32

[10] Sultana S and Radecka K 2011 Reversible implementation of square-root circuit *18th IEEE Int. Conf. on Electronics, Circuits and Systems* (Piscataway, NJ: IEEE) pp 141–4

[11] Sultana S, Radecka K and Pang Y 2011 A study on relating redundancy removal in classical circuits to reversible mapping *IEEE 29th Int. Conf. on Computer Design* (Piscataway, NJ: IEEE) pp 206–11

Chapter 12

Quantum latches and counter circuits

A latch is a circuit that has two stable states and can be used to store state information. It is the basic storage element in sequential logic. A counter is a register which produces a specified output sequence or counts the clock pulses arriving at a clock input. It has a wide variety of applications in almost all fields, such as industrial, domestic, security, surveillance, and communication, to name just a few. The properties along with the designs for quantum latches and quantum counters are discussed in this chapter.

12.1 Properties of latches

The current model of computing is based on the concept of a state machine. Such a machine exhibits different behaviors depending on the value of its current state. Thus some structure is required to keep track of this state. The basic building block used in the memory of current computers is the flip–flop, which themselves are built out of latches. There are various types of latches, but the basic type upon which other designs are generally based is the set–reset (SR) latch. The primary inputs of such a structure are *S* (set) and *R* (reset). A diagram illustrating a NOR-gate implementation of an SR latch is given in figure 12.1.

A variety of quantum latches has been introduced by several research works. This chapter examines the behavior of some of these latches. Quantum realizations of the latches are also included in this chapter. These quantum latches may be used to emulate a traditional SR latch. The behavior of the SR latch is characterized by the truth table given in table 12.1.

12.2 The design of the quantum latches

In this section, some quantum latches are shown. Subsections 12.2.1–12.2.4 illustrate the quantum SR latch, quantum D latch, quantum T latch, and quantum J-K latch, respectively.

doi:10.1088/978-0-7503-2747-3ch12

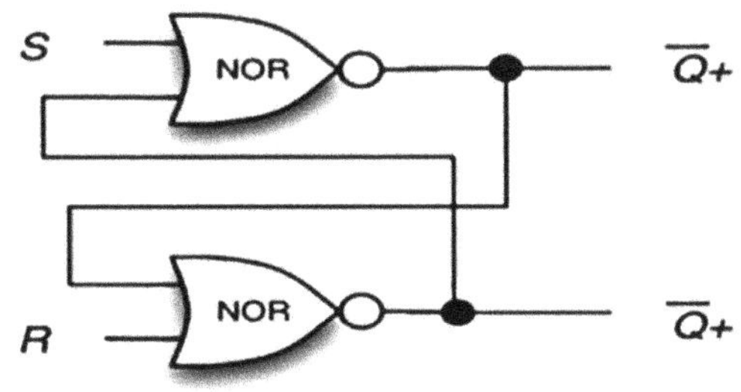

Figure 12.1. The SR latch (built from traditional logic gates). Reproduced with permission from [3]. Copyright 2013 IEEE.

Table 12.1. The next state values for the SR latch.

Inputs		Next state	
S	R	Q^+	$\overline{Q^+}$
0	0	Q	$\bar{Q}$
0	1	0	1
1	0	0	1
1	1	Not permitted	

12.2.1 The quantum SR latch

The SR latch can be designed using two cross-coupled NOR gates or two cross-couple NAND gates. The S input sets the latch to 1, while the R input resets the latch to 0. The cross-coupled NOR-gate strategy was used to design the Fredkin gate based SR latch and the cross-coupled NAND gate strategy was used to design the quantum Toffoli gate based SR latch. The modified Toffoli gate (MTG) based SR latch was also introduced. These quantum SR latch designs do not have the enable signal (clock).

The characteristic equation of the SR latch can be written as $Q^+ = S + \bar{R} \cdot Q$. From the characteristic equation we observe that for five-input combinations ($S = 0$, $R = 0$, $Q = 1$), ($S = 1$, $R = 0$, $Q = 0$), ($S = 1$, $R = 0$, $Q = 1$), ($S = 1$, $R = 1$, $Q = 0$), and ($S = 1$, $R = 1$, $Q = 1$), we have $Q^+ = 1$. In order to design a quantum SR latch that can work for all input combinations, we modify the output Q for the selective input combinations as follows. When ($S = 1$, $R = 1$, $Q = 0$) the output Q^+ is assigned the value of 0, and when ($S = 1$, $R = 1$, $Q = 1$) the output Q^+ is assigned the value of 1. Thus, when $S = 1$ and $R = 1$, we have $Q^+ = Q$. After the modifications in the output for these two selective input combinations, we observe that now for only four input combinations ($S = 0$, $R = 0$, $Q = 1$), ($S = 1$, $R = 0$, $Q = 0$), ($S = 1$, $R = 0$, $Q = 1$), and ($S = 1$, $R = 1$, $Q = 1$), do we have $Q^+ = 1$.

The modified truth table of the SR latch is shown in table 12.2. From table 12.1 a new characteristic equation of $Q^+ = (S \oplus Q) \cdot (S \oplus R) \oplus Q$ is derived for the quantum SR latch. The quantum SR latch is shown in figure 12.2.

Table 12.2. The modified truth table of the SR latch.

S	R	Q	Q^+
0	0	0	0
0	0	1	1
0	1	0	0
0	1	1	0
1	0	0	1
1	0	1	1
1	1	0	0
1	1	1	1

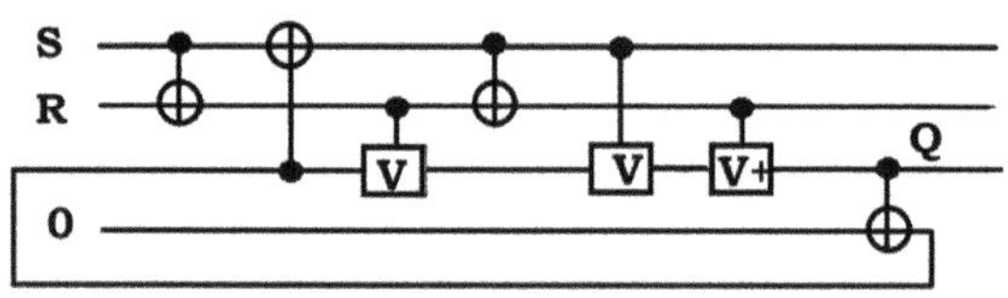

Figure 12.2. The quantum SR latch.

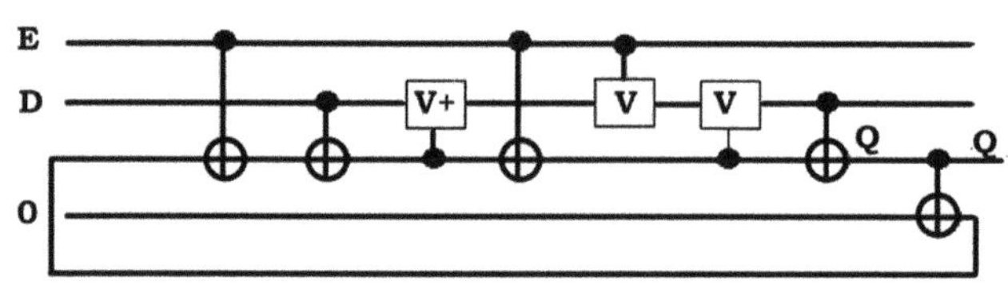

Figure 12.3. The quantum D latch.

12.2.2 The quantum D latch

The characteristic equation of the D latch can be written as $Q^+ = D \cdot E + \bar{E} \cdot Q$. When the enable signal (clock) is 1, the value of the input D is reflected at the output, that is $Q^+ = D$. When $E = 0$ the latch maintains its previous state, that is $Q^+ = Q$.

This can be understood as follows. From the characteristic equation of the D latch $Q^+ = D \cdot E + \bar{E} \cdot Q$, we can see that for four input combinations ($E = 0$, $D = 0$, $Q = 0$), ($E = 0$, $D = 1$, $Q = 0$), ($E = 1$, $D = 0$, $Q = 0$), and ($E = 1$, $D = 0$, $Q = 1$), the output Q^+ is 0. The addition of one garbage output can resolve only two output positions since one bit can produce only two distinct output combinations. The quantum D latch is shown in figure 12.3.

12.2.3 The quantum T latch

The characteristic equation of the T (for toggle) latch can be written as $Q^+ = (T \cdot Q) \cdot E + \bar{E} \cdot Q$. However, the same result can also be obtained from $Q^+ = (T \cdot E) \oplus Q$. The T latch is a complementing latch which complements its

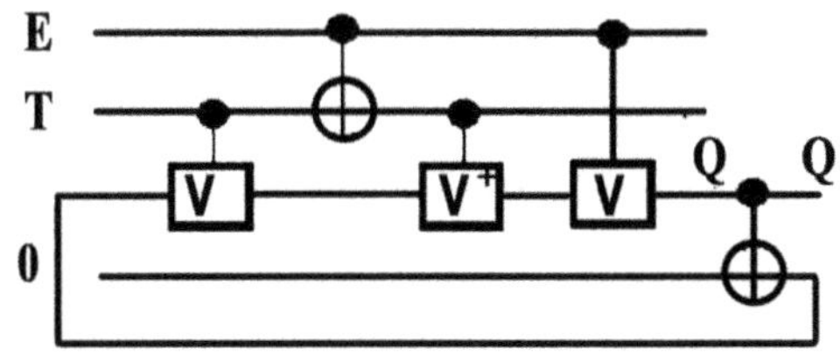

Figure 12.4. The quantum T latch. Reproduced with permission from [4]. Copyright 2010 IEEE.

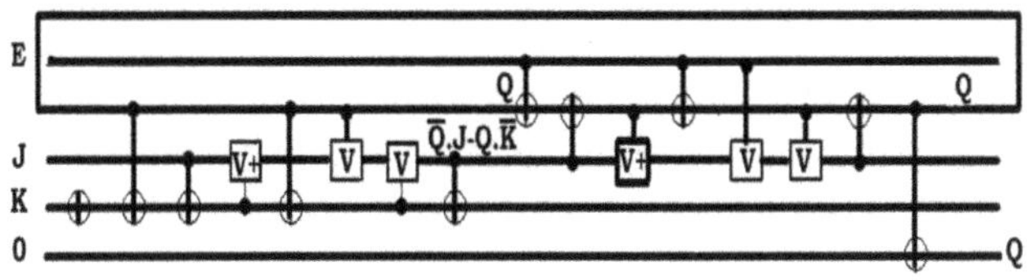

Figure 12.5. The quantum J-K latch.

value when $T = 1$, that is when $T = 1$ and $E = 1$ we have $Q^{+} = \bar{Q}$. When $T = 0$, the T latch maintains its state and we have no change in the output. The T latch characteristic equation can be directly mapped to the quantum Peres gate and the fan-out at output Q can be avoided by cascading the quantum Feynman gate (we can see that $T \cdot E \oplus Q$ matches the template of the quantum Peres gate). The design is shown in figure 12.4. We can see that for the four input combinations ($E = 0, T = 0, Q = 0$), ($E = 0, T = 1, Q = 0$), ($E = 1, D = 0, Q = 0$), and ($E = 1, T = 1, Q = 1$), the output Q^{+} is 0.

The design of the quantum T latch has both Q output and the complementary $\bar{Q}$ output. In this section a T latch is introduced based on quantum Peres and quantum Feynman gates with the outputs Q and $\bar{Q}$. The quantum T latch is shown in figure 12.5.

12.2.4 The quantum J-K latch

From the characteristic equation of the J-K latch for the eight input combinations ($E = 0, J = 0, K = 0, Q = 0$), ($E = 0, J = 0, K = 1, Q = 0$), ($E = 0, J = 1, K = 0, Q = 0$), ($E = 0, J = 1, K = 1, Q = 0$), ($E = 1, J = 0, K = 0, Q = 0$), ($E = 1, J = 0, K = 1, Q = 0$), ($E = 1, J = 0, K = 1, Q = 1$), and ($E = 1, J = 1, K = 1, Q = 1$), the output Q^{+} is 0.

The J-K latch has the capability of setting the output, resetting the output or complementing the output depending on the value of J and K. When E (the clock) is 1, the J input can set the output to 1 (when $J = 1$), the reset input can reset the output to 0 (when $K = 1$), and when both J and K are set to 1 the output Q is complemented.

The J-K latch does not produce the complemented output $\bar{Q}$. Thus we illustrate a quantum realization of the J-K latch in figure 12.5 which produces both the output Q and its complement $\bar{Q}$. In this design the Feynman gate is used to generate the complement of the output Q.

12.3 Properties of counter circuits

The important feature of the counter circuit is that different subsystems of a digital system can be controlled by the sequence generated; timing signals can be generated and clocks of different frequencies can be generated. A variety of counters are available. They can be classified as synchronous or asynchronous counters based on the clock input, as positive edged or negative edged based on the clock trigger, as a binary or decade counter based on the counts, as an up, down, or up/down counter based on the direction of the count, and as J-K, T, or D based on the latches used in the counter circuit.

The primitive logic blocks used in counters are latches and combinational gates. Asynchronous or ripple counters do not use a common clock and the output of a stage affects the clock input of the next stage. In a synchronous counter the clock is simultaneously given to all the latches. The total bits of the counter do not change simultaneously but synchronously at the rising or falling edge of the clock, depending on the type of edge triggered latches used in the counter.

12.4 The design of the quantum counters

In this section the designs for synchronous and asynchronous quantum counter circuits are shown. Subsections 12.4.1 and 12.4.2 describe the quantum asynchronous counter and quantum synchronous counter, respectively.

12.4.1 The quantum asynchronous counter

In an asynchronous counter the quantum T latches are arranged in such a way that the output of one latch is connected to the clock input of the next higher order latch. The output of a latch triggers the next latch. The latch holding the least significant bit receives the incoming count pulse. A four-bit quantum asynchronous counter is shown in figure 12.6.

12.4.2 The quantum synchronous counter

In a synchronous counter, the clock pulses are applied to the inputs of all the latches at a time. A latch is complemented depending on the input value T and the clock pulse. The latch in the least significant position is completed with every clock pulse. A latch in the other position is complemented only when all the outputs of preceding latches produce 1, and the same strategy is followed here to implement the quantum synchronous counter. Figure 12.7 shows the four-bit quantum synchronous counter.

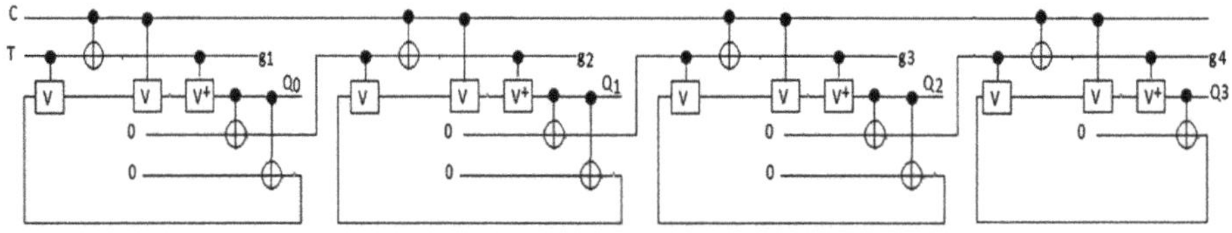

Figure 12.6. The quantum asynchronous counter.

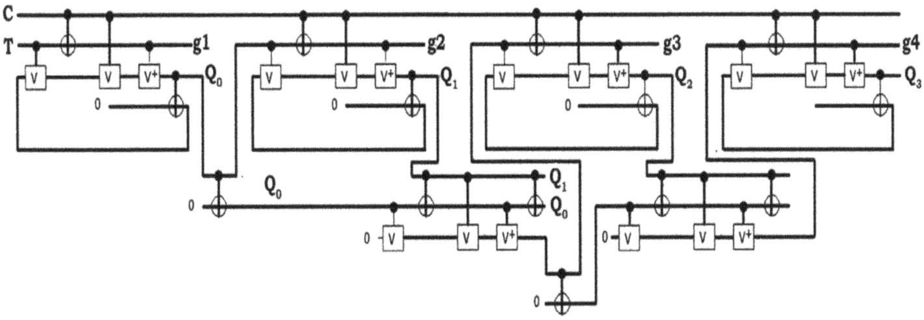

Figure 12.7. The four-bit quantum synchronous counter.

12.5 Summary

This chapter presents the design methodologies of four quantum latches (quantum SR latch, quantum D latch, quantum T latch, and quantum J-K latch) and two quantum counter circuits (quantum asynchronous counter and quantum synchronous counter). A further application of this chapter would be to use the latches towards the design of complex quantum sequential circuits such as flip-flops, storage registers, shift registers, etc. In addition, the quantum counters can also be used to build quantum sequential circuits of higher order and complex designs of quantum computers. These designs form the basis of a quantum arithmetic circuit design and thus contribute to reducing power dissipation.

Further reading

[1] Chuang M-L and Wang C-Y 2008 Synthesis of reversible sequential elements *ACM J. Emerg. Tech. Com. Sys.* **3** 4

[2] Picton P 1996 Multi-valued sequential logic design using Fredkin gates *Mult-Valued Log. J.* **1** 241–51

[3] Rice J E 2008 An introduction to reversible latches *Comput. J.* **51** 700–9

[4] Thapliyal H and Ranganathan N 2010 Design of reversible sequential circuits optimizing quantum cost, delay, and garbage outputs *ACM J. Emerg. Tech. Com. Sys.* **6** 14

[5] Thapliyal H and Vinod A P 2007 Design of reversible sequential elements with feasibility of transistor implementation *IEEE Int. Symp. on Circuits and Systems* (Piscataway, NJ: IEEE) pp 625–8

[6] Thapliyal H and Zwolinski M 2006 Reversible logic to cryptographic hardware: a new paradigm *49th IEEE Int. Midwest Symp. on Circuits and Systems* vol 1 (Piscataway, NJ: IEEE) pp 342–6

Chapter 13

The quantum controlled ternary barrel shifter

Most mathematical operations in quantum computing need shifting and rotation operations. Some of those applications are convolution, correlation, multiplication, etc. This requirement has made the use of shifters an important functional part of modern computers. To make these shifters faster, barrel shifters are being designed. Barrel shifters can shift data in one single clock cycle and do not have an inbuilt memory, and this makes the response of the systems faster when compared to other flip-flop shifters.

A barrel shifter is an n input and n output quantum logic circuit in which k select lines control the shift operation. A barrel shifter can be unidirectional, allowing data to be shifted only to the left (or the right), or bidirectional, which provides data to be rotated or shifted in both directions. A barrel shifter with n inputs and k select lines is called an (n, k) barrel shifter. Among the different designs of barrel shifter the logarithmic barrel shifter is the most widely used because of its simple design, smaller area, and the elimination of the decoder circuitry. The conventional irreversible design of a logarithmic barrel shifter is shown in figure 13.1. An n-bit logarithmic barrel shifter contains an $log_2(n)$ stage where the ith stage either shifts over 2^i bits or leaves the data unchanged. Each stage of a logarithmic barrel shifter is controlled by a control bit. If the control bit is set to one, the input data will be shifted in the associated stage otherwise it remains unchanged.

13.1 Ternary quantum gates

The elementary logical operation of quantum computing can be performed using conditional two-qubit logic gates, which are called quantum Muthukrishnan–Stroud gates (quantum M-S gates). A generalized quantum ternary gate (GQTG) is built on top of quantum M-S gates. The ternary quantum logics are $|0\rangle$, $|1\rangle$, and $|2\rangle$. A quantum M-S gate for ternary logic is presented in figure 13.2, where the input A acts as the control input and input B works as the controlled input. The output of the circuit P is equal to the input A. If $A = 2$, the other output Q is the Z-transform,

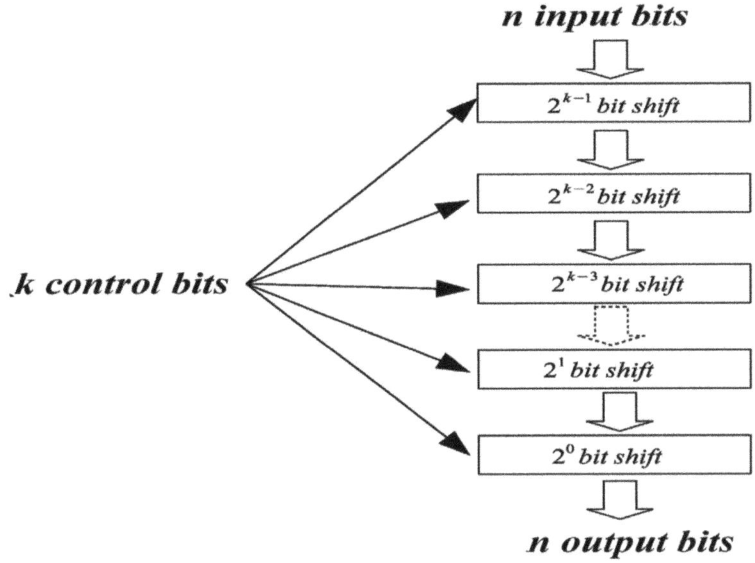

Figure 13.1. Block diagram of an (*n*, *k*) logarithmic barrel shifter.

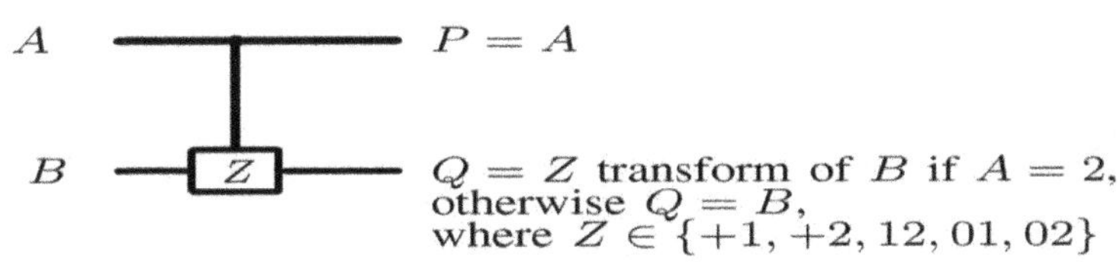

Figure 13.2. The quantum M-S gates.

where $Z \in \{+1, +2, 12, 01, 21\}$ of the input B, otherwise $Q = B$. The Z-transforms are as follows: +1 and +2 indicate increasing by one and two, respectively, and 12, 01, and 02 stand for swap the last, first, and middle two values.

The diagram of a GQTG and its representation by an quantum M-S gate are illustrated in figure 13.3. The controlling input A controls an abstract ternary multiplexer (a conditional gate) which operates on the controlled input B. From the value of A, a ternary shift operation is employed on B.

13.1.1 The quantum ternary Peres gate

The quantum Peres gate is an important concept in universal computing and the ternary quantum version of this gate is also important for ternary quantum logic synthesis. The quantum ternary Peres gate (QTPG) is 3 × 3 quantum gate, mapping (A, B, C) to $(P = A, Q = A + B, R = AB + C)$, where A, B, C are the inputs and P, Q, R are the respective outputs. The QTPG is implemented using seven quantum M-S gates and six quantum shift gates. Thus the number of quantum gates is 13. The quantum M-S and quantum shift gates based on the QTPG are shown in figure 13.4.

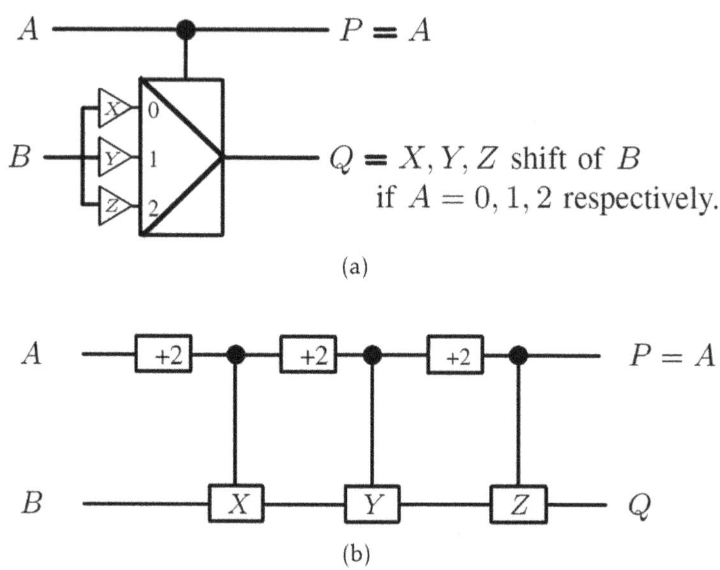

Figure 13.3. The GQTG using quantum M-S gates. (a) The GQTG. (b) Quantum realization of an M-S gate.

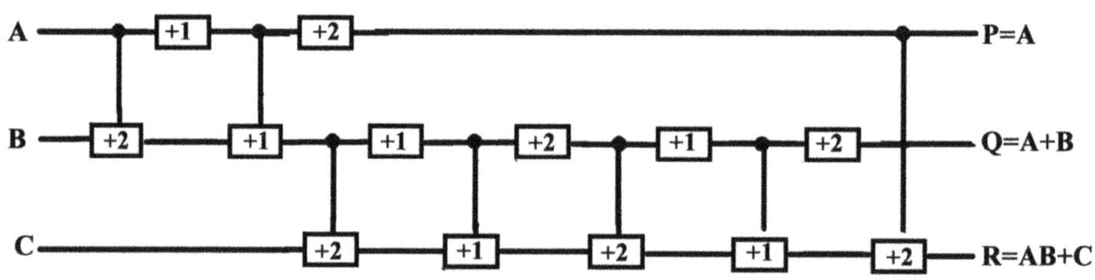

Figure 13.4. The realization of a QTPG using quantum M-S gates.

13.1.2 The quantum ternary modified Fredkin gate

The quantum Fredkin gate (see figure 2.4 in subsection 2.2.3) is a binary gate, and the quantum ternary modified Fredkin gate (QTMFG) is the multi-valued sequential logic gate. The QTMFG is the universal gate for quantum multi-valued sequential logic. Figure 13.5 presents the realization of the QTMFG using quantum M-S gates.

13.2 Properties of ternary quantum circuits

The ternary quantum gate calculation complexity refers to the number of ternary quantum gates (quantum M-S gates, quantum Shift gates) used to synthesize the given circuit. Let ϵ be the quantum M-S gate calculation complexity and γ be the quantum Shift gate calculation complexity. For example, the quantum ternary Feynman gate (QTFG) (figure 2.2 in subsection 2.2.1 shows the quantum Feynman gate) has two quantum M-S gates and two quantum Shift gates. Therefore, the quantum gate calculation complexity is $2\epsilon + 2\gamma$, which is depicted in figure 13.6.

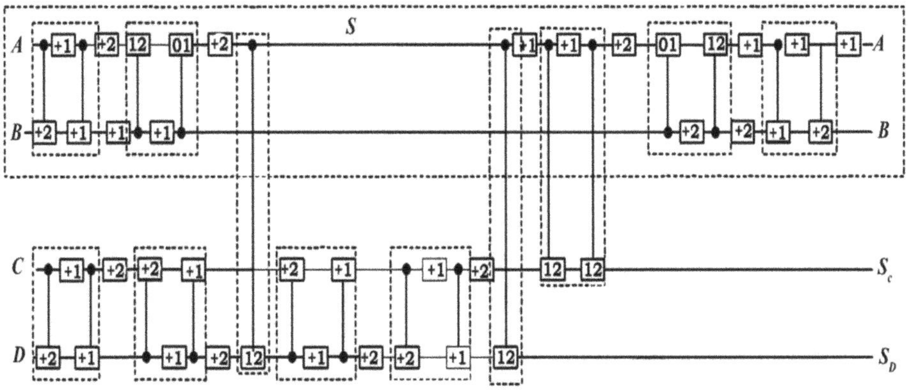

Figure 13.5. The QTMFG realization using quantum M-S gates.

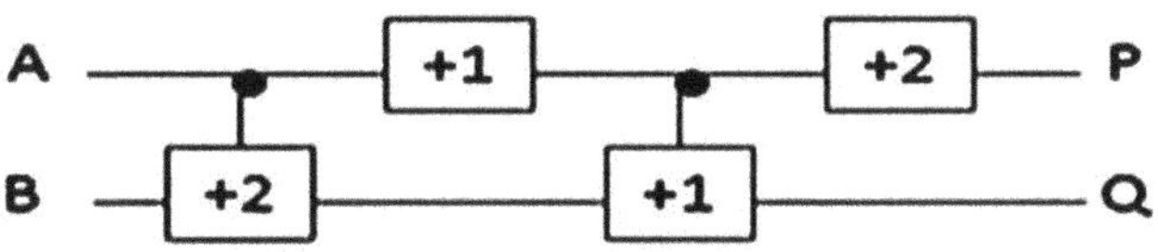

Figure 13.6. The QTFG realization using quantum M-S gates.

Table 13.1. Operations of an eight-bit barrel shifter with a three-bit shift value.

Operation performed	X (final output[a])
Logical right shift	$000i_7i_6i_5i_4i_3$
Arithmetic right shift	$i_7i_7i_7i_7i_6i_5i_5i_3$
Right rotation	$i_2i_1i_0i_7i_6i_5i_4i_3$
Logical left shift	$i_4i_3i_2i_1i_0000$
Arithmetic left shift	$i_7i_3i_2i_1i_0000$
Left rotation	$i_4i_3i_2i_1i_0i_7i_6i_5$

[a] The input data are $i_7i_6i_5i_4i_3i_2i_1i_0$.

13.3 The quantum barrel shifter

As shown in figure 13.1, a barrel shifter has an n input and n output logic circuit in which k select lines control the bit shift operation. The design of a quantum barrel shifter presents the design methodologies which can perform six operations: logical right shift, arithmetic right shift, right rotate, logical left shift, arithmetic left shift, and left rotate. For illustration, all of these operations are shown in table 13.1 for an eight-bit logarithmic barrel shifter where the eight-bit input data are denoted as i_7, i_6, i_5, i_4, i_3, i_2, i_1, i_0, where i_7 is the sign bit, and the shift or rotate operation is performed by three bits, and X denotes the shifted result.

All the operations that can be performed by a logarithmic barrel shifter, as shown in table 13.1, are as follows.

13.3.1 Logical right shift

As illustrated in table 13.1, a three-bit logical right shift operation right shifts the input data by three bits and sets the left-most three-bit to zero. Thus the output will be $000i_7i_6i_5i_4i_3$.

13.3.2 Arithmetic right shift

A three-bit arithmetic right shift operation right shifts the input data bit by three bits and sets the left-most three-bit to the sign bit (i_7). Thus the output will be $i_7i_7i_7i_7i_6i_5i_4i_3$.

13.3.3 Right rotation

A three-bit right rotation operation performs a right shift operation on input data by three bits. Further, the left-most three-bit is set to the right-most three-bit of the original input data. As illustrated in table 13.1, the final output after the right rotation operation will be $i_2i_1i_0i_7i_6i_5i_4i_3$ as the input data $i_7i_6i_5i_4i_3i_2i_1i_0$ are shifted three times to the right and the three left-most bits ($i_7i_6i_5$) are set to the three right-most bits ($i_2i_1i_0$) of the original data.

13.3.4 Logical left shift

A three-bit logical left shift operation left shifts the input data by three bits and sets the right-most three-bit to zero. Thus the final output will be $i_4i_3i_2i_1i_0000$.

13.3.5 Arithmetic left shift

In the arithmetic left shift operation, the sign bit of the input data remains intact and the remaining bits are logically left shifted by three bits. As illustrated in table 13.1 the output will be $i_7i_3i_2i_1i_0000$ as the input data $i_7i_6i_5i_4i_3i_2i_1i_0$ is logically left shifted by three bits and the sign bit (i_7) remains intact.

13.3.6 Left rotation

A three-bit left rotation operation performs a left shift operation on input data by three bits. Further the right-most three-bit is set to the left-most three-bit of the original input data. As illustrated in table 13.1, the final output after the left rotation operation will be $i_4i_3i_2i_1i_0i_7i_6i_5$ as the input data $i_7i_6i_5i_4i_3i_2i_1i_0$ are shifted three times to the left and the three right-most bits ($i_2i_1i_0$) are set to the three left-most bits ($i_7i_6i_5$) of the original data.

13.4 The design of a quantum ternary barrel shifter

In this section two new designs for a quantum ternary barrel shifter are shown. The design of the quantum ternary barrel shifter and the properties of the designed circuits are shown in detail.

13.4.1 The optimized quantum ternary barrel shifter

Two new designs for a quantum ternary barrel shifter are shown in this subsection. Unidirectional and bidirectional quantum barrel shifters are designed. A unidirectional barrel shifter shifts data only to the left (or right). A four-bit quantum ternary unidirectional barrel shifter's block diagram, shown in figure 13.7, has two stages (k_i, $i = 0, 1$) and it is constructed using QTFGs, QTMFGs, and QTPGs. A bidirectional logarithmic logical shifter is a non-rotating barrel shifter which can shift input data to the left or right. It has a control signal (K) for determining the direction of the shift. If the K signal is set to zero, the logical shifter will work as a logical right shifter, otherwise it will work as a logical left shifter. For example, a four-bit quantum ternary bidirectional barrel shifter's block diagram, illustrated in figure 13.8, has two stages (k_i, $i = 0, 1$) and it is constructed by QTFGs, QTMFGs, and QTPGs. The QTFG is utilized to avoid fan-out. Moreover, the QTMFG and QTPG are applied as 2-to-1 multiplexer and AND gates, respectively. The (4, 2) quantum ternary unidirectional and bidirectional logical barrel shifters, which are depicted in figures 13.7 and 13.8, respectively, take n_3, n_2, n_1, n_0 as data inputs and k_1, k_0 as select inputs. The (4, 2) quantum ternary unidirectional and bidirectional logical barrel shifters can be generalized for quantum (n, k) ternary unidirectional and bidirectional logical barrel shifters, respectively. The ternary gate representation of the (4, 2) quantum ternary bidirectional barrel shifter is depicted in figures 13.9–13.12 for each step.

13.4.2 Properties of the designed circuit

In this subsection the properties of an (n, k) quantum ternary barrel shifter are presented.

Property 13.1. QTPG is utilized for AND operation in the design. Every stage in the barrel shifter is denoted by i. Thus the number of required QTPGs in each stage is equal to $2^i - 1$, $i = 0, 1, \ldots, k - 1$. Thus in the design, k stages will have a total of $N_{\mathrm{QTPG.TU}} = (\sum_{i=0}^{k-1} 2^i) - 1 = 2^k - 2$ QTPGs. The QTMFGs are responsible for

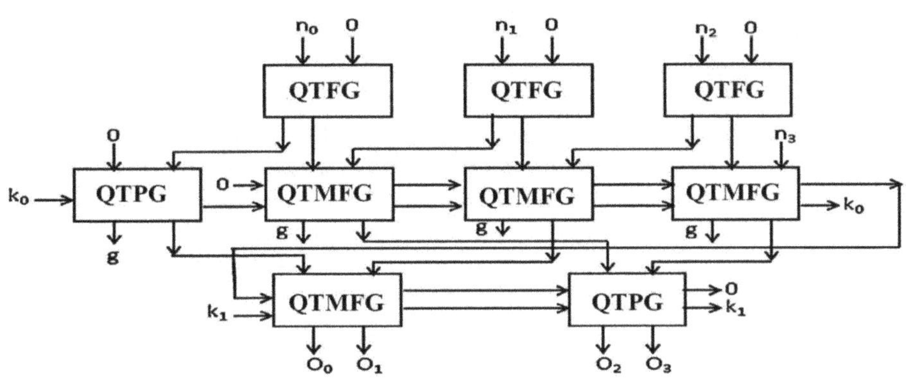

Figure 13.7. Block diagram of a quantum ternary unidirectional barrel shifter.

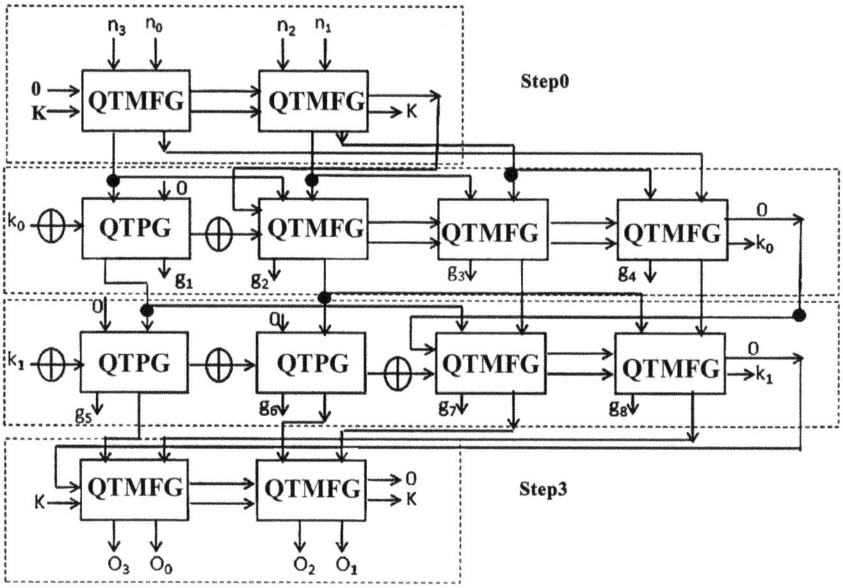

Figure 13.8. Block diagram of a (4, 2) quantum ternary bidirectional barrel shifter.

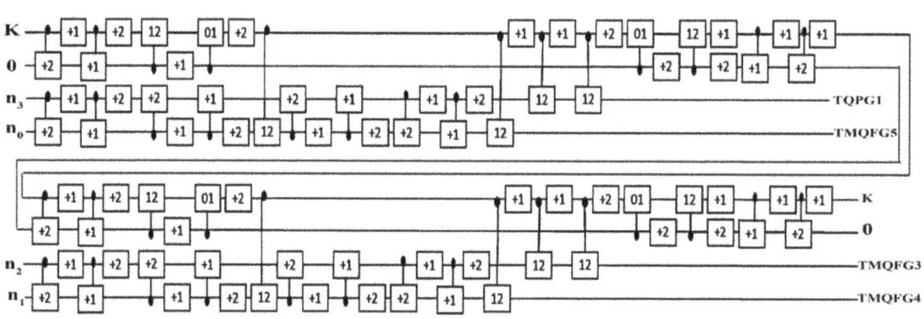

Figure 13.9. The (4, 2) quantum ternary bidirectional barrel shifter (step 0) realization using quantum M-S gates. Reproduced with permission from [3]. Copyright 2015 IEEE

shifting/rotating data input bits. Each stage has n QTMFGs, except for the last stage. The last stage requires only $n/2$ QTMFGs. Thus there are $k - 1$ stages which require n QTMFGs and the last stage requires $n/2$ QTMFGs, but there are $2^k - 2$ QTMFGs which are replaced by QTPGs. Hence in the design the total number of QTMFGs is $N_{\text{QTMFG.TU}} = ((n(k - 1) + n/2) - (2^k - 2))$. QTFGs are used for copying signals. The number of QTFGs in the input stage is $(n - 1)$. Hence the design can be realized with $N_{\text{QTFG.TU}} = (n - 1)$ QTFGs. Therefore the design requires $N_{\text{QTPG.TU}} = 2^k - 2$, $N_{\text{QTMFG.TU}} = ((n(k - 1) + n/2) - (2^k - 2))$, and $N_{\text{QTFG.TU}} = n - 1$ ternary quantum gates, where n is the number of bits, k is the total number of stages of the circuit, and $N_{\text{QTPG.TU}}$, $N_{\text{QTMFG.TU}}$, and $N_{\text{QTFG.TU}}$ are the total numbers of ternary quantum Peres, ternary modified quantum Fredkin, and quantum ternary Feynman gates, respectively.

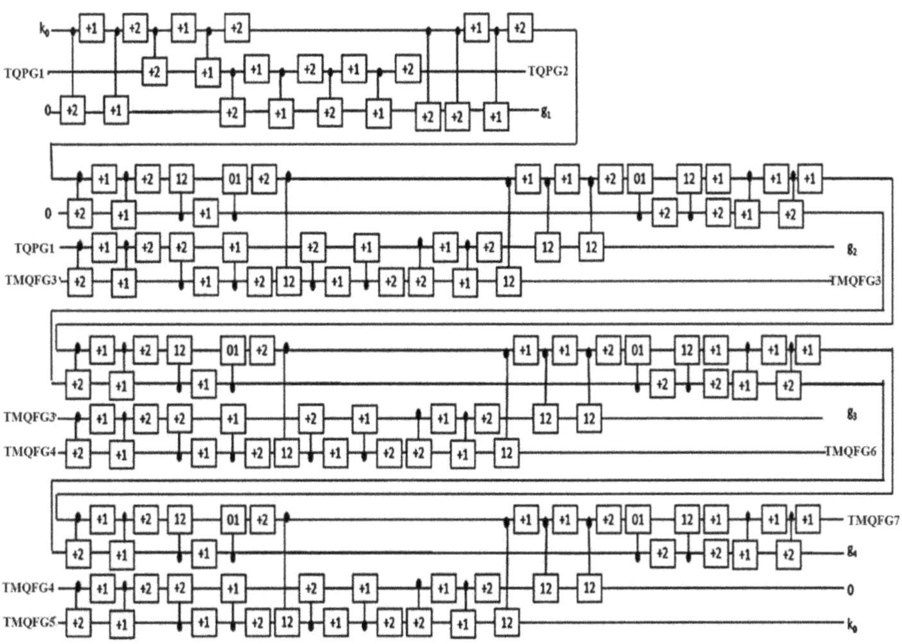

Figure 13.10. The (4, 2) quantum ternary bidirectional barrel shifter (step 1) realization using quantum M-S gates. Reproduced with permission from [3]. Copyright 2015 IEEE.

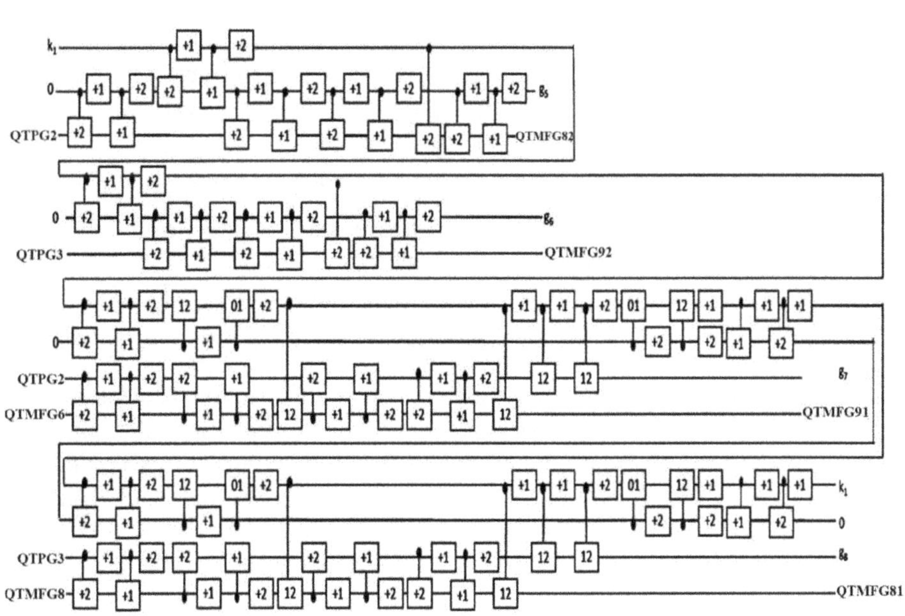

Figure 13.11. A (4, 2) quantum ternary bidirectional barrel shifter (step 2) realization using quantum M-S gates. Reproduced with permission from [3]. Copyright 2015 IEEE.

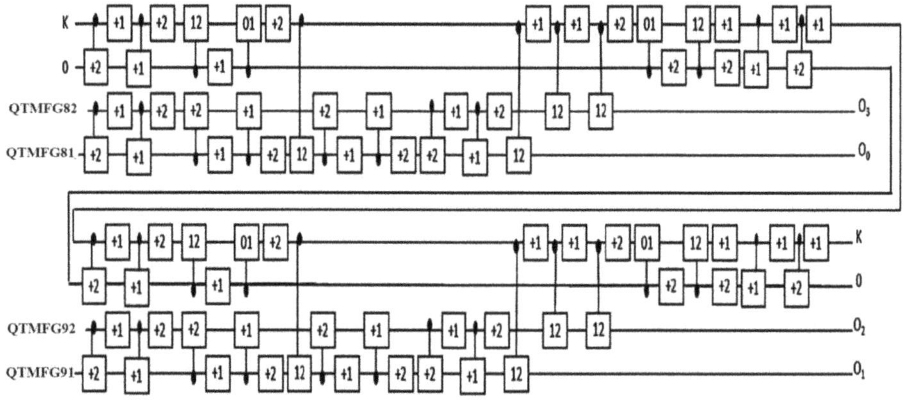

Figure 13.12. A (4, 2) quantum ternary bidirectional barrel shifter (step 3) realization using quantum M-S gates. Reproduced with permission from [3]. Copyright 2015 IEEE.

Property 13.2. The QTPG is utilized as an AND gate in the design. Every stage in the barrel shifter is denoted by i. Thus the number of required QTPGs in each stage is equal to 2^i, $i = 0, 1, \ldots, k - 1$. Thus in the design k stages will have a total of $N_{\mathrm{QTPG.TB}} = \sum_{i=0}^{k-1} 2^i = 2^k - 1$ QTPGs. The input stage and the output stage of the circuit need $n/2$ QTMFGs and the barrel shifter requires $(n - 2^i)$ QTMFGs for the rest of the stages. Thus an n-bit bidirectional barrel shifter with k stages will have a total of $N_{\mathrm{QTMFG.TB}} = \sum_{i=0}^{k-1}(\frac{n}{2} + \frac{n}{2} + n - 2^i) = \sum_{i=0}^{k-1}(n + n - 2^i) = (k + 1)n - \sum_{i=0}^{k-1}(2^i) = (k + 1)n - (2^k - 1)$ MQFGs. Fan-out gates must be used for copying the signal in the quantum circuit. One of the fan-out gates is a QTFG. The number of QTFGs in each stage is $(n - 2^i)$. Hence the circuit can be realized with $N_{\mathrm{QTFG.TB}} = \sum_{i=0}^{k-1}(n - 2^i)$ QTFGs. Therefore, the design requires $N_{\mathrm{QTPG.TB}} = 2^k - 1$, $N_{\mathrm{QTMFG.TB}} = (k + 1)n - (2^k - 1)$, and $N_{\mathrm{QTFG.TB}} = \sum_{i=0}^{k-1}(n - 2^i)$ ternary quantum gates, where n is the number of bits, k is the total number of stages of the circuit, and $N_{\mathrm{QTPG.TB}}$, $N_{\mathrm{QTMFG.TB}}$, and $N_{\mathrm{QTFG.TB}}$ are the total numbers of ternary quantum Peres, ternary modified quantum Fredkin, and quantum ternary Feynman gates, respectively.

Property 13.3. The ternary quantum unidirectional and bidirectional barrel shifters are realized by QTPGs, QTFGs, and QTMFGs with the number of quantum gates being 13, 4, and 41, respectively. Thus the optimized (n, k) quantum ternary unidirectional and bidirectional barrel shifters require $(13N_{\mathrm{QTPG.TU}} + 41N_{\mathrm{QTMFG.TU}} + 4N_{\mathrm{QTFG.TU}})$ and $(13N_{\mathrm{QTPG.TB}} + 41N_{\mathrm{QTMFG.TB}} + 41N_{\mathrm{QTFG.TB}})$ quantum gates, respectively, where n is the number of bits and k is the total number of stages of the circuit.

Property 13.4. The ternary quantum unidirectional and bidirectional barrel shifters are realized using QTPGs, QTFGs, and QTMFGs. A QTPG has 7ϵ quantum M-S gate calculation complexity and 6γ quantum Shift gate calculation complexity, a QTFG has

2ϵ quantum M-S gate calculation complexity and 2γ quantum Shift gate calculation complexity, and a QTMFG has 20ϵ quantum M-S gate calculation complexity and 21γ quantum Shift gate calculation complexity. Thus the total quantum gate calculation complexities are $(7N_{QTPG.TU\epsilon} + 6N_{QTPG.TU\gamma} + 20N_{QTMFG.TU\epsilon} + 21N_{QTMFG.TU\gamma} + 2N_{QTFG.TU\epsilon} + 2N_{QTFG.TU\gamma})$ and $(7N_{QTPG.TB\epsilon} + 6N_{QTPG.TB\gamma} + 20N_{QTMFG.TB\epsilon} + 21N_{QTMFG.TB\gamma} + 2N_{QTFG.TB\epsilon} + 2N_{QTFG.TB\gamma})$ for the ternary quantum unidirectional and bidirectional barrel shifters, respectively, where ϵ is the M-S quantum gate calculation complexity and γ is the Shift quantum gate calculation complexity.

Property 13.5. The optimized (n, k) ternary unidirectional and bidirectional barrel shifters produce at least $n(k - 1)$ and nk garbage outputs, respectively, where n is the number of bits and k is the total number of stages of the circuit.

Proof. The above statement has been proved by the method of contradiction. Suppose (n, k) ternary quantum unidirectional and bidirectional barrel shifters do not produce at least $n(k - 1)$ and nk garbage outputs, respectively, where n is the number of bits and k is the total number of stages of the circuit.

In every stage of the ternary quantum unidirectional and bidirectional barrel shifters, QTPGs and QTMFGs produce one garbage output, except for the first and last stages. In other words, in each row the quantum barrel shifter produces at least n garbage outputs. There are $k - 1$ and k stages in the unidirectional and bidirectional quantum barrel shifters, respectively. Thus $(k - 1)$ stages of a unidirectional quantum barrel shifter will have a total of $n(k - 1)$ garbage outputs, whereas k stages of a bidirectional quantum barrel shifter produce nk garbage outputs. Thus the optimized (n, k) unidirectional and bidirectional quantum ternary barrel shifters produce at least $n(k - 1)$ and nk garbage outputs, respectively.

This contradicts the supposition that (n, k) unidirectional and bidirectional quantum ternary barrel shifters do not produce at least $n(k - 1)$ and nk garbage outputs, respectively. Hence the supposition is false and the property 13.5 is true and this completes the proof.

13.5 Summary

In this chapter two new design methodologies of a quantum ternary barrel shifter are presented. The optimized shifters are designed using quantum ternary Feynman gates, quantum ternary modified Fredkin gates, and quantum ternary Peres gates. The (n, k) bidirectional quantum ternary barrel shifter requires $13N_{QTPG.TB} + 41N_{QTMFG.TB} + 4N_{QTFG.TB})$ quantum gates, nk garbage outputs, and $(7N_{QTPG.TB\epsilon} + 6N_{QTPG.TB\gamma} + 20N_{QTMFG.TB\epsilon} + 21N_{QTMFG.TB\gamma} + 2N_{QTFG.TB\epsilon} + 2N_{QTFG.TB\gamma})$ quantum gate calculation complexity, where n is the number of bits, k is the total number of stages of the circuit, $N_{QTPG.TB}$, $N_{QTMFG.TB}$, and $N_{QTFG.TB}$ are the total numbers of quantum ternary Peres gates, quantum ternary modified Fredkin gates, and quantum ternary Feynman gates, respectively, ϵ is the quantum M-S gate calculation complexity, and γ is the quantum Shift gate calculation complexity. The quantum ternary barrel shifters will be useful in a wide variety of applications of signal processing systems. A general use of barrel shifters is in the hardware employment of floating-point arithmetic. For a floating-point add or subtract operation, the significants of the two numbers must be associated, which requires shifting the smaller number to the right and increasing its exponent until it matches the exponent of the larger number. This is performed by subtracting the exponents and using the barrel shifter to shift the smaller number to the right by the difference in one cycle. If a humble shifter were used, shifting by n-bit positions would need n clock cycles.

Further reading

[1] Khan A I, Nusrat N, Khan S M, Hasan M and Khan M H A 2007 Quantum realization of some ternary circuits using Muthukrishnan–Stroud gates *37th Int. Symp. on Multiple-Valued Logic* (Piscataway, NJ: IEEE) p 20

[2] Kotiyal S, Thapliyal H and Ranganathan N 2010 Design of a ternary barrel shifter using multiple-valued reversible logic *10th IEEE Conf. on Nanotechnology* (Piscataway, NJ: IEEE) pp 1104–8

[3] Lisa N J and Babu H M H 2015 A compact representation of a quantum controlled ternary barrel shifter *2015 IEEE Int. Symp. on Circuits and Systems (ISCAS)* (Piscataway, NJ: IEEE) pp 2145–8

[4] Muthukrishnan A and Stroud C R Jr 2000 Multivalued logic gates for quantum computation *Phys. Rev.* A **62** 052309

[5] Nia N H 2012 Design of an optimized reversible ternary and binary bidirectional and normalization barrel shifters for floating point arithmetic *Life Sci. J.* **9** 1904–15

[6] Asher P 1985 Reversible logic and quantum computers *Phys. Rev.* A **32** 3266

IOP Publishing

Chapter 14

Quantum random access memory

Random access memory (RAM) is a highly versatile device for storing and accessing information. It consists of an array of memory cells (write-enabled master–slave flip-flop) where information is stored in the form of bits. Each cell is associated with a unique address, i.e. a number which identifies the location of the cell. The main characteristic of RAM is that each memory cell can be separately addressed as well and hence the designation is 'random access'. To access a cell, its address must be provided by a decoder. The device will then output the content of the memory cell.

If the decoder has n selection bits, the RAM is capable of addressing 2^n different memory locations: when given an n-bit address k, RAM returns the bit string f_k which is stored in the memory slot of the database labeled k. Figure 14.1 presents the general architecture of a $2^n \times m$ RAM, where n is the selection bits of the decoder and m is the number of bits of input data.

This chapter focuses on the construction procedure and the complexities of quantum RAM, and each of the underlying components in the quantum circuitry used to build quantum RAM is described.

14.1 The quantum n-to-2^n decoder

A decoder is a collection of logic gates set up in such a way that, for an input combination, all output terms are low except one. These terms are the mintersm which use a variable only once. Thus, when an input combination changes, two outputs will change. Let there be n inputs, then the number of outputs will be 2^n. In other words, there will be one line at the output for each possible input. Figure 14.2 presents the general architecture of an n-to-2^n decoder. Table 14.1 presents the truth table of a 2-to-4 decoder circuit.

A quantum decoder can be realized using the quantum Feynman double gate (QF2G) and quantum Fredkin gate (QFRG). Figure 14.3 illustrates the QRFG and QF2G. Figure 14.3(a) presents the quantum 1-to-2 decoder and figure 14.3(b)

doi:10.1088/978-0-7503-2747-3ch14

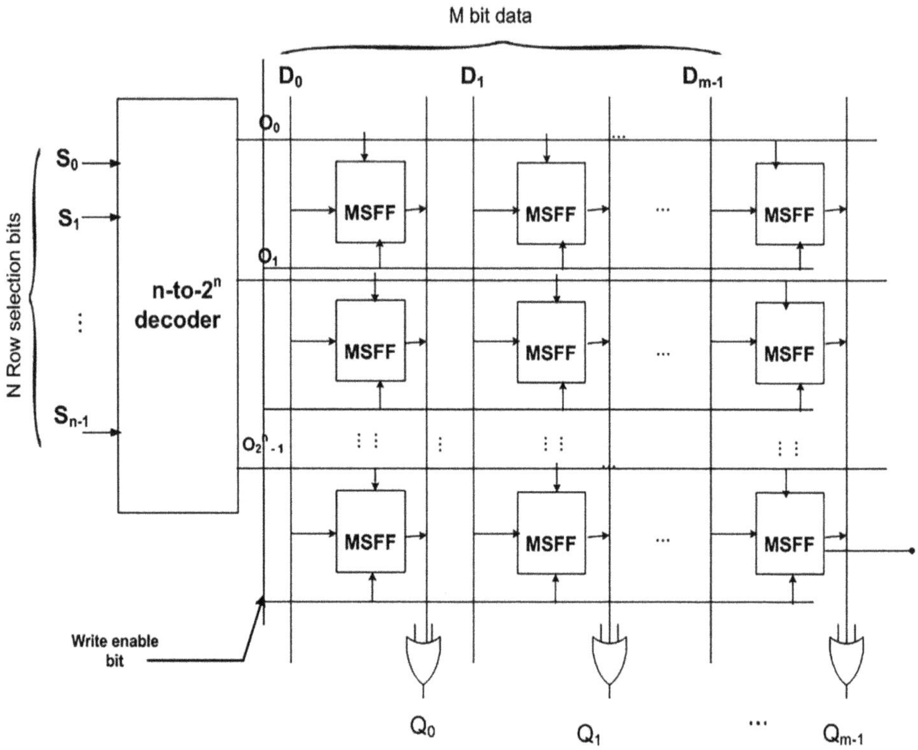

Figure 14.1. General architecture of a RAM.

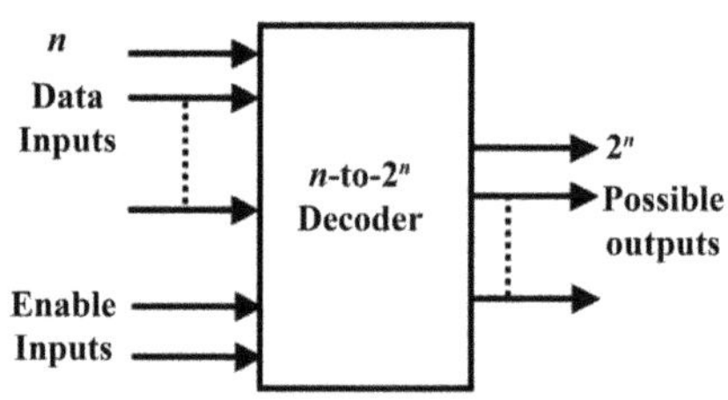

Figure 14.2. Block diagram of a generalized n-to-2^n decoder.

Table 14.1. The truth table of the 2-to-4 decoder.

Input			Output			
EN	A	B	Y_3	Y_2	Y_1	Y_0
0	x	x	0	0	0	0
1	0	0	0	1	0	1
1	0	1	0	0	1	0
1	1	0	0	1	0	0
1	1	1	1	0	0	0

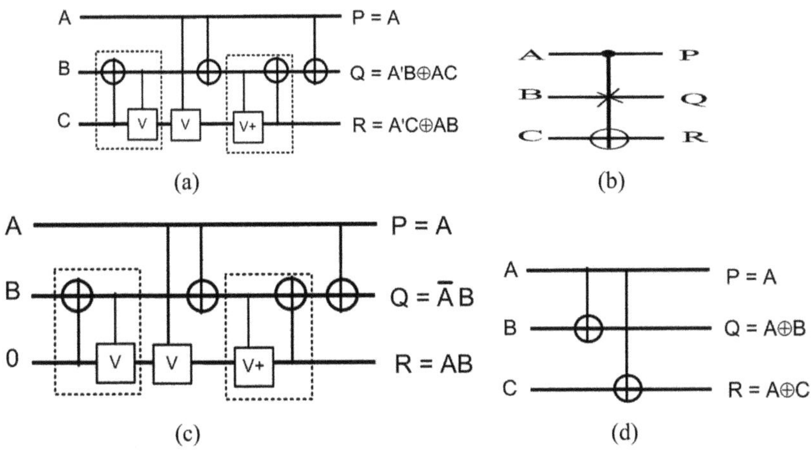

Figure 14.3. The quantum Fredkin gate and Feynman double gate. (a) A quantum Fredkin gate. (b) Symbol of a quantum Fredkin gate. (c) The simultaneous generation of two terms. (d) A quantum Feynman double gate.

presents the quantum 2-to-4 decoder. The quantum cost of a 2-to-4 decoder circuit is 12.

A higher order quantum decoder, such as the 3-to-8 shown in figure 14.4, can be designed with a 2-to-4 quantum decoder and an additional four QFRGs, as each quantum Fredkin gate is able to generate two different terms, shown in figure 14.3(c). In the same way an n-to-2^n quantum decoder can be designed with an $(n - 1)$-to-2^{n-1} quantum decoder and an additional 2^{n-1} quantum Fredkin gates, as shown in figure 14.5. Algorithm 14.1 presents the construction of an n-to-2^n quantum decoder.

Algorithm 14.1 The quantum decoder (S, QF2G, QFRG).

Input: $S_0S_1, \cdots, S_{n-1}$, QF2G, QFRG
Output: n-to-2^n quantum decoder circuit
START
i = input, o = output
for $j = 0$ to $n - 1$ **do**
    if $j = 0$ **then**
        $S_j \rightarrow$ first.i.QF2G, 1 $\rightarrow$ second.i.QF2G, 0 $\rightarrow$ third.i.QF2G
    else
        $S_j \rightarrow$ first.i.QFRG, 0 $\rightarrow$ second.i.QFRG
        if $j = 2$ **then**
            third.o.QF2G $\rightarrow$ third.i.QFRG
        else

(*Continued*)

```
        call Quantum Decoder(j – 1), Quantum Decoder.o.j →third.i.QFRG_j
      end if
    end if
end for
return QFRG.o.3 and QFRG.o.2 → desired output, and the remaining QF2G.o and QFRG.o
  → garbage output
END
```

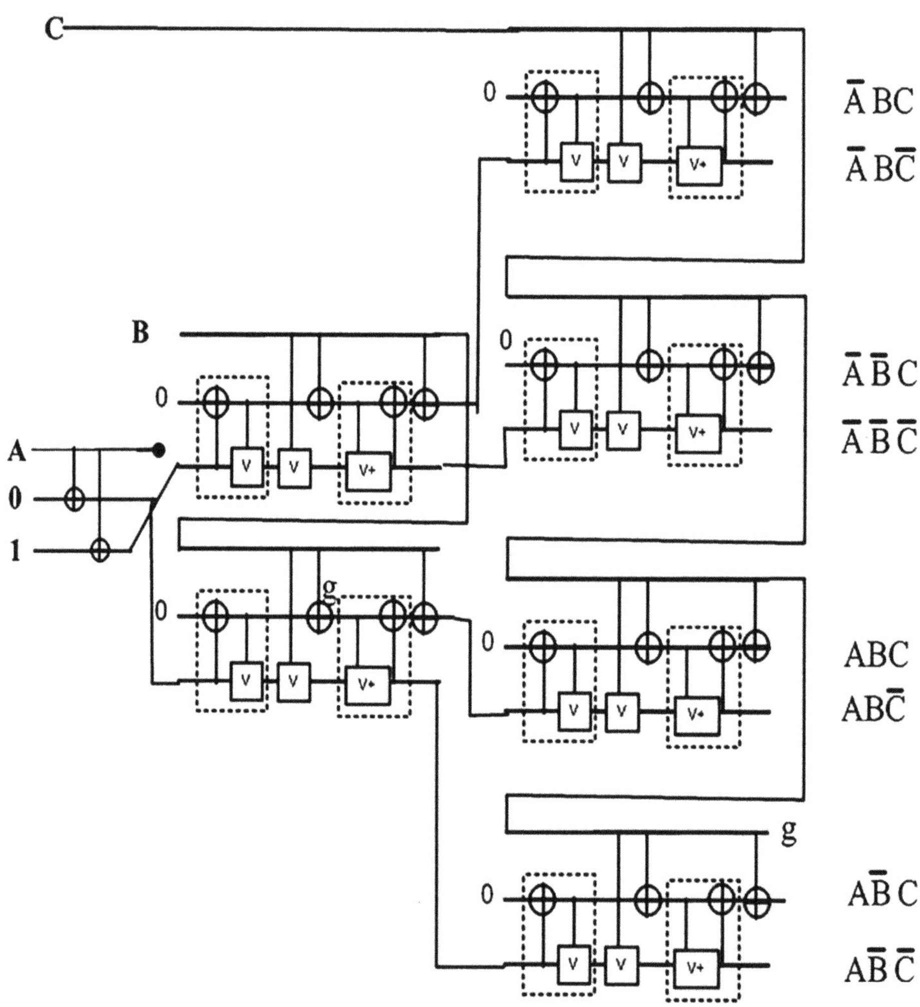

Figure 14.4. The quantum 3-to-8 decoder.

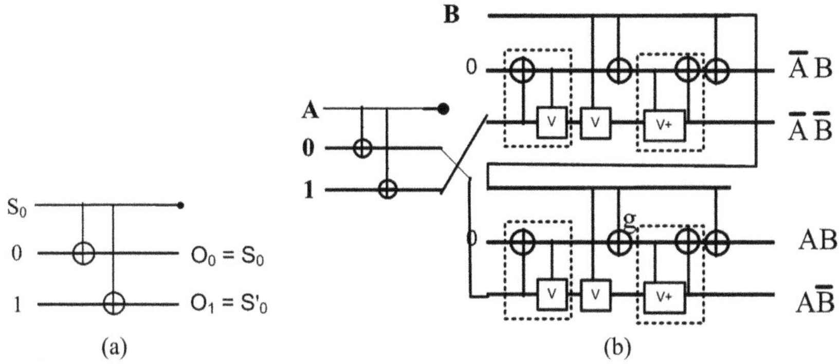

Figure 14.5. Quantum decoders for $n = 1$ and $n = 2$. (a) Quantum 1-to-2 decoder. (b) Quantum 2-to-4 decoder.

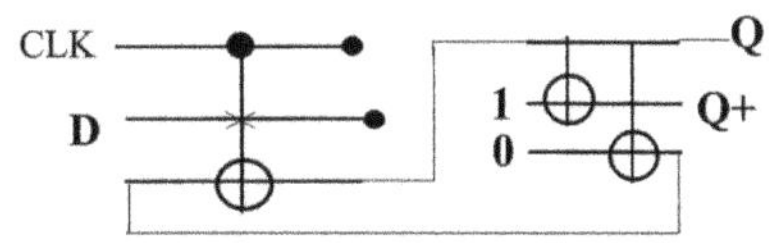

Figure 14.6. The quantum D latch.

Line 5 of the algorithm assigns the input to the QF2G for the first control bit (S_0), whereas line 7 assigns the first two inputs to the QFRG for all the remaining control bits. Lines 8–9 assign the third input to the FRG for $n = 2$, while lines 10–12 assign the third input to the QFRG through a recursive call to a previous quantum decoder for $n = 2$. Line 15 returns the outputs. The complexity of this algorithm is $O(n)$, where n is the number of data bits. The decoder is used to address the row of the memory cells of the quantum RAM.

14.2 The quantum memory unit

In quantum RAM, there are write-enabled master–slave flip-flops. To design a write-enabled master–slave flip-flop, first a D latch is needed.

Figure 14.6 presents the quantum D latch, and it can be seen that a QFRG and a QF2G are used in the circuit. The QFRG is needed to produce Q, whereas QF2G produces Q^+ and Q. The equations for the latch are $Q_n = D.\ \mathrm{CLK} + \mathrm{CLK}'.\ Q_{n-1}$, and $Q^+ = Q_0$ (n is time varying), which are used to capture the logic level when the data are present and the clock is SET. A quantum D latch can be realized with seven quantum gates, two garbage outputs, and a quantum cost of 7.

After designing the quantum D latch, the quantum write-enabled master–slave flip-flop (QMSFF) can be designed with two D latches and additional QFRGs and QF2Gs. Figure 14.7 illustrates the architecture of a QMSFF.

The QMSFF works on two modes (read and write), since the data are both read from and written into the QRAM. The CLK signal provides the clock pulse, if CLK = 1, the D input is stored in the first latch, i.e. the master flip-flop, but the

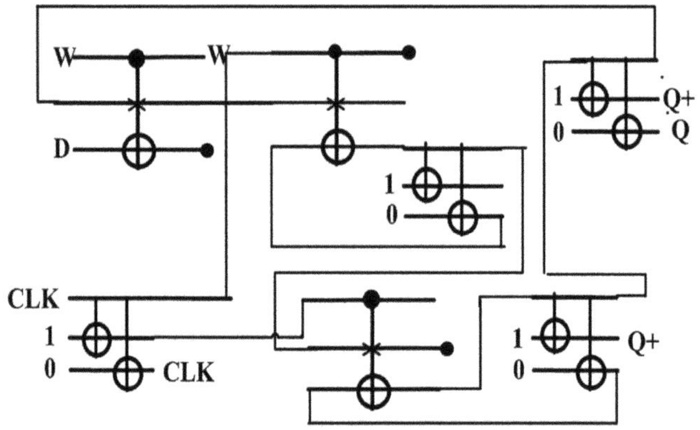

Figure 14.7. The QMSFF.

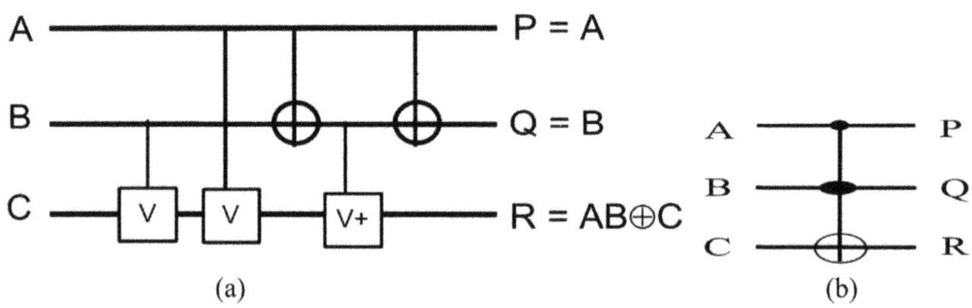

Figure 14.8. (a) The QTG. (b) Symbol of the QTG.

second latch cannot change state. When CLK = 0, the first latch's output is stored in the second latch, i.e. the slave flip-flop, but the first latch cannot change state. The CLK signal is maintained by the first QF2G. In figure 14.7, W works as the write enable bit. It determines whether a read or a write operation is performed. When $W = 1$, a write operation is performed. The W and CLK outputs are propagated to the next MSFF if required and Q indicates the output of the MSFF.

The circuit consists of two D flip-flops connected together. When the clock is high, it can be found from the quantum circuit that the QMSFF consists of a QFRG, two QF2Gs, and two quantum D latches (the D latch is presented in figure 12.3 of subsection 12.2.1).

Considering the unused clock pulses as garbage outputs, this design has six garbage outputs and it has a quantum cost of 23.

14.3 The construction procedure of the quantum RAM

To design the quantum RAM, the addressable components, i.e. the decoder and memory component MSFF, are replaced with the quantum decoder and QMSFF. Then the OR gates are replaced with Ex-OR gates. Additionally, quantum Toffoli gates (QTGs), as shown in figure 14.8, are used to propagating both the CLK and

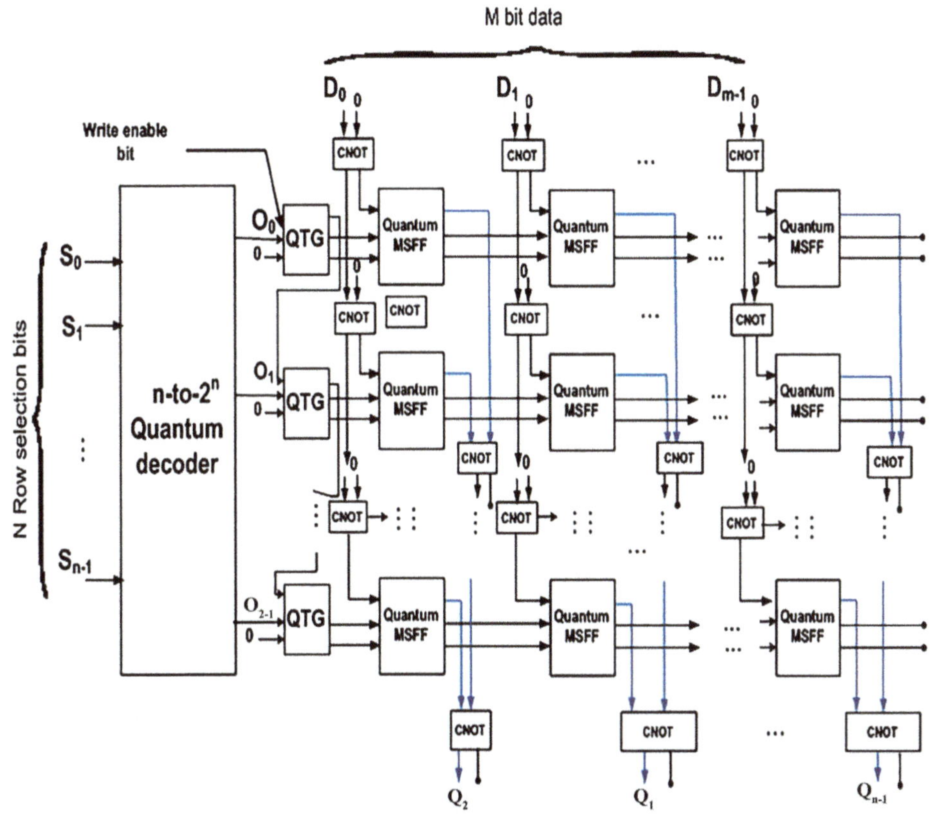

Figure 14.9. Block diagram of the $2^n \times m$ quantum RAM.

write-enabled bit W while diminishing the fan-out problem. Figure 14.9 presents the block diagram of a $2^n \times m$ quantum RAM. The working principles of the quantum RAM are as follows.

In the $2^n \times m$ RAM, n is the number of control bits of the quantum decoder and 2^n decoded outputs are the row selection bits. For a particular combination of n input variables of the configured quantum decoder, only one of the 2^n outputs is active at a time. This active output is propagated through the second output of the QTG, which acts as the CLK and hence selects only one row of the array of memory units as the CLK propagate through the row of the QMSFF.

The write bit W specifies whether there is a read or a write operation. When $W = 1$ or high, then the m data-inputs D_0 to D_{m-1} are written in the m flip-flops of the selected row as W is propagated by the third output of QTG throughout the selected row.

When $W = 0$ or low, then the m outputs Q_0 to Q_{m-1} are the previously stored bits in the flip-flops, and they are read from the m flip-flops of the selected row. The use of a quantum CNOT gate (Ex-OR gate) eliminates the fan-out problem. These gates also perform the Ex-OR operations of the outputs of the flip-flops in a column.

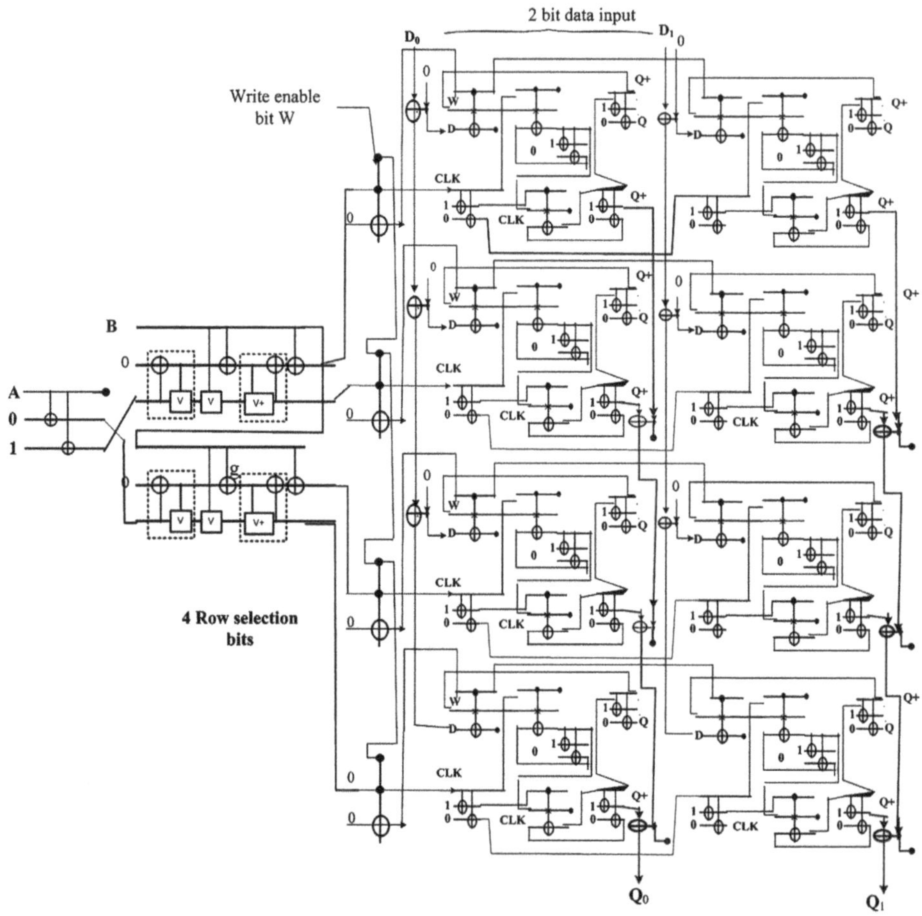

Figure 14.10. The 4 × 2 quantum RAM.

In addition, for simplicity the quantum circuit of the 4 × 2 quantum RAM is given in figure 14.10. This circuit is developed by accumulating the quantum 2-to-4 decoder (figure 14.5(b)), eight quantum MSFFs (figure 14.7), four QTGs (figure 14.8), and twelve quantum CNOT gates. In figure 14.10 the red line indicates the output and the blue lines indicate the clock pulse.

14.4 Summary

Quantum RAM is an essential component in quantum memory devices. This chapter focuses on the design procedure for quantum RAM. This chapter covers the design procedures of the quantum decoder step by step to address the RAM and the quantum MSFF (the unit memory component of RAM). A composite quantum realization of a 4 × 2 quantum RAM is also given at the end of the chapter which will help the reader to better understand quantum RAM.

Further reading

[1] Asfestani M N and Heikalabad S R 2017 A novel multiplexer-based structure for random access memory cell in quantum-dot cellular automata *Physica* B **521** 162–7

[2] Fredkin E and Toffoli T 2001 Conservative logic *Collision-based Computing* (Berlin: Springer) pp 47–81

[3] Giovannetti V, Lloyd S and Maccone L 2008 Architectures for a quantum random access memory *Phys. Rev.* A **78** 52310

[4] Mahammad S N and Veezhinathan K 2010 Constructing online testable circuits using reversible logic *IEEE Trans. Instrum. Meas.* **59** 101–9

[5] Sayem A S M and Mitra S K 2011 Efficient approach to design low power reversible logic blocks for field programmable gate arrays *2011 IEEE Int. Conf. on Computer Science and Automation Engineering* vol 4 pp 251–5

[6] Shamsujjoha M, Babu H M H and Jamal L 2013 Design of a compact reversible fault tolerant field programmable gate array: a novel approach in reversible logic synthesis *Microelectron. J.* **44** 519–37

[7] Sharmin F, Polash M M A, Shamsujjoha M, Jamal L and Babu H H 2011 Design of a compact reversible random access memory *4th IEEE Int. Conf. on Computer Science and Information Technology* vol 10 pp 103–7

[8] Tanaka M, Sato R, Hatanaka Y and Fujimaki A 2016 High-density shift-register-based rapid single-flux-quantum memory system for bit-serial microprocessors *IEEE Trans. Appl. Supercond.* **26** 1–5

[9] Tayari M and Eshghi M 2011 Design of 3-input reversible programmable logic array *J. Circuit. Syst. Comp.* **20** 283–97

[10] Thapliyal H and Ranganathan N 2010 Design of reversible sequential circuits optimizing quantum cost, delay, and garbage outputs *ACM J. Emerg. Tech. Com. Sys.* **6** 14

Chapter 15

The quantum arithmetic logic unit

An arithmetic logic unit (ALU) is a major component of the central processing unit of a computer system. It performs all processes related to arithmetic and logic operations that need to be performed on instruction words. An ALU executes one of the different micro-operations for performing the instructions on two operands A and B depending on the control signal inputs, and is suitable for quantum technology and embedded processors. The arithmetic logic unit can carry out Boolean functions, namely Ex-OR, AND, OR, NAND, NOR, and MUX, as well as basic arithmetic operations, such as addition (ADD) and subtraction (SUB). In this chapter three different designs of quantum ALUs are presented. The main basic operations that are required to realize ALU are as follows:

- ADD (addition): $(A, B) \rightarrow (A, A + B)$
- SUB (subtraction): $(A, B) \rightarrow (A, A - B)$
- Ex-OR (bitwise exclusive-OR): $(A, B) \rightarrow (A, A \oplus B)$
- AND (bitwise-AND): $(A, B) \rightarrow (A, A \wedge B)$
- OR (bitwise-OR): $(A, B) \rightarrow (A, A \vee B)$
- NOP (no-operation): $(A) \rightarrow (A)$

15.1 The design of a quantum ALU

The component of the computer system that executes the majority of data-processing operations is called the central processing unit (CPU). The CPU is constructed with three major components, consisting of the control unit, ALU, and processor unit. The ALU has an important role and performs the required micro-operations for executing instructions. In order to design an effective quantum ALU, the number of logical and arithmetic calculations produced on the outputs must be considered in addition to the quantum cost and other evaluation parameters. The quantum ALU circuit produces a large number of calculations at the least quantum cost.

doi:10.1088/978-0-7503-2747-3ch15

15.1.1 The first approach

The quantum ALU has eight inputs and eight outputs. The simple quantum gates employed in the implementation of all ALUs are covered in this chapter. In the first approach, the six logical calculations of the resulting output are AND, NOR, NAND, OR, ADD, and SUB. The inputs consist of three data inputs (A, B, and C_{in}), three select lines (S_0, S_1, S_2), and two constant inputs. The circuit can generate all selector inputs on the outputs (S_0, S_1, S_2). The eight outputs are S_0, S_1, and S_2 propagated to the output, C_{out}/borrow, result, and three garbage outputs. It produces two arithmetic and four logical operations.

The first approach for designing a one-bit quantum ALU is illustrated in figure 15.1 which shows five dotted rectangles, four V, two V+, and eight CNOT gates. A dotted rectangle is equivalent to a CNOT gate with the cost of 1. The total quantum cost of the circuit is 19. The functions produced by the first approach are presented in table 15.1. From table 15.1, depending on the values in select inputs, the circuit produces six functions, namely ADD and SUB, which are arithmetic functions, and AND, NAND, OR, and NOR, which are logical functions.

15.1.2 The second approach

The quantum ALU has nine inputs and nine outputs and is illustrated in figure 15.2. The quantum ALU logic gates have used two dotted rectangles, eight V, four V^+, and six CNOT gates. From the ALU, three input values (A, B, and C_{in}), three select lines (S_0, S_1, S_2, S_3, S_4), and one constant input are selected. The principal outputs are F (as the product output) and C_{out}.

In the second approach the total quantum cost is 24. There are two garbage values and one constant input. It has nine circuit lines and five selection lines. The presented design can be implemented with seven gates. Here it is clear that the inputs A, B, and C_{in} can be controlled depending on the values of the selection and carry input lines. The product output is F (as a final output) which will be obtained from 12 different operations, either arithmetic or logical, depending upon the values of C_{in} (as the carry input) and selection lines. The other primacy output is carry out (C_{out}). The 12 operations generated in the second approach are given in table 15.2. A special function is selected through (S_0, S_1, S_2, S_3, S_4). The presented circuit has

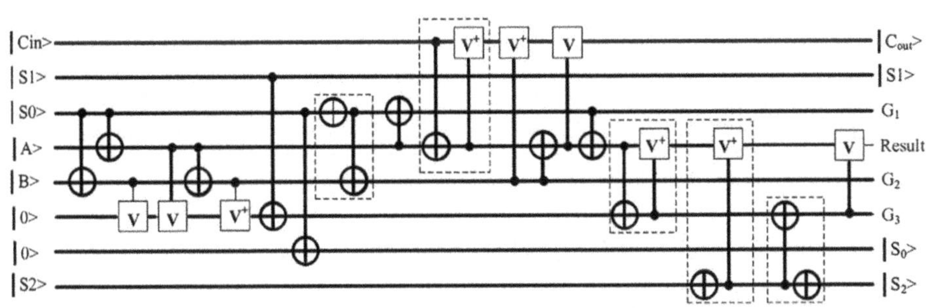

Figure 15.1. The quantum arithmetic logic unit (first approach). Reproduced with permission from [4]. Copyright 2016 Springer (GmbH).

Table 15.1. Different functions of the quantum ALU (the first approach).

S_2	S_1	S_0	Result	Functions
0	0	0	(AB)	AND
0	0	1	$(A + B)$	NOR
0	1	0	(AB)	NAND
0	1	1	$(A + B)$	OR
1	0	0	$A \oplus B \oplus C_{\text{in}}$	ADD
1	0	1	$A \oplus B \oplus C_{\text{in}}$	SUB

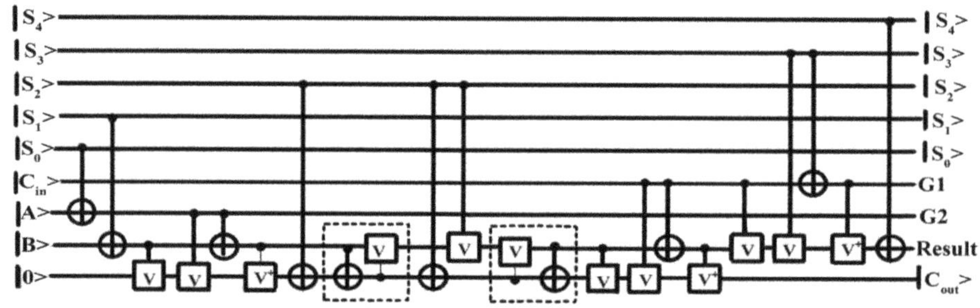

Figure 15.2. Quantum ALU (the second approach). Reproduced with permission from [4]. Copyright 2016 Springer (GmbH).

Table 15.2. Simple operations of the quantum ALU (the second approach).

Result	C_{in}	S_0	S_1	S_2	S_3	S_4	Transformations	Description
$A + B$	0	0	0	0	0	0	$F = A + B + 0$	ADD without carry
$A + B + 1$	1	0	0	0	0	0	$F = A + B + 1$	ADD with carry
$A - B$	1	0	1	0	0	0	$F = A + B' + 1$	SUB
$A \oplus B$	0	0	0	0	1	0	$F = A \oplus B$	Exclusive-OR
$A \wedge B$	0	0	0	1	1	0	$F = A \wedge B$	Bitwise-AND
$A \vee B$	0	1	1	1	1	1	$F = (A' \wedge B')'$	Bitwise-OR
$(A \oplus B)'$	0	0	0	0	1	1	$F = (A \oplus B)'$	Exclusive-NOR
$A + 1$	1	0	0	0	0	0	$F = A + 0 + 1$	Self-increment of A
$A - 1$	1	0	1	0	0	0	$F = A + 1' + 1$	Self-decrement of A
A'	0	1	0	0	0	0	$F = A' + 0 + 0$	Bitwise negation of A
A'	1	1	0	0	0	0	$F = A' + 0 + 1$	Complement of A
A	0	0	0	0	0	0	$F = A + 0 + 0$	NOP (no-operation)

two main outputs (result and C_{out}). This circuit has five select control signals with a provision to realize 12 logical and arithmetic operations. The total elementary operations of the second ALU design are shown in table 15.2. By transforming the expressions based on the second approach, an extension of ALU can perform other operations (such as NAND and NOR), which are the universal functions. Therefore, the second approach for the ALU circuit can implement several universal and arithmetic operations, namely ADD, SUB, AND, NAND, OR, NOR, Ex-OR, and Ex-NOR. Moreover, an n-bit ALU can be constructed by cascading n of these one-bit ALU circuits.

15.1.3 The third approach

The quantum ALU has seven inputs and seven outputs. The third approach for a one-bit quantum ALU is illustrated in figure 15.3. One V, two V^+, and six CNOT gates are employed to construct the ALU. From the ALU, two input values (A, B) and five selectors (S_0, S_1, S_2, S_3, S_4) are used for the input and the primary outputs.

In the third approach the total quantum cost is 12. There are no garbage values or constant inputs. The product output (result) will be obtained from the ten different operations presented in table 15.3. After transforming the expressions which are based on the third approach, an extension of ALU can perform basic arithmetic and logical operations (ADD, SUB, NSUB, Ex-OR, and NOP). The third approach is a garbage-free quantum ALU, which has no constant inputs.

15.2 Summary

In this chapter three approaches are explained for one-bit quantum arithmetic logic unit design, constructed using a combination of simple quantum gates. All the ALUs can handle many simple logical and arithmetic operations, namely addition, subtraction, bitwise-AND, bitwise-OR, NAND, NOR, exclusive-OR, exclusive-NOR, negation, and complement. In order to design a cost-effective quantum ALU, the number of logical and arithmetic calculations produced on the outputs must be considered. The ALU designs are evaluated in terms of the number of circuit lines, garbage outputs, and constant inputs, and the quantum cost. Moreover, the ALU designs are versatile approaches for the realization of quantum computing, both having remarkably low power consumption and nano-scaling. Therefore, the designs are implemented with simple quantum gates which have the scope to be realized in nanotechnology based systems, such as quantum cellular automata,

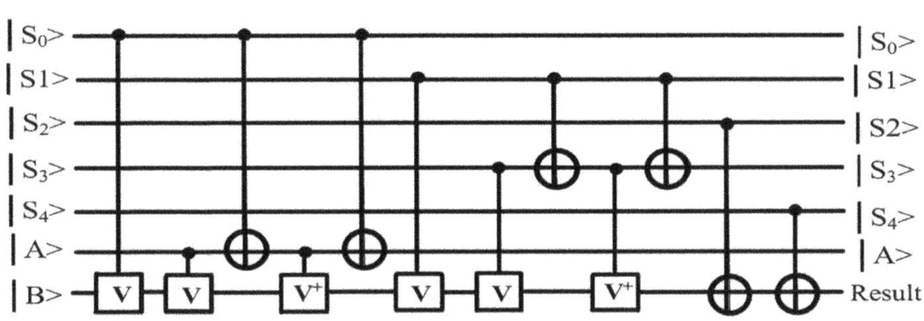

Figure 15.3. The quantum ALU (the third approach). Reproduced with permission from [4]. Copyright 2016 Springer (GmbH).

Table 15.3. Basic operations of the quantum ALU (the third approach).

S_0	S_1	S_2	S_3	S_4	ALU operations	Description
1	1	0	0	0	$F = A + B$	ADD
1	1	1	0	1	$F = B - A$	SUB
1	1	0	1	1	$F = A - B$	NSUB
1	0	0	0	0	$F = A \oplus B$	Ex-OR
0	0	0	0	0	$F = B$	NOP
1	1	0	1	0	$F = B + A + 1$	ADD with carry
1	1	1	1	1	$F = B - A - 1$	SUB with borrow
1	1	0	0	1	$F = A - B - 1$	NSUB with borrow
1	0	1	0	0	$F = A \oplus B$	Ex-NOR
0	0	1	0	0	$F = B$	Negation of *B*

quantum processor and different NANO-metric architectures. The described circuits can be used in a wide variety of large quantum systems which enhance the performance of quantum computers. Furthermore, an optimized ALU is a requirement in computer systems and digital signal processors for the design of the instruction set of more complex systems and for low power VLSI design in nanotechnology, quantum computers, and programmable computing devices.

Further reading

[1] Bennett C H 1973 Logical reversibility of computation *IBM J. Res. Dev.* **17** 525–32

[2] Dirac P A M 1981 *The Principles of Quantum Mechanics* (Oxford: Oxford University Press)

[3] Große D, Wille R, Dueck G W and Drechsler R 2008 Exact synthesis of elementary quantum gate circuits for reversible functions with don't cares *38th Int. Symp. on Multiple Valued Logic* pp 214–9

[4] Haghparast M and Bolhassani A 2016 Optimization approaches for designing quantum reversible arithmetic logic unit *Int. J. Theor. Phys.* 55 1423–37

[5] Kaye P *et al* 2007 *An Introduction to Quantum Computing* (Oxford: Oxford University Press)

[6] Landauer R 1961 Irreversibility and heat generation in the computing process *IBM J. Res. Dev.* **5** 183–91

[7] Lim J, KIM D-G and Chae S-I 1999 Reversible energy recovery logic circuits and its 8-phase clocked power generator for ultra-low-power applications *IEICE Trans. Electron.* **82** 646–53

[8] Morrison M and Ranganathan N 2011 Design of a reversible ALU based on novel programmable reversible logic gate structures *IEEE Computer Society Annual Symp. on VLSI* pp 126–31

[9] Morrison M and Ranganathan N 2013 A novel optimization method for reversible logic circuit minimization *IEEE Computer Society Annual Symp. on VLSI* pp 182–7

[10] Morrison M A 2012 Design of a reversible ALU based on novel reversible logic structures *Thesis* University of South Florida

[11] Nielsen M A and Chuang I 2002 *Quantum Computation and Quantum Information* 10th edn (Leiden: Cambridge University Press)

[12] Moore G E 2000 Cramming more components onto integrated circuits *Readings in Computer Architecture* (San Francisco, CA: Morgan Kaufmann) p 56

[13] Thomsen M K, Glück R and Axelsen H B 2010 Reversible arithmetic logic unit for quantum arithmetic *J. Phys. A: Math. Theor.* **43** 382002

[14] Toffoli T 1980 Reversible computing *Int. Coll. on Automata, Languages, and Programming* (Berlin: Springer) pp 632–44

[15] Weste N H and Harris D 2015 *CMOS VLSI Design: A Circuits and Systems Perspective* (India: Pearson Education)

IOP Publishing

Quantum Computing
A pathway to quantum logic design
Hafiz Md Hasan Babu

Chapter 16

Applications of quantum computing technology

There are many potential future uses for quantum computing technology. This is an important phenomenon which could change the total level of computing in the coming decade and the computers we know today will become museum pieces. In this chapter we discuss some real world applications of quantum computers/circuits. The basic structure of a quantum computer is shown in figure 16.1.

16.1 Optimization

Optimization problems are some of the most difficult problems to solve.

Imagine that you are building a house and have a list of things you need to have in your home. However, you cannot afford the cost of everything on your list since you have limited finances. What you truly need to work out is the combination of things which will give you the best value for your money.

This is a case of an optimization issue, where you are attempting to locate the best combination of things given a few requirements. While issues with only a couple of decisions are simple, as the number of decisions increases, they rapidly become difficult to settle ideally. With only 270 on/off switches, there are more conceivable combinations than there are molecules in the Universe! These types of advancement issues exist in various areas, for example framework structures, mission arrangements, aircraft planning, money related investigations, web searches, and malignant growth radiotherapy. These are some of the most complex issues on the planet, with tremendous potential advantages for organizations, individuals, and science if ideal solutions can be achieved quickly.

It is easily understandable how quantum computing can help in these types of scenarios. Some of these optimization problems are examined below, where quantum computers are playing a role in solving them.

doi:10.1088/978-0-7503-2747-3ch16

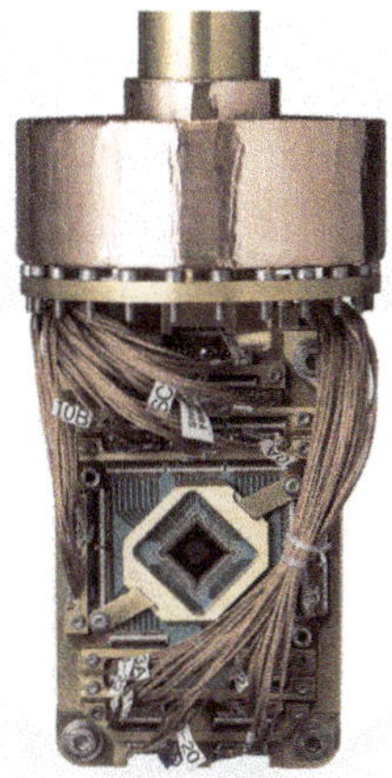

Figure 16.1. A quantum computer. Reproduced courtesy of D-Wave Systems Inc.

16.1.1 Roswell Park Cancer Institute

Enhancement techniques are a basic requirement to supply enough radiation to kill malignant cells while avoiding non-carcinogenic adjacent cells. Researchers at Roswell Park Cancer Institute utilized a D-Wave framework for use in quantum annealing for beamlet intensity optimization in intensity-modulated radiation therapy.

16.1.2 Volkswagen Group

Volkswagen was the primary vehicle maker to utilize a quantum PC to determine traffic streams, as shown in figure 16.2. Their researchers carried out an explorative study of traffic stream improvement. Utilizing information from 10,000 taxi cabs in Beijing, they modified a calculation to improve the movement time of taxi cabs in the city.

16.1.3 Recruit Communications

Recruit Communications and D-Wave are working together to apply quantum processing to showcasing, promoting, and correspondence. Its first task was to streamline the efficiency of coordinating notices to clients for web publicizing.

16.2 Machine learning

When you look at a photograph it is simple for you to identify the distinct element in the picture: trees, mountains, and so forth. This assignment is relatively easy for humans, however, it is a tremendously difficult job for PCs to accomplish. This is because software engineers do not know how to characterize the essence of a 'tree' in PC code.

Machine learning is the best way to deal with this issue, using which developers can devise algorithms that naturally determine how to perceive the 'substance' of objects by recognizing repeating designs in enormous quantities of information. In

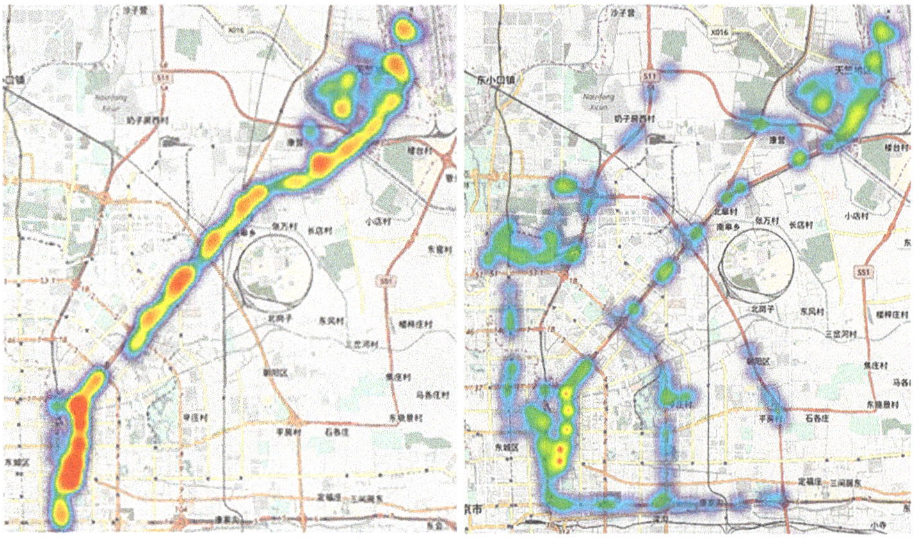

Figure 16.2. Non-optimized and optimized traffic flow research using quantum computers. Reproduced with permission from [11].

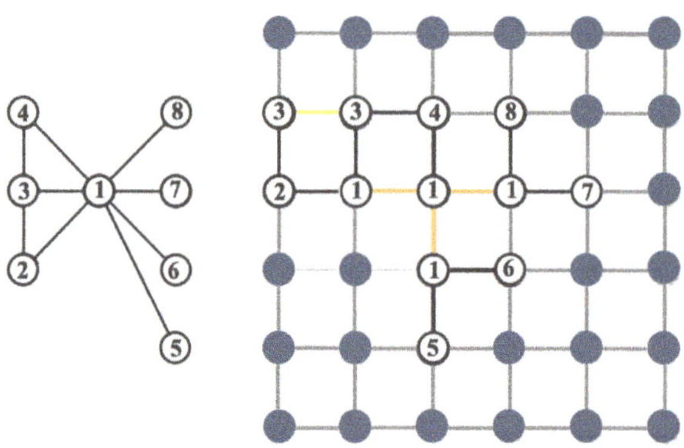

Figure 16.3. An example of embedding problem into a quantum computation model.

view of the amount of information involved in this procedure, and the vast number of potential combinations of information components, this is a computationally costly issue. Similarly to other streamlining issues, these types of tasks can be mapped to the local ability of the D-Wave quantum computer.

16.2.1 QxBranch

Foreseeing the result of an election is a challenge. Consolidating quantum computing with neural system innovations could enhance such conjectures, as indicated by the work carried out by QxBranch. They utilized such a system in a simple manner to predict the 2016 US presidential race. An example of embedding a problem into a quantum computation model is shown in figure 16.3.

16.2.2 Los Alamos National Laboratory

Utilizing a set of 2429 pictures of faces, Los Alamos National Laboratory researchers utilized unsupervised machine learning techniques by means of framework factorization to break down large datasets. This is one of the numerous quick reaction extensions that Los Alamos scientists have carried out on the D-Wave quantum computing framework, first introduced at Los Alamos, which is shown in figure 16.4.

16.2.3 NASA

Machine learning depends strongly on inspecting complex likelihood calculations. NASA researchers effectively prepared the D-Wave 2X framework for picture information collection in a generative unsupervised learning setting, which stands out as one of the most challenging ideal models in machine learning.

16.3 Biomedical simulations

With quantum computers we can mimic and model sub-atomic structures. In 2012, analysts at Harvard University utilized a D-Wave One quantum PC, as shown in figure 16.5, to understand the riddle of how a few proteins overlap.

'The model comprised of scientific portrayals of amino acids in a cross section, associated by various communication qualities' wrote Geoffrey Brumfiel in a news article for *Nature* about the Harvard specialists' protein collapsing models, noting that '[t]he D-Wave PC found the least setups of amino acids and associations, which

Figure 16.4. Te National Security Sciences Building at Los Alamos National Laboratory. Photo courtesy of Los Alamos National Laboratory.

Figure 16.5. D-Wave quantum computers. Reproduced courtesy of D-Wave Systems Inc.

relates to the most efficient collapsing of the proteins.' While the innovation was not even close to complete—as the researchers wrote, '10 000 estimations utilizing an 81-qubit rendition of the examination gave the right answer only multiple times'—it had the ability to achieve a remarkable accomplishment by displaying the behavior of protein collapsing with some level of exactness.

16.4 Financial services

Quantum processing is now on its way. D-Wave, an organization supported by Goldman Sachs and Bezos Expeditions, among others, introduced its first business quantum computer, the D-Wave 2000Q, a quantum enhancing framework with 2000 qubits and propelled including controls.

Notwithstanding their limitations, these computers are being used in specialist fields as demonstrated by the Harvard team's utilization of D-Wave's first model in 2012. These frameworks could be utilized for complex budgetary displays. Quantum computing could be utilized to discover 'better approaches to show budgetary information' and exclude 'key worldwide hazard factors', as stated by IBM. A development architecture of a quantum computing network by IBM is shown in figure 16.6.

'It is extraordinary to assemble frameworks to help Wall Street better oversee hazard utilizing this sort of innovation', D-Wave Systems President and CEO Vern Brownell told Bloomberg News: '[t]hey spend a ton on figuring power to oversee risk'.

16.5 Computational chemistry

There are numerous avenues in materials science that could accomplish tremendous results if we could simply locate the correct catalyst or procedure to construct a new material or to produce a current material more productively. Currently, notable efforts are being made to utilize traditional PCs to mimic mixture interactions, yet much of the time these problems prove too difficult to solve

Figure 16.6. An IBM expanded quantum computing network. Image courtesy of International Business Machines Corporation. Copyright International Business Machines Corporation.

traditionally. Thus, the first thought, introduced by Richard Feynman, is why not utilize a quantum PC to reproduce the quantum mechanical procedures that occur. The following are only a few of the myriad of critical issues that could see significant solutions if only we could understand them:

1. Supplanting the Haber procedure to deliver ammonium nitrate for use in fertilizers.
2. Finding new materials that can be used to achieve a room-temperature superconductor.
3. Discovering a catalyst that can enhance the efficiency of carbon capture.
4. Devising a new battery science that can improve performance compared to current lithium-ion batteries.

16.6 Logistics and scheduling

Numerous standard improvements made in industry can be categorized under coordination and planning. Consider an aircraft coordination supervisor who needs to make sense of how to organize aircraft to achieve the best management at minimal cost, or a manufacturing plant supervisor who has a regularly changing combination of machines, stock, generation requests, and individuals, and requirements to limit cost and throughput times while improving yield. Consider, on the other hand, a chief assessor in a car organization who needs to make sense of the optimal costs of a considerable variety of vehicle choices to enhance consumer loyalty and benefit. Although traditional calculations are utilized extensively to perform such tasks, some questions might be too difficult for traditional processing while a quantum approach might have the capacity to achieve such a task.

16.7 Cyber security

Digital security is becoming a more significant problem these days as threats around the globe are expanding their capabilities and we are becoming progressively more powerless as we increase our reliance on advanced digital frameworks. Different strategies to combat digital security risks can be created utilizing quantum machine learning approaches to perceive the dangers ahead of time and alleviate the harm that they may cause.

16.8 Circuit, software, and system fault simulation

When one develops expansive programming with a huge number of lines of code or substantial ASIC chips that have billions of transistors, it can become very difficult and costly to assess them for correctness. There can be billions or trillions of different states and it is unthinkable for a traditional PC to check each and every one in simulation. In addition to wanting to comprehend what will happen when a framework is working normally, one also needs to understand what occurs if there is an equipment or other error. Will the framework identify this error and does it have a recovery system to alleviate any conceivable issues? The costs of an error can be high, as some of these frameworks may be utilized in situations where lives or a huge amount of money may depend on their being error-free. By utilizing quantum computing to help in these simulations, one may be able to provide a vastly improved assessment of the performance in a much shorter time.

16.9 Weather forecasting

National Oceanic and Atmospheric Administration (NOAA) Chief Economist Rodney F Weiher has claimed that almost 30% of the US GDP ($6 trillion) is specifically or in a oblique way influenced by the weather, affecting food production, transportation, and retail, among others. The capacity to more accurately predict the weather would have huge advantages for numerous fields, while also providing better opportunities to protect people from extreme weather events.

While this has been an objective of researchers for quite some time, the conditions underlying weather and the climate include numerous factors, making traditional simulations protracted. As quantum specialist Seth Lloyd pointed out, '[u]tilizing a traditional PC to perform such investigation may take longer than it takes the real climate to advance!' This inspired Lloyd and co-workers at MIT to demonstrate that the conditions that determine the climate have a concealed wave nature which is amenable to be addressed by a quantum computer.

The executive of design at Google, Hartmut Neven, likewise noticed that quantum computers allowed better atmosphere models that could provide us with a greater understanding of how humans are affecting the natural world. These models are what we base our estimation of future global warming on, and what will enable us to determine what steps need to be taken now to prevent future calamities. The United Kingdom's national weather service, the Met Office, has recently started putting resources into such developments to meet the energy and adaptability requirements they will be facing from 2020 onward, and discharged its very own account necessities for bigger scale computing.

Thus we can easily understand that quantum computers are the future of computing. All the types of quantum computers are very important for various types of computational problems. We only discuss some of it greater impacts, but there are many other areas where we will see the effects of the future of computing—quantum computing.

16.10 Summary

This chapter describes different applications of quantum computing. Quantum computing is leading to a paradigm shift in many ways. Quantum computing provides access to new levels of computational ability and it also inspires new ways of thinking. As we look at problems in new ways, this can, in turn, contribute new ideas for how we could advance classical computation as well. With more and more individuals conceiving problems from different angles, more and more ideas and results will be available. As we dive into the world of quantum evolution, we will exercise the ability to think about problems in new ways, become familiar with programming quantum computers, and even simulate this work so that we can obtain useful quantum computers.

Further reading

[1] Jackson M 2017 6 things quantum computers will be incredibly useful for *SingularityHub* https://singularityhub.com/2017/06/25/6-things-quantum-computers-will-be-incredibly-useful-for/#sm.0000trlmgo7c0f9mwo3107yechmv7 (Accessed: 27 December 2018)

[2] Volkswagen 2018 Avoiding traffic jams and surviving tsunamis *Volkswagen* https://www.volkswagenag.com/en/news/stories/2018/06/avoiding-traffic-jams-and-surviving-tsunamis.html (Accessed: 27 December 2018)

[3] D-Wave D-Wave: practical quantum computing *D-Wave* https://www.dwavesys.com/home (Accessed: 27 December 2018)

[4] D-Wave 2012 Harvard researchers use D-Wave quantum computer to fold proteins *D-Wave* https://www.dwavesys.com/news/harvard-researchers-use-d-wave-quantum-computer-fold-proteins (Accessed: 27 December 2018)

[5] NASA Quantum computing *NASA Advanced Supercomputing Division* https://www.nas.nasa.gov/projects/quantum.html (Accessed: 27 December 2018)

[6] Microsoft Quantum computing applications for innovation and impact https://www.microsoft.com/en-us/quantum/what-is-quantum-computing (Accessed: 27 December 2018)

[7] LANL Quantum Institute *Los Alamos National Laboratory* https://www.lanl.gov/collaboration/research-opportunities/quantum-institute.php (Accessed: 27 December 2018)

[8] QxBranch QxBranch builds software and applications for quantum computers *QxBranch* https://www.qxbranch.com/quantum-computing/ (Accessed: 27 December 2018)

[9] Recruit communications and D-Wave collaborate to apply quantum computing to marketing, advertising, and communications optimization https://www.dwavesys.com/press-releases/recruit-communications-and-d-wave-collaborate-apply-quantum-computing-marketing (Accessed: 27 December 2018)

[10] Roswell park https://www.roswellpark.org/ (Accessed: 27 December 2018)

[11] Neukart F, Compostella G, Seidel C, Dollen D V, Yarkoni S and Parney B 2017 Traffic flow optimization using a quantum annealer *Front. ICT* **4** 29

Part II

Quantum fault tolerance

An overview of quantum fault-tolerant circuits

A fault is an error that occurs in a system which forces the system to deviate from its normal behavior. The purpose of fault tolerance is to enable reliable quantum computations when the computer's basic components are unreliable. To achieve this, the qubits in the computer are encoded in blocks of a quantum error-correcting code, which allows the state to be corrected even when some qubits are wrong. A fault-tolerant protocol prevents catastrophic error propagation by ensuring that a single faulty gate or time step produces only a single error in each block of the quantum error-correcting code. The fault models vary according to the type of description that is being considered, which in turn varies according to the level of abstraction.

Fault tolerance is the property that enables a system to operate accurately in the presence of the failure of one or more of its components. Fault tolerance in quantum circuits reflects the robustness of the system. Fault-tolerant systems are capable of detecting and correcting faults. If the logic circuit itself is made of fault-tolerant components, then the detection and correction of faults in the circuit become cheaper, easier, and more simple. To achieve fault tolerance, the first step is to identify the occurrence of a fault. To detect the occurrence of a fault, parity-preserving techniques are considered as one type of solution. Any fault that affects only one signal is detectable at the circuit's primary outputs in fault-tolerant quantum circuits.

A conventional circuit dissipates energy to reload missing information because of overlapped mapping between input and output vectors. Quantum computing recovers this energy loss and prevents bit errors by including fault-tolerant mechanisms. A fault-tolerant quantum circuit is capable of preventing error at the outputs. The goal of any program to produce a fault-tolerant logic gate is to implement a universal set of gates with the properties of fault tolerance. Essentially, the goal is to implement a set of gates that will allow any arbitrary quantum computation.

In practice, quantum computers are difficult to implement. The reason for this is that they require the qubits they are constructed of to be isolated from the environment and remain completely unaltered by any processes other than the logical gates which are applied to them. In reality, such isolation is extremely difficult to achieve. Undesired interactions with the qubits' surroundings, and even imperfect executions of quantum gates and measurements all introduce the possibility of error. These errors are by nature random, so in the modeling it is considered that faults are probabilistic events that occur throughout the evolution of a quantum circuit. Here, the ramifications of error rates and circuit designs are considered, which will inform the basis for fault-tolerant quantum computations.

Quantum computers appear to be capable, at least in principle, of solving certain problems far faster than any conceivable classical computer. In practice, quantum computing technology is still in its infancy. While a practical and useful quantum computer may eventually be constructed, at present we cannot clearly envision what the hardware of that machine will be like. Nevertheless, one can be quite confident that any practical quantum computer will incorporate some types of error correction in its operation. Quantum computers are far more susceptible to making errors than conventional digital computers, and some method of controlling and correcting such errors will be needed to prevent a quantum computer from crashing.

The basic idea of error-correcting codes is that, to store and transmit data without said data being compromised, it is necessary to build some kind of larger mathematical structure to serve as a shield for the structure containing the actual data. This shield must be effective, namely, it should be able to both survive one or more errors and should be able to correct those errors over time by restoring the original state.

This part of the book describes some background and preliminary studies of quantum fault-tolerant circuits in chapter 17. The approach for designing a fault-tolerant quantum adder (full-adder) is given in this chapter. A design for a fault-tolerant multiplier (signed multiplier) is also shown here. Finally, the design procedure of for a quantum fault-tolerant integer division circuit is introduced in this chapter. In this design, a restoring division algorithm as well as quantum restoring integer division circuitry have been introduced.

Chapter 17

Quantum fault-tolerant circuits

The discovery of quantum error correction has greatly improved the long-term prospects of quantum computing technology. Encoded quantum information can be protected from the errors that arise due to uncontrolled interactions with the environment, or due to imperfect implementations of quantum logical operations. Recovery from errors can work effectively even if occasional mistakes occur during the recovery procedure.

17.1 The need for quantum fault-tolerant circuits

The discovery of quantum effects alone is not sufficient to ensure that a quantum computer can perform reliably. To carry out a quantum error-correction protocol we must first encode the quantum information we want to protect and then repeatedly perform recovery operations that reverse the errors that accumulate. However, encoding and recovery are themselves complex quantum computations, and errors will inevitably occur while we perform these operations. Thus we need to find methods for recovering from errors that are sufficiently robust to succeed with high reliability even when we make some errors during the recovery step.

Furthermore, to operate a quantum computer, we must process the information of two or more encoded qubits that interact with one another and if an error occurs in one qubit, then that qubit interacts with another quantum gate and the error is likely to propagate to the second qubit. We must design our gates to minimize the propagation of error.

Incorporating quantum error correction will surely complicate the operation of a quantum computer. To establish the redundancy needed to protect against errors, the number of elementary qubits will have to increase. The encoded information on performing gates, and inserting periodic error-recovery steps, will slow down the computation.

doi:10.1088/978-0-7503-2747-3ch17

A device that works perfectly even when its primary components are imperfect is said to be fault-tolerant. With the development of fault-tolerant methods, it will be possible, in principle, for the operator of a quantum computer to actively intervene to stabilize the device against errors in a chaotic environment. In the long run, fault tolerance might be achieved in practical quantum computers by a rather different route—with intrinsically fault-tolerant hardware. Such hardware, designed to be impervious to localized influences, could be operated relatively unsupervised, yet could still store and process quantum information robustly.

Fault-tolerant quantum computing falls under the framework of proposed ideas that allow qubits to be protected from quantum errors introduced by the inadequate control of environmental interactions, called quantum error correction (QEC). Moreover, the appropriate innovations in quantum circuits will be implemented in both QEC and encoded logic operations in such a way as to avoid these errors cascading through quantum circuits. By avoiding a cascade of errors, there comes a point (when the fundamental accuracy of individual qubits is high enough), when the QEC is correcting more errors than are being created. Once this threshold has been achieved, the expansion of the size of the protective quantum will exponentially decrease the failure of encoded information and allows us to achieve algorithms implemented with noisy devices.

The parity-preserving quantum circuits are a class of quantum logic with the additional property that the parity of the input is the same as the parity of the output. A quantum circuit will be parity-preserving if the Ex-OR of the inputs matches the Ex-OR of the outputs, i.e. the parity of the input and the output remains the same.

17.2 The fault-tolerant quantum adder

A quantum adder is a quantum circuit that performs the addition operation. In a full-adder, two binary numbers and a carry or overflow bit might input to complete the addition operation. The output of the circuit will be the sum and another carry bit. Full-adders are commonly connected to each other to add bits to an arbitrary length of bits, such as 32 or 64 bits.

Fault tolerance is the ability of a system to hold its normal operation without failure when some part of the system fails to operate properly. It increases the wear-out time for any system at a cost of increased hardware. Fault-tolerant approaches must be incorporated in any safety-critical system to continuing functioning without failure even if an error occurs in the system.

The adder is an important circuit in quantum circuit design. To design a fault-tolerant quantum full-adder some basic circuits are necessary, which are discussed in the next section.

17.2.1 The fault-tolerant full-adder

In this section, the conventional full-adder, as presented in figure 17.1, is discussed and a fault-tolerant full-adder is designed. Finally, the quantum fault-tolerant full-adder is presented. A classic full-adder, as shown in figure 17.1, will have two inputs

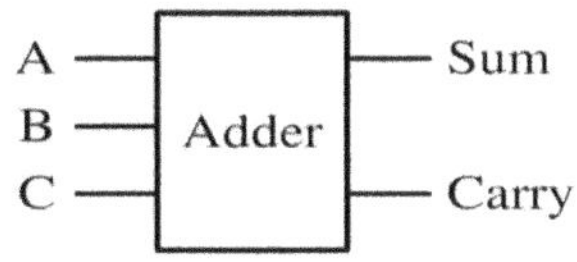

Figure 17.1. A conventional full-adder.

Table 17.1. The truth table of a conventional full-adder.

A	*B*	*C*	Sum	Carry
0	0	0	0	0
0	0	1	1	0
0	1	0	1	0
0	1	1	0	1
1	0	0	1	0
1	0	1	0	1
1	1	0	0	1
1	1	1	1	1

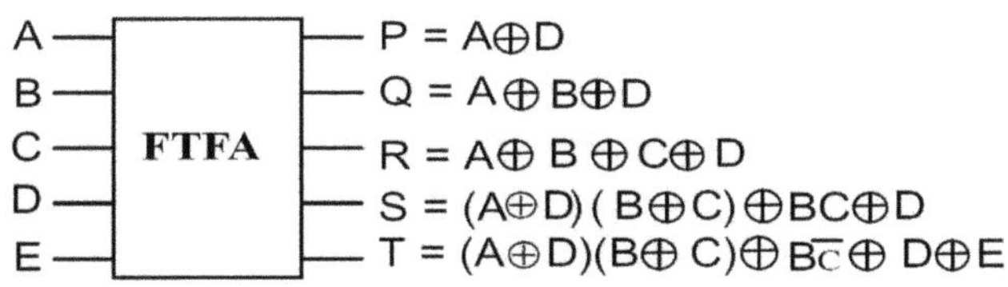

Figure 17.2. Block diagram of a FTFA.

and one carry or overflow bit and from the truth table (table 17.1) the following expressions can be achieved:

$$\text{Sum} = A \oplus B \oplus C \tag{17.1}$$

$$\text{Carry} = (A \oplus B)C \oplus AB. \tag{17.2}$$

Based on equations (17.1) and (17.2), a parity-preserving logic gate called the fault-tolerant full-adder (FTFA) is implemented and is presented in figure 17.2. In order to verify that the implemented gate has worked as a fault-tolerant quantum adder, the corresponding truth table of the gate is presented in table 17.2. It can be seen from the truth table that the input and output vectors show one-to-one mapping and at the same time the condition of the input parity $A \oplus B \oplus C \oplus D \oplus E$ is equal to the output parity $P \oplus Q \oplus R \oplus S \oplus T$.

From figure 17.2 it is discovered that if the inputs $D = 0$ and $E = 0$, the extended fault-tolerant full-adder (EFTFA), generates three garbage outputs, which is the minimum (see figure 17.3).

Table 17.2. The truth table of a quantum FTFA.

Input					Output				
A	*B*	*C*	*D*	*E*	*P*	*Q*	*R*	*S*	*T*
0	0	0	0	0	0	0	0	0	0
0	0	0	0	1	0	0	0	0	1
0	0	0	1	0	1	1	1	1	1
0	0	0	1	1	1	1	1	1	0
0	0	1	0	0	0	0	1	0	0
0	0	1	0	1	0	0	1	0	1
0	0	1	1	0	1	1	0	0	0
0	0	1	1	1	1	1	0	0	1
0	1	0	0	0	0	1	1	0	1
0	1	0	0	1	0	1	1	0	0
0	1	0	1	0	1	0	0	0	1
0	1	0	1	1	1	0	0	0	0
0	1	1	0	0	0	1	0	1	0
0	1	1	0	1	0	1	0	1	1
0	1	1	1	0	1	0	1	0	1
0	1	1	1	1	1	0	1	0	0
1	0	0	0	0	1	1	1	0	0
1	0	0	0	1	1	1	1	0	1
1	0	0	1	0	0	0	0	1	1
1	0	0	1	1	0	0	0	1	0
1	0	1	0	0	1	1	0	1	1
1	0	1	0	1	1	1	0	1	0
1	0	1	1	0	0	0	1	1	1
1	0	1	1	1	0	0	1	1	0
1	1	0	0	0	1	0	0	1	0
1	1	0	0	1	1	0	0	1	1
1	1	0	1	0	0	1	1	1	0
1	1	0	1	1	0	1	1	1	1
1	1	1	0	0	1	0	1	1	0
1	1	1	0	1	1	0	1	1	1
1	1	1	1	0	0	1	0	0	1
1	1	1	1	1	0	1	0	0	0

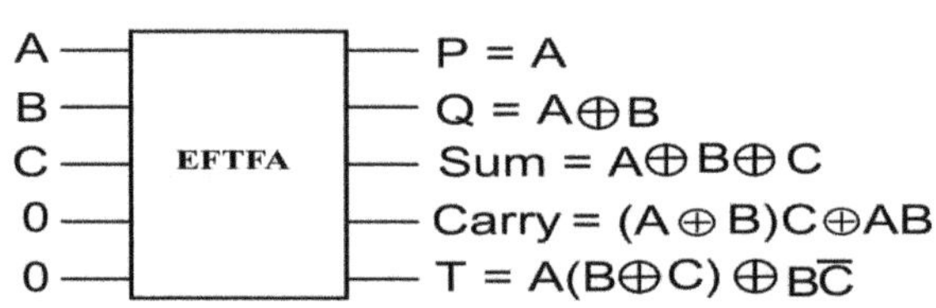

Figure 17.3. Block diagram of an extended fault-tolerant full-adder, when $D = 0$ and $E = 0$.

The equivalent quantum circuit of the EFTFA is illustrated in figure 17.4 and the dotted rectangle in the figure is equivalent to a 2 × 2 CNOT gate. Therefore, the quantum cost of the fault-tolerant full-adder is 8.

17.3 The fault-tolerant multiplier

Fault-tolerant multiplier circuits have special importance because of the fact that they are the integral components of computer systems, mobile devices, and most audio/video components. To implement a fault-tolerant multiplier, we need a quantum representation of the Ex-OR, NAND, and basic logic operations. The parity-preserving quantum fault-tolerant operations are as follows:

A. *Quantum fault-tolerant Ex-OR operation*: figure 17.5 presents the block diagram (block A) and the quantum realization for the XOR operation, where the inputs are A, B, and C, and the outputs are $P = A$, $Q = A \oplus B$, and $R = A \oplus C$.

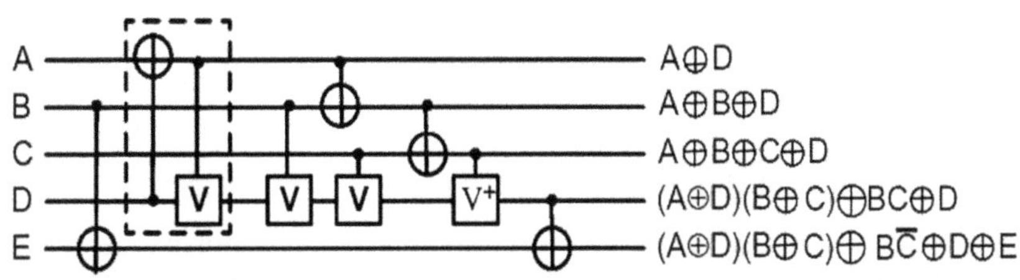

Figure 17.4. Quantum implementation of the extended fault-tolerant full-adder. Reproduced with permission from [9]. Copyright 2014 Taylor and Francis.

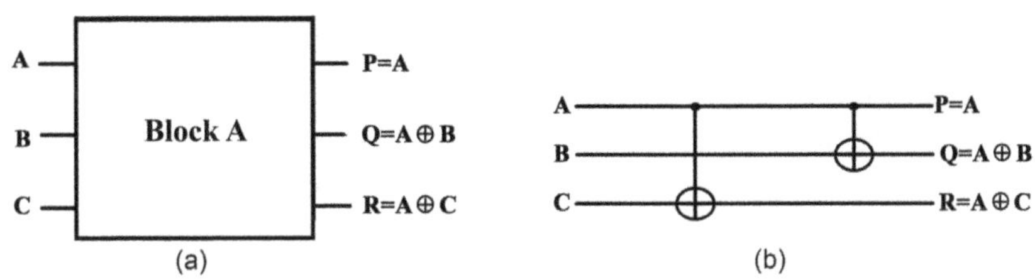

Figure 17.5. The fault-tolerant quantum realization of Ex-OR operation. (a) Block diagram. (b) Quantum implementation of block A.

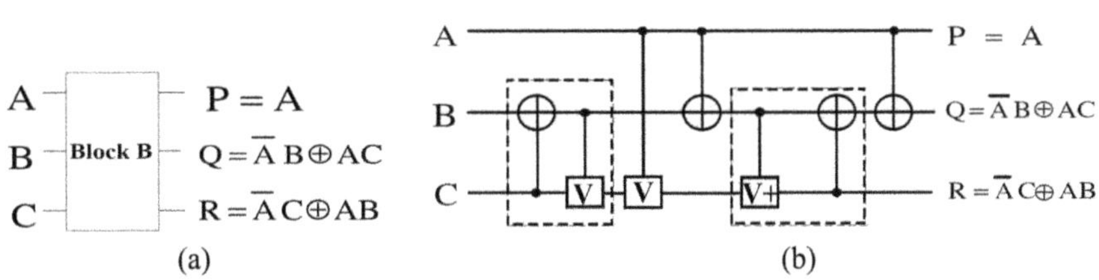

Figure 17.6. The fault-tolerant quantum realization of basic logic operations. (a) Block diagram. (b) Quantum implementation of block B. Reproduced with permission from [9]. Copyright Taylor and Francis.

B. *Quantum fault-tolerant basic logic operations*: figure 17.6 presents the block diagram (block B) and the quantum realization for basic logic operations, where the inputs are A, B, and C, and the outputs are $P = A$, $Q = \bar{A}B \oplus AC$, and $R = \bar{A}C \oplus AB$.
C. *Quantum fault-tolerant NAND, Ex-OR, and basic logic operations*: figure 17.7 presents the block diagram (block C) and the quantum realization for the NAND, XOR, and basic logic operations, where the inputs are A, B, and C, and the outputs are $P = A\bar{C} \oplus BC$, $Q = A \oplus B$, and $R = A\bar{C} \oplus \bar{B}C$.
D. *Quantum fault-tolerant half-adder operation*: figure 17.8 presents the block diagram (block D) of the half-adder operation and its quantum realization, where the inputs are A, B, C, and D and the outputs are $P = A, Q = A \oplus B$, $R = AB \oplus C$, and $S = A\bar{B} \oplus D$.
E. *Quantum fault-tolerant full-adder operation*: figure 17.9 presents the block diagram (block E) of the full-adder operation and the quantum realization, where the inputs are A, B, C, D, and E and the outputs are $P = AC \oplus B\bar{C}$, $Q = A \oplus B$, $R = A \oplus B \oplus C$, $S = (A \oplus B)C \oplus AB \oplus D$, and $T = A\bar{B} \oplus E$. The working principles of block E are presented in table 17.3.

17.3.1 The fault-tolerant signed multiplier

The speed of the multiplier plays an important role in the performance of a system. In general the speed of a multiplier depends on the algorithm and structure of the circuits. The Baugh–Wooley algorithm is adopted to perform the multiplication process, as shown in figure 17.10. Baugh–Wooley algorithms can be directly used in the complement multiplier, improving the speed of the multiplier. It only needs one kind of full-adder which requires supplements and it can be achieved as shown in figure 17.9(a).

In the signed multiplier, a Wallace tree structure is applied to implement one addition for n operands. In order to reduce the carry transfer delay of the partial product in the adder array of the multiplier, it is necessary to design its Wallace tree. In other words, it improves the operating speed in the whole adder array. In the Wallace tree, the bits with the same weight in the corresponding partial products are added together rather than one by one, as shown in figure 17.11. Usually, a full-adder is used to accomplish the addition of bits with the same weight. In the layers of

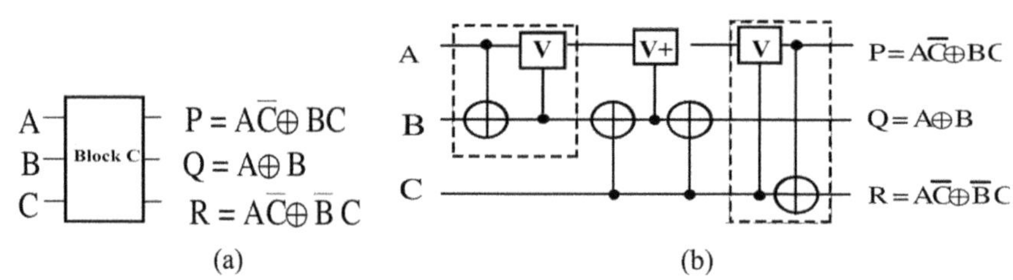

Figure 17.7. The fault-tolerant quantum realization of NAND, Ex-OR, and basic logic operations. (a) Block diagram. (b) Quantum implementation of block C.

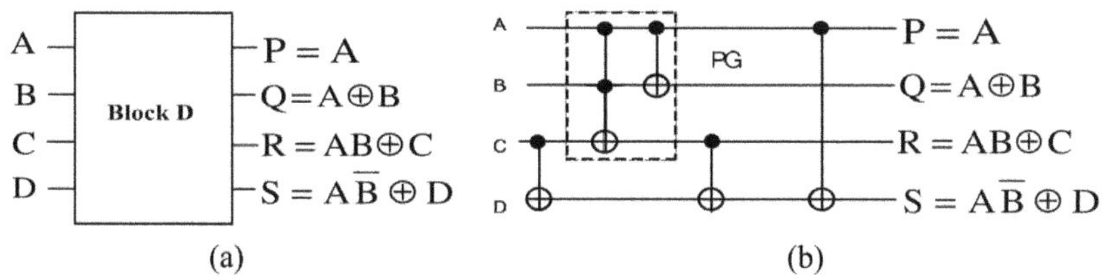

Figure 17.8. The fault-tolerant quantum realization of half-adder operation. (a) Block diagram. (b) Quantum implementation of block D.

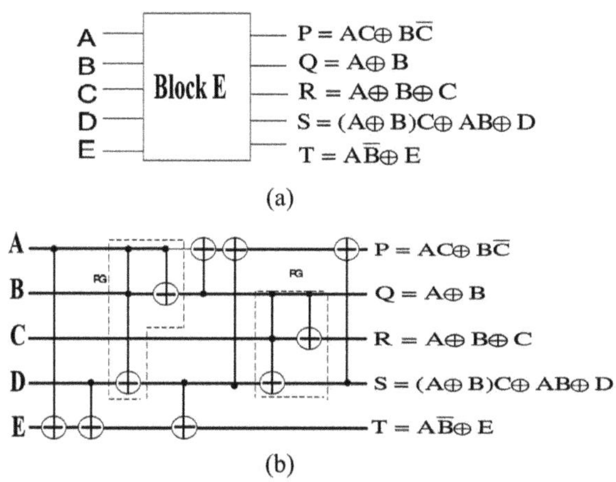

Figure 17.9. The fault-tolerant quantum realization of full-adder operation. (a) Block diagram. (b) Quantum implementation of block E.

the Wallace tree, the number of partial product vectors can be decreased according to the proportion 3:2, e.g. every three figures produces one sum bit and one carry bit. The fault-tolerant signed multiplier is shown in figure 17.12.

The fault-tolerant quantum signed multiplier is composed of block D and block E, which are the quantum realizations of the half-adder and full-adder circuits, respectively.

17.4 The quantum fault-tolerant integer divider

Dividers are one of the major computational units in quantum arithmetic. Integer division has applications in the circuit design of quantum algorithms, computation of power series, and trigonometric functions. Quantum computers of many qubits are extremely difficult to realize. Thus the number of qubits in quantum circuits needs to be minimized. The fabrication constraint of realizing quantum circuits with a large number of qubits has the objective of optimizing the number of ancillary qubits in a quantum circuit. Designing a scalable and reliable quantum computer is necessary now as well as in the future. Hence fault-tolerant quantum circuits are being explored based on Cliford and T gates.

Cliford+T gates based on a quantum circuit design of integer division are presented with n ancillary qubits, where n is the number of qubits in the operands.

Table 17.3. The truth table for block E.

Input					Output				
A	*B*	*C*	*D*	*E*	*P*	*Q*	*R*	*S*	*T*
0	0	0	0	0	0	0	0	0	0
0	0	0	0	1	0	0	0	0	1
0	0	0	1	0	0	0	0	1	0
0	0	0	1	1	0	0	0	1	1
0	0	1	0	0	0	0	1	0	0
0	0	1	0	1	0	0	1	0	1
0	0	1	1	0	0	0	1	1	0
0	0	1	1	1	0	0	1	1	1
0	1	0	0	0	1	1	1	0	0
0	1	0	0	1	1	1	1	0	1
0	1	0	1	0	1	1	1	1	0
0	1	0	1	1	1	1	1	1	1
0	1	1	0	0	0	1	0	1	0
0	1	1	0	1	0	1	0	1	1
0	1	1	1	0	0	1	0	0	0
0	1	1	1	1	0	1	0	0	1
1	0	0	0	0	0	1	1	0	1
1	0	0	0	1	0	1	1	0	0
1	0	0	1	0	0	1	1	1	1
1	0	0	1	1	0	1	1	1	0
1	0	1	0	0	1	1	0	1	1
1	0	1	0	1	1	1	0	1	0
1	0	1	1	0	1	1	0	0	1
1	0	1	1	1	1	1	0	0	0
1	1	0	0	0	1	0	0	1	0
1	1	0	0	1	1	0	0	1	1
1	1	0	1	0	1	0	0	0	0
1	1	0	1	1	1	0	0	0	1
1	1	1	0	0	1	0	1	1	0
1	1	1	0	1	1	0	1	1	1
1	1	1	1	0	1	0	1	0	0
1	1	1	1	1	1	0	1	0	1

The design has no garbage outputs and the quantum circuit is based on the restoring division algorithm. It employs optimized quantum designs of a conditional integer ADD operation and integer subtraction.

17.4.1 The restoring division algorithm

The restoring division algorithm for quantum circuits is shown in algorithm 17.1. In the algorithm, the inputs are: (a) ($|Q_{[0:n-1]}\rangle$, n-qubit register in which the dividend is

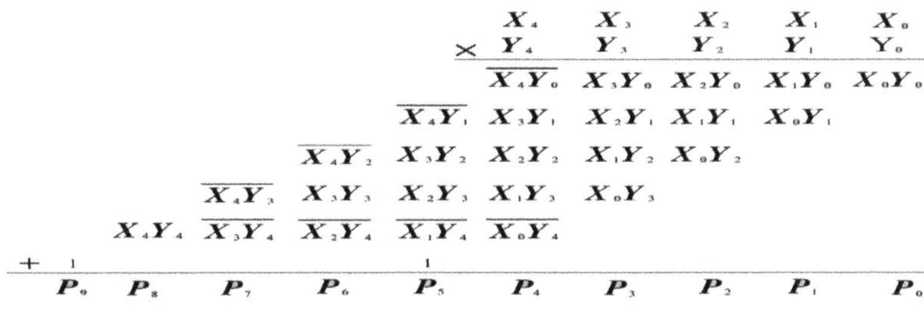

Figure 17.10. The modified Baugh–Wooley signed multiplier.

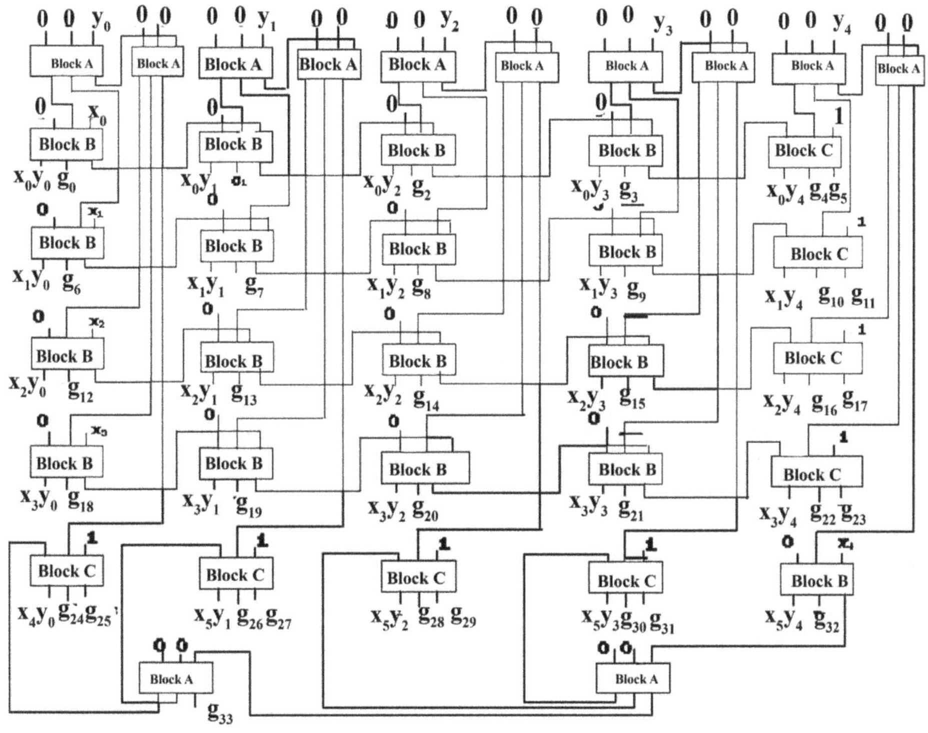

Figure 17.11. Block diagram of a fault-tolerant quantum circuit for partial product generation.

stored; (b) $|D_{[0:n-1]}\rangle$, n-qubit register in which the divisor is loaded; (c) $|R_{[0:n-1]}$, n-qubit remainder register which is initiated to 0 at the start. Therefore, for initiating $|R_{[0:n-1]}\rangle$, n ancillary qubits are required. The algorithm executes repeatedly and from the algorithm it can be observed that at the end of n iterations the quotient is $(|Q_{[0:n-1]}\rangle$ and the remainder is $|R_{[0:n-1]}\rangle$. The divisor is retained at the output.

The quantum circuits that are required to develop the hardware implementation of the restoring division algorithm are: (i) left shift operation circuitry; (ii) an n-qubit quantum subtractor; and (iii) conditional ADD operation circuitry.

Combining $|R_{[0:n-2]}\rangle$ and $(|Q_{[n-1]}\rangle$ to form an n-qubit register is actually equal to performing a left shift operation. By aggregating the qubits in this way, a separate left shift operation circuitry is not necessary.

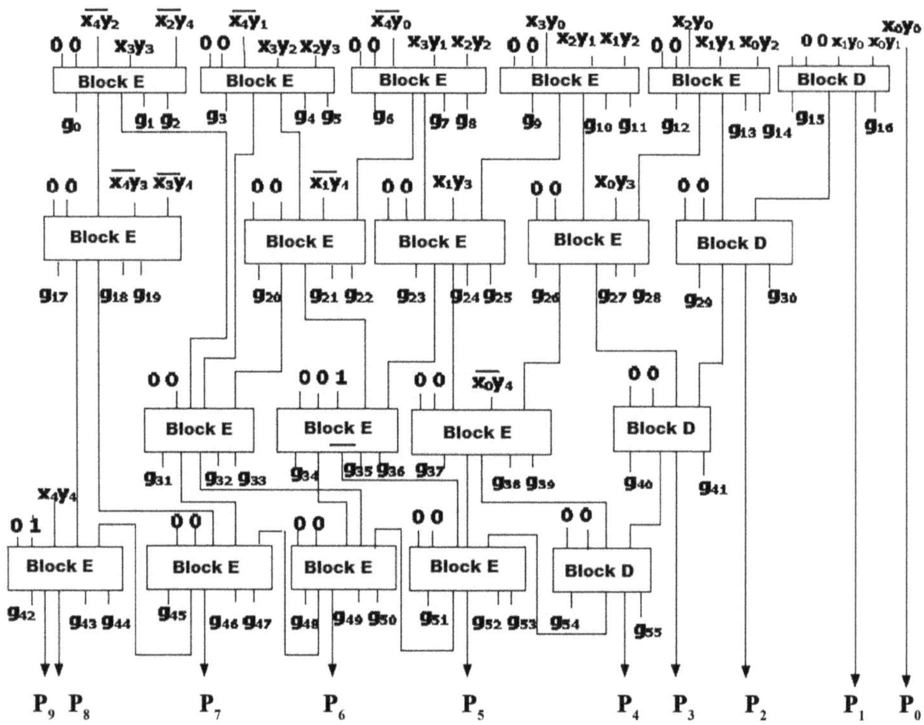

Figure 17.12. Block diagram of a fault-tolerant quantum circuit for signed multiplier generation.

Algorithm 17.1. Restoring division algorithm.

START
Restore ($|Q_n\rangle$, $|R_n\rangle$, $D_n\rangle$)
for $i = 0$ to $n - 1$ **do**
  $(|Q_{[1:n-1]}\rangle|R_{[0:n-1]}\rangle) = \text{LEFTSHIFT}(|Q_{[0:n-1]}\rangle), |R_{[0:n-1]}\rangle;$
  $|R - D_{[0:n-1]}\rangle = |R_{[0:n-1]}\rangle - |D_{[0:n-1]}\rangle;$
  if $(|R_{[0:n-1]}\rangle > 0)$ **then**
    $|Q_{[0]}\rangle{=}1$
    $R_{[0:n-1]}\rangle = |R - D_{[0:n-1]}\rangle;$
  else
    $|Q_{[0]}\rangle = 0;$
    $|R_{[0:n-1]}\rangle = |R - D_{0:n-1}\rangle + |D_{[0:n-1]}\rangle;$
  end if
end for
return R;
END

17.4.2 The subtractor module

Figure 17.13 shows the symbol and working principle of the quantum subtractor circuitry. The subtractor circuitry takes two n-qubit inputs $a_{[0:n-1]}$ and $b_{[0:n-1]}$. The input a is regenerated at the output. The n-qubit output $s_{[0:n-1]}$ has the result of the subtraction of b and a, i.e. $b - a$. Figure 17.14 presents the circuit design of an N-qubit subtractor based on an N-qubit quantum ripple carry adder. As demonstrated in figure 17.14, a quantum ripple carry adder is required to develop a quantum subtractor circuitry. The approach that is followed for developing the quantum subtractor circuitry is $b - a = \overline{(\bar{b} + a)}$. Both inputs are passed through the quantum ripple carry adder. The input qubits $b_{[0:n-1]}$ are complemented at the start and at the end. The qubits $a_{[0:n-1]}$ are just passed through the quantum ripple carry adder.

17.4.3 The conditional addition operation module

Figure 17.15 presents the graphic symbol of the quantum conditional ADD operation circuit. The quantum controlled ADD operation circuitry operates as follows: (i) when the input labeled 'ctrl' is high, the circuit output is $|P\rangle = |B + A\rangle$, and (ii) when the 'ctrl' input is low, the circuit output is $|P\rangle = |B\rangle$. The complete working circuit of quantum conditional ADD operation circuitry is demonstrated in figure 17.16 for four-qubit operands. The quantum conditional ADD circuit uses a modified version of the ripple carry adder. The addition architecture uses Peres gates to perform the addition operation. The Peres gate can be decomposed into a quantum Feynman gate and a Toffoli gate. By replacing the Feynman gates with three input quantum Toffoli gates, it will be possible to use the control line (ctrl) to perform addition or no operation. It is possible to extend the four-qubit operands to any operand size.

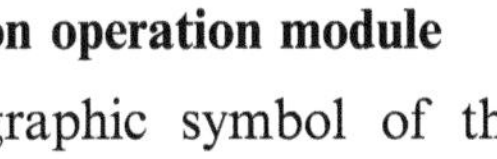

Figure 17.13. Graphical representation of a quantum subtractor.

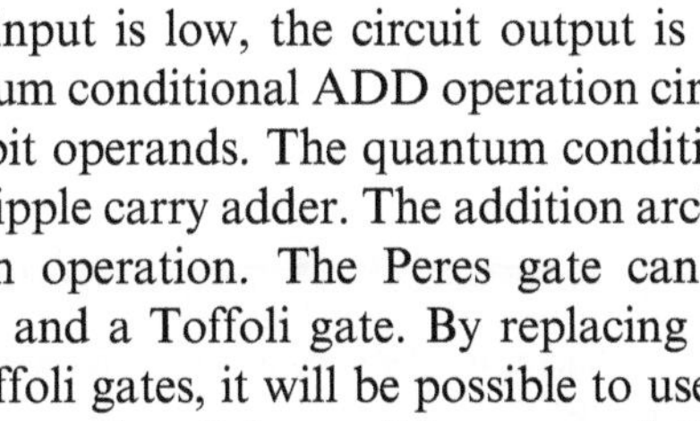

Figure 17.14. The quantum subtractor based on an N-qubit ripple carry adder.

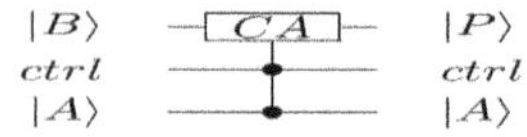

Figure 17.15. Graphical representation of the conditional ADD operation.

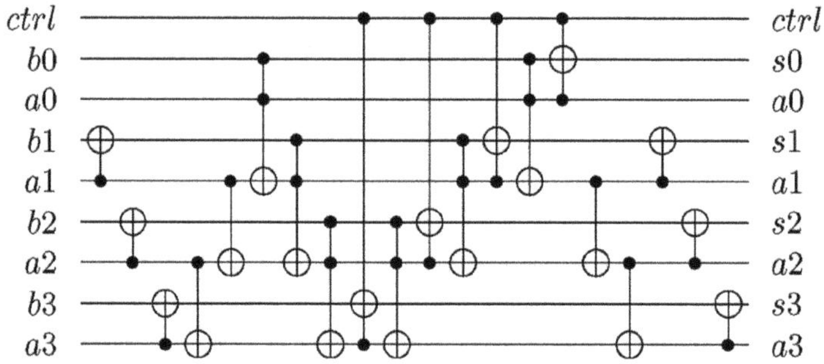

Figure 17.16. Circuit design of a quantum conditional ADD operation. Reproduced with permission from [10]. Copyright 2017 IEEE.

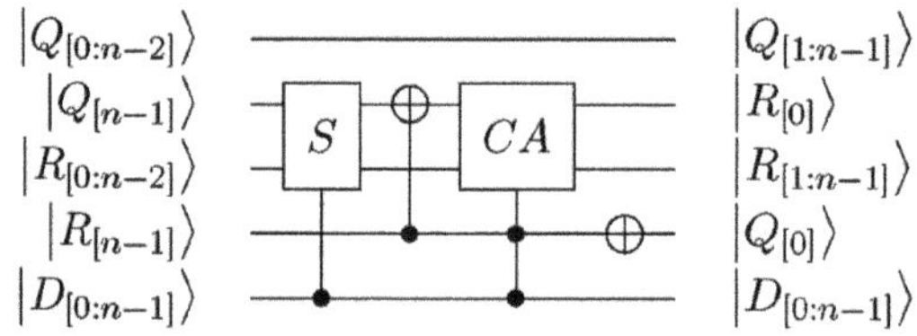

Figure 17.17. The quantum restoring integer divider circuitry design for a single iteration. Reproduced with permission from [10]. Copyright 2017 IEEE.

17.4.4 Quantum restoring integer division circuitry

Figure 17.17 presents the quantum circuit of restoring division for one iteration. The details of the movement of the circuit are as follows:

Step 1. The $|D_{[0:n-1]}\rangle$ keeps the divisor, $|R_{[0:n-1]}\rangle$ initialized to zero, and $|Q_{[0:n-1]}\rangle$ holds the dividend.

Step 2. Let $|Q_{[n-1]}\rangle$ and $|R_{[0:n-2]}\rangle$ be one combined register. This allows us not to use a left shifting circuit thus saving quantum resources.

Step 3. The combined register mentioned above in step 2 and $|D_{[0:n-1]}\rangle$ are given as inputs to the quantum subtractor circuitry. Register $|D_{[0:n-1]}\rangle$ emerges unchanged. The combined register now holds the result of subtraction of the R and D registers. Let us call this result $|R - D_{[0:n-1]}\rangle$.

Step 4. Qubits $|R - D_{[n-1]}\rangle$ and $|R_{[n-1]}\rangle$ are supplied to a CNOT gate. $|R - D_{[n-1]}\rangle$ is the control qubit and $|R_{[n-1]}\rangle$ is the target qubit. The target now holds the value of $|R - D_{[n-1]}\rangle$ because $|R - D_{[n-1]}\rangle$ is always zero throughout the computation.

Step 5. The $|R - D_{[n-1]}\rangle$ computed in step 4 now acts as the control qubit to the conditional ADD circuit. $|R - D_{[0:n-1]}\rangle$ and $|D_{[0:n-1]}\rangle$ are two n-qubit inputs to

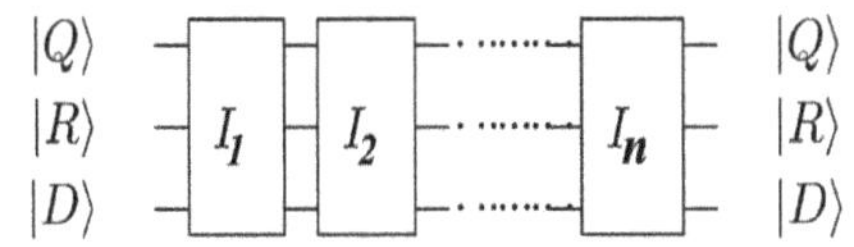

Figure 17.18. The quantum restoring integer divider circuitry design (for n iterations). Reproduced with permission from [10]. Copyright 2017 IEEE.

the conditional ADD operation circuit. The outputs of conditional ADD operation are collected. $|R - D_{[n-1]}\rangle$ is complemented.

Step 6. All the above operations constitute the first iteration. From algorithm 17.1 it can be seen that the whole circuit is iterated n times. Hence the circuit in figure 17.17 is also iterated n times. This is performed by using the outputs of the first iteration which will be used as inputs for the next iteration.

Step 7. This process continues for n iterations. In figure 17.18, I_1 represents the first iteration, I_2 represents the second iteration, and I_n represents the nth iteration. The steps 1 through 6 have to go through each iteration until it reaches n iterations.

Step 8. At the end of n iterations, the quotient is $|Q_{[0:n-1]}\rangle$, the remainder is $|R_{[0:n-1]}\rangle$, and the divisors are retained.

17.5 Summary

In this chapter fault-tolerant adder, multiplier, and division circuits are described with illustrations. The fault-tolerant quantum half-adder and full-adder are discussed and we show how circuits function as parity-preserving. The fault-tolerant quantum adder circuits, namely FTFA and EFTFA, are described and implemented by the quantum gates, and it is shown that a fault-tolerant adder can be used for implementing all Boolean functions. In the fault-tolerant quantum multiplier design a Wallace tree structure is explained and demonstrated. A resource efficient design of the quantum circuitry of integer division is presented by optimizing the quantum circuit modules, and knitting them together efficiently. It is observed that a non-restoring division algorithm can be an attractive choice to design a quantum integer division circuit when minimizing the number of qubits is of primary concern. The designs can be integrated in larger data path subsystem designs to provide resource efficient implementation of quantum algorithms.

Further reading

[1] Amy M, Maslov D, Mosca M and Roetteler M 2013 A meet-in-the-middle algorithm for fast synthesis of depth-optimal quantum circuits *IEEE Trans. Comput.-Aided Des. Integr. Circuits Syst.* **32** 818–30

[2] Baugh C R and Wooley B A 1973 A two's complement parallel array multiplication algorithm *IEEE Trans. Comput.* **100** 1045–7

[3] Khosropour A, Aghababa H and Forouzandeh B 2011 Quantum division circuit based on restoring division algorithm *2011 Eighth Int. Conf. on Information Technology: New Generations* (Piscataway, NJ: IEEE) pp 1037–40

[4] Kliuchnikov V, Maslov D and Mosca M 2012 Fast and efficient exact synthesis of single qubit unitaries generated by Clifford and T gates (arXiv:1206.5236)

[5] Li J, Peng X, Du J and Suter D 2012 An efficient exact quantum algorithm for the integer square-free decomposition problem *Sci. Rep.* **2** 260

[6] Qi X, Chen F, Zuo K, Guo L, Luo Y and Hu M 2012 Design of fast fault tolerant reversible signed multiplier *Int. J. Phys. Sci.* **7** 2506–14

[7] Thapliyal H and Ranganathan N 2013 Design of efficient reversible logic-based binary and BCD adder circuits *ACM J. Emerg. Technol. Comput. Syst.* **9** 17

[8] Thapliyal H, Varun T S S and Munoz-Coreas E 2016 Quantum circuit design of integer division optimizing ancillary qubits and T-count (arXiv:1609.01241)

[9] Zhou R-G, Li Y-C and Zhang M-Q 2014 Novel designs for fault tolerant reversible binary coded decimal adders *Int. J. Electron.* **101** 1336–56

[10] Thapliyal H, Varun T S S, Munoz-Coreas E, Britt K A and Humble T S 2017 *IEEE International Symposium on Nanoelectronic and Information Systems (iNIS)* pp 123–8

Part III

Quantum-dot cellular automata

An overview of quantum-dot cellular automata

Quantum-dot cellular automata (QCA) are not transistor-based, they are a type of nanotechnology. In QCA, the basic element is a cell (containing multiple quantum dots). The interaction between cells is purely Coulombic and there is no physical transport of charge. QCA are very promising for a number of reasons, among which their low power consumption and high density are two important features. Therefore, they seem to be appropriate for designing circuits for general-purpose computation as well as for embedded applications. Digital design in QCA is the focus of this part of the book.

QCA technology was introduced in 1993. It consists of an array of cells. Each cell contains several quantum dots, each of which is a nanoparticle or a crystal made of semiconducting materials such as silicon, cadmium, selenide, etc. A cell typically contains four quantum dots at the corners of a square and one central dot. Due to Coulomb interactions, electrons may occupy two 'diametrically' opposite quantum dots (or two antipodal sites). Labeling the quantum dots at the corners from 1 to 4 (in order), it can be observed that electrons may occupy dots 1 and 3 (or 2 and 4). Each cell has bi-stable behavior, facilitating its use in large scale cellular arrays. The physical interaction between neighboring cells is used to implement various logic functions. The key aspect of QCA is that the interaction between cells is purely Coulombic and there is no transport of charge between cells. The notion of cellular automata is due to the fact that state of a given cell at a particular time depends on the state of its neighbors during the previous clock cycle.

Arrays can, in general, be regular or irregular and can lead to the realization of various logic functions. Early efforts were created on a single QCA device using different approaches. Semiconductor, molecular, and magnetic implementations have been pursued. One of the earliest logic functions realized was the three-input majority gate. This has been used for computation of other basic logic functions, such as AND and OR. Combining a majority gate with an inverter has facilitated the realization of NAND, NOR, and other logic elements. All circuits in QCA technology are clocked. Clocking serves as a means of synchronization. Further, circuits for the realization of arithmetic functions may involve some types of wires crossing in the plane.

Contemporary microprocessors and application-specific integrated circuits are largely based on complementary metal oxide semiconductor (CMOS) technology. It is believed that the performance of various circuits in current CMOS-based architectures are close to reaching the limit. When the feature size of transistors is reduced to a nanometer, quantum effects such as tunneling take place. Further, when device scaling takes place, the interconnections do not scale automatically due to the effects of wire resistance and capacitance. This part of the book introduces the reader to QCA, an emerging nanotechnology and an alternative to CMOS. We also examine the problem of designing efficient circuits in QCA.

The changes that QCA devices have experienced have been evolutionary, but the most advanced chips still use the same computing paradigm as their predecessors. There is a lot of expectation that new paradigms will be developed for the processing of information. CMOS technology uses current switching, whereas QCA deal with encoding the binary information as it is the array of cells and every cell has quantum dots. The interaction of the quantum mechanics with the Coulomb force in every cell determines the cell state.

In the near future, the era of 'beyond CMOS' will begin, as the scaling of the current CMOS technology will reach the fundamental limit. QCA are a transistor-less computation paradigm and a viable candidate for 'beyond CMOS' device technology. The continued and fast-dimensional scaling of CMOS will eventually approach the fundamental limit. Also, short channel effects, high power dissipation, and quantum effects are limiting the further scaling of current CMOS technology devices. Emerging device technology could overcome the scaling limitation in current CMOS technology. Single electron transistors (SETs), QCA, and resonant tunneling diodes (RTD) are some of the 'beyond CMOS' technologies. Among these evolving nanotechnologies, QCA are the most favorable technology. QCA can achieve a device density of 1012 devices/cm^2 and an operating speed of THz. The QCA device paradigm can replace FET based logic and exploits the quantum effects of small size. QCA are a means of representing binary information on cells through which no current flows, and achieving device performance by the coupling of those cells.

This part of the book starts with some fundamentals of QCA circuits, which are provided in chapter 18. Many of the QCA gates are included in this chapter. The designs for a QCA cell and QCA clock are described here. Chapter 19 presents the characteristics and design procedure of QCA adder and subtractor circuits. The designs for a QCA half-adder and subtractor as well as a QCA full-adder and subtractor are also provided here. The designs for a QCA multiplier and divider circuit are shown in chapter 20. In this chapter, the technique of the non-restoring binary divider is described. QCA asynchronous and synchronous counters are introduced in chapter 21. The design of a dual-edge triggered J-K flip-flop is also discussed in this chapter. The design procedures for the QCA decoder and encoder circuit are introduced in chapter 22. In this chapter, a QCA 2-to-4 decoder and QCA 3-to-8 decoder are shown. In addition, the design of an encoder with single-feedback and multi-feedback is also given in this chapter. Chapter 23 describes the design for a QCA multiplexer and demultiplexer circuit. In this chapter, a QCA 2-to-1 multiplexer and QCA 4-to-1 multiplexer are presented. In addition, a QCA 1-to-2 demultiplexer and QCA 1-to-4 demultiplexer are also provided in this chapter. Chapter 24 presents the design for a QCA random access memory (RAM) cell, read only memory (ROM), and processor. In addition, the design procedures for instruction memory, data memory, arithmetic logic units (ALUs), and integrated processors are also described in this chapter.

Chapter 18

Quantum-dot cellular automata

Quantum-dot cellular automata (QCA) are a new computational paradigm which encodes binary information by charge configuration within a cell instead of using the conventional current switches. As there is no current flow, the Coulombic interaction is sufficient for computation. This revolutionary paradigm provides a possible solution for transistorless computation at the nanoscale. QCA allow operating frequencies in a high range with a small feature size and ultra-low power consumption. Currently, devices are becoming smaller and consume low power with a higher frequency of operations. For this reason, optimizing the overall cost is becoming more difficult day by day.

For circuit layout and functionality checking, a simulation tool for QCA circuits, QCA Designer, is used. This tool allows users to create a custom layout and then verify QCA circuit functionality using simulations. It includes two different simulation engines, such as a bi-stable approximation and a coherence vector.

18.1 Fundamentals of QCA circuits

The basic properties of QCA circuits are area, delay, power, and overall cost. The descriptions of basic components are given below.

Area

The area of a QCA circuit is the area of the base rectangular block on which the circuit is plotted. The area of a QCA circuit is determined by multiplying the vertical length and horizontal length of the circuit. Thus, the area (A) of a QCA circuit is calculated as

$$A = C_h \times C_w \times L^2, \tag{18.1}$$

where C_h = maximum number of cells in height, C_w = maximum number of cells in width, and L = length of a cell. The units to measure the area are m^2, nm^2, and μm^2.

doi:10.1088/978-0-7503-2747-3ch18

Example 18.1. If a circuit has 22 cells in the maximum horizontal chain, 25 cells in the maximum vertical chain, and the length of a cell is 20 μm, then the area of the circuit is $A = 0.22\ \mu\text{m}^2$.

Delay

Delay represents the number of clock cycles to generate the output from an input. For a QCA circuit, the delay is the value with the maximum clock cycle among all the clock cycles of different input paths to the same output of the QCA circuit. The delay (D) is determined as

$$D = \max_{D_i \in D_a}(D_i), \tag{18.2}$$

where D_a = set of minimum clock cycles of each path, D_i = ith element of the set D_a, and the measuring unit of delay is the clock cycle.

Example 18.2. If a QCA circuit contains three different input paths such as D_1, D_2 and D_3 to the same output and the paths need 12, 11, and 13 clock cycles to reach the same corresponding output, then the delay of the QCA circuit is 13 clock cycles.

Kink energy

The kink energy E_{kink} represents the energy difference between two horizontally or vertically adjacent polarized cells with the same or the opposite polarization and electrostatics. Coulomb's law gives the means to calculate its value. According to Coulomb's law we have the following equation:

$$E_{\text{kink}} = F_q \times d, \tag{18.3}$$

where q = charge of an electron and d = distance between two repulsive electrons.

Power

We know that in a tightly coupled environment and for an irreversible operation, each cell will dissipate E_{kink} for each cycle. If the delay for each cycle is constant, then the power dissipation is proportional to E_{kink}. The power dissipation of any QCA circuit is directly related to the number of cells needed to design the circuit. Thus the power (P) is measured as

$$P = N \times P_r, \tag{18.4}$$

where N = number of cells of the circuit and P_r = relative power of a cell. The measuring unit of power is a relative value. Thus it has no unit. The power needed for a four-dot QCA is taken as 1 unit.

If a circuit needs 2345 cells and the relative power of the QCA cell is 0.4, then the power of the QCA cell is $P = 938$.

Overall cost

Cost is the parameter of a circuit which defines its efficiency. Cost in a QCA circuit is the measurement in terms of area, power, and delay. The overall cost of a circuit can be formulated in different ways. The overall cost of a circuit can be calculated as

$$\text{Cost} = N \times P_r \times A \times D, \tag{18.5}$$

where N = number of cells (i.e. complexity implies to power), Pr = relative power of a cell (which equals 1 for a standard four-dot QCA cell), A = area of the circuit, and D = delay of the circuit.

Thus the derivation of the overall cost for a standard four-dot QCA cell is Cost = Area × Delay × Power.

18.2 The QCA cell

In general, an electronics-based transistor functions through the transportation of electrons, whereas QCA apply the adjustment of electrons in the fixed area of tiny square nanometers. QCA are implemented in a small quadratic region which is called the QCA cell. Each small square cell contains exactly four potential dots to hold the extra electrons. Electrons can only reside in the potential dots. The QCA innovation depends on the association of bi-stable QCA cell outlines from four quantum dots. A schematic chart of a basic cell structure is shown in figure 18.1. The cell is charged by two free electrons by burrowing between nearby spots. These electrons have a tendency to possess antipodal locales as an after-effect of their shared electrostatic Columbic repulsion.

The potential dots are connected with electron tunnel junctions. They can be opened for the electrons to travel through them under a particular condition by a clock signal. Without any interaction from outside, the two electrons will try to separate from each other as far as possible due to the Coulomb force that acts between them. As a result they will reside in diagonally located potential dots, because the diagonal is the largest possible distance apart for them to reside (figure 18.2).

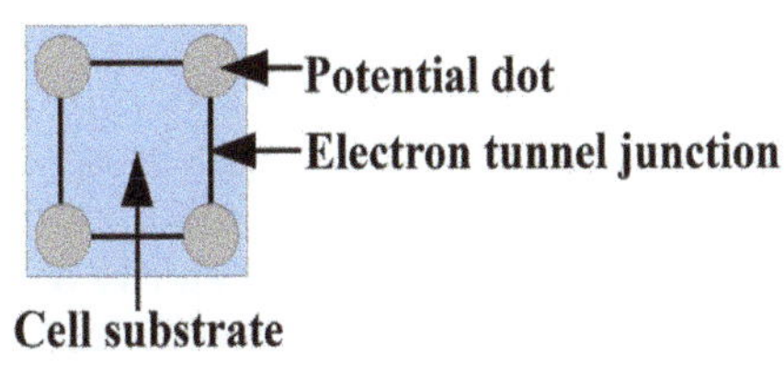

Figure 18.1. The structure of a QCA cell.

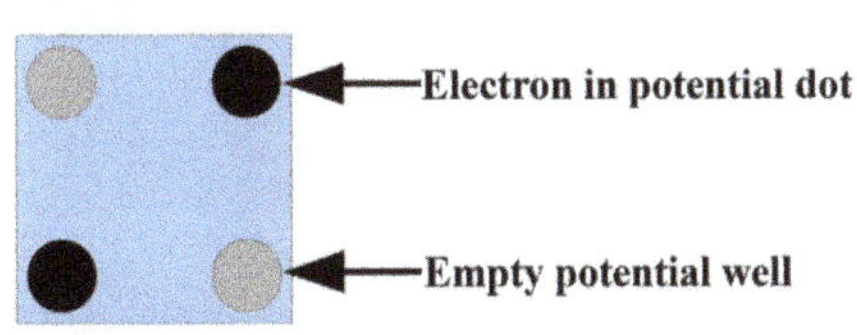

Figure 18.2. Electrons in potential dots.

There are two diagonals in a square, which means the electrons can reside in exactly two possible variants in the QCA cell. Regarding these two arrangements, they are interpreted as a binary 0 and binary 1, i.e. each cell can be in two states. The state 0 and the state 1 are shown in figure 18.2. A binary system is something familiar, as Boolean logic is already used in today's computers. There, a high voltage is often interpreted as binary 1 and a low voltage as binary 0.

18.3 Information and data propagation

If two QCA cells are placed next to each other, it is possible to exchange their states, i.e. the arrangements of the electrons in them. The QCA cell that should transfer its state to a neighboring cell must have its tunnel junctions closed, and the tunnel junctions in the neighboring cell have to be open to allow the electrons to travel through the tunnel junctions between the potential wells. As soon as they open, the electrons in the neighboring cell are pushed by the Coulomb force of the original cell as far away as possible. As they are also pushed away from each other, they will travel into the same potential wells as in the original cell. As soon as the tunnel junctions are closed again, the transfer of the state is completed.

The state of a cell can also be transferred to multiple neighboring cells. This works the same way as for a single neighboring cell, but the tunnel junctions of all the sequentially neighboring cells should be opened at the same time, which makes the transfer much faster than transferring the state cell by cell. This allows us to build 'wires' made of QCA cells to transport information over larger distances.

18.4 Basic QCA elements and gates

So far, we know how to interpret and transport information with QCA cells, as shown in figure 18.3, but as yet we lack the possibility of computations. For QCA cells, the basic gate is a three-input majority voter. It is built from five cells, arranged as a cross.

18.4.1 The QCA majority voter

It is known that the Coulomb forces of several electrons are summed up and the majority voter takes advantage of this effect. The cells at top, at the left, and at the bottom work as input connection cells. As the Coulomb forces of the electrons of all input cells sum up, the middle cell adjusts to the majority of adjustments of the input connection cells. Finally the output cell adjusts to the middle cell and the resulting

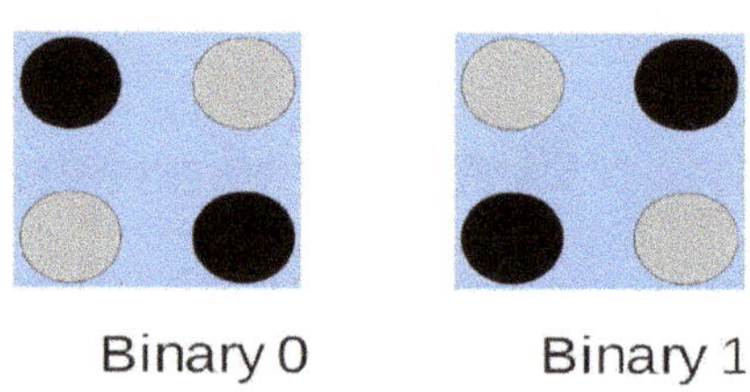

Figure 18.3. Binary interpretation of adjustments.

state of the majority vote, as shown in figure 18.4, can be obtained from the output cell.

18.4.2 The QCA AND gate

As we work in the field of QCA with known binary representation, it is preferable to have further logic gates we are already familiar with. With a slight modification, it is possible to turn the majority voter into an AND gate.

The Boolean AND outputs are 1 if all inputs are 1, otherwise 0. Regarding two inputs of the majority voter, as the inputs of an AND gate and the voter should not output 1 if only one of the two inputs is 1, a fixed cell is added as the third input that is always in the 0 state. If both AND inputs are 1, the two ones sum up to a stronger Coulomb force than the single fixed 0 cell and the majority voter is turned into a two-input AND gate, as shown in figure 18.5. The fixed cell can be obtained by setting it to the 0 state and never opening the electron tunnel junctions.

18.4.3 The QCA OR gate

The OR gate is built almost exactly like the AND gate, but instead of a fixed 0, a fixed 1 QCA cell must be attached as one input. The fixed 1 cell sums up to a stronger Coulomb force with a single other input being adjusted to 1, so that the OR gate will output 1, if one of the free inputs is 1.

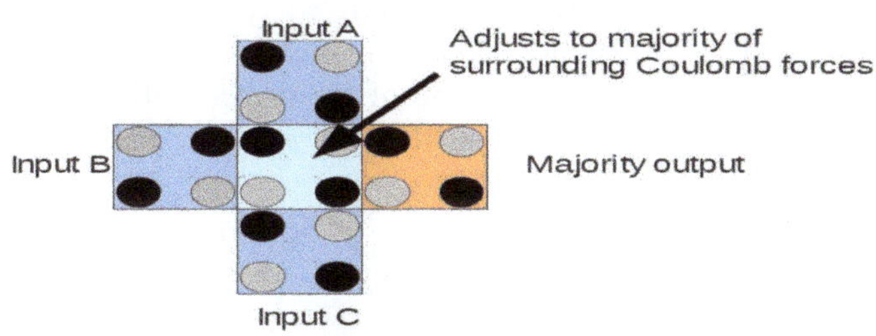

Figure 18.4. The QCA majority voter.

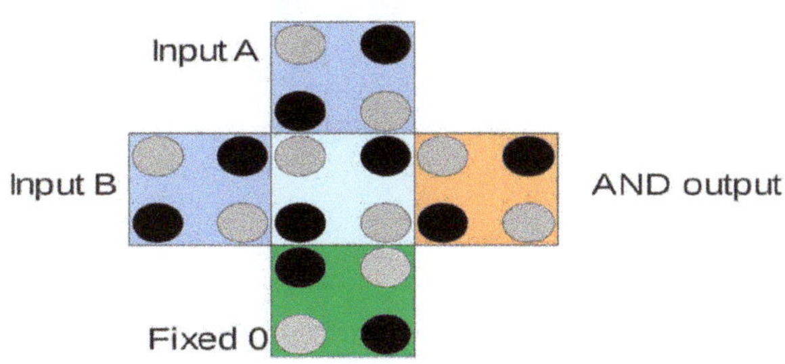

Figure 18.5. The QCA AND gate.

18.4.4 The QCA NOT gate

It is also possible to build a QCA NOT gate, as shown in figure 18.6. The implementation in QCA takes advantage of the geometry of cell adjustments. One QCA wire is forked to two wires. The switch of the cell adjustment takes place by putting the output cell next to the forked wires so that only the corners touch. Since only cell corners are touching the right of the fork and the cells at the end of the fork will have the same adjustment, and the cell on the right of the fork will not adjust with an electron close to an electron at the corner at the end of the fork, the adjustment on the right of the fork will be inverted. This creates a 1 at the input an a 0 at the output and vice versa.

18.4.5 The QCA wire

A string of QCA cells acts like a wire. A representation of a QCA wire is shown in figure 18.7. In each clock cycle, half of the wire is active for signal transmission, while the other half remains in a steady state. In the middle of the following clock cycle, half of the past active clock zone is deactivated and the remaining active zone cells trigger the recently initiated cells to be polarized. In this way the signal propagates from one clock zone to the next.

18.5 The QCA clock

A QCA cell has four clock zones, where each clock zone has four stages: switch, hold, release, and relax. Figure 18.8 demonstrates its operation process. In the switch stage the QCA cells start as unpolarized and their inter-dot potential boundaries are low. The obstructions are then raised in the middle stage and the QCA can be polarized as per the condition of the driver of the QCA gate (i.e. the information cell). It is in this clock stage that the real calculation (or exchanging)

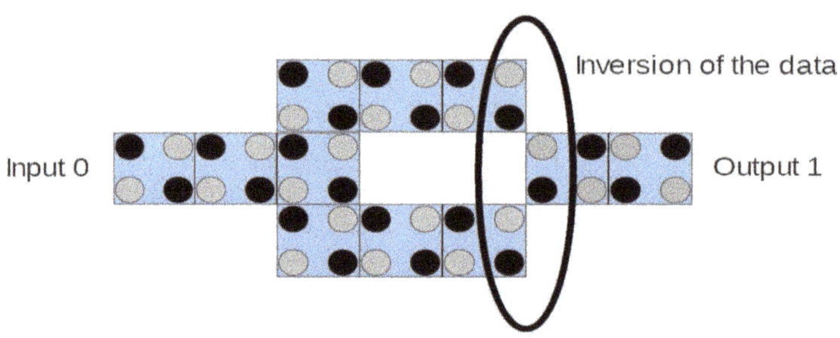

Figure 18.6. The QCA NOT gate.

Figure 18.7. A QCA wire that propagates the signal across the circuit.

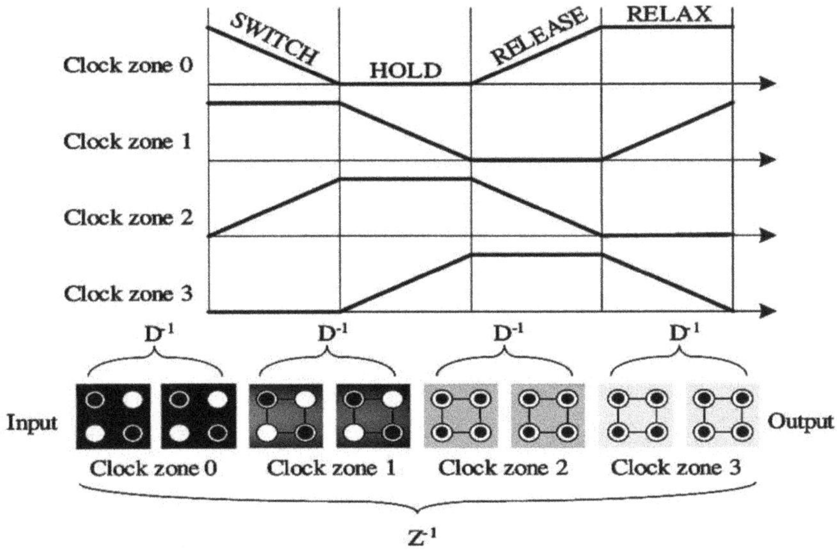

Figure 18.8. The four phases of a QCA clocking mechanism.

happens. Before the end of this clock stage, boundaries are sufficiently high to stifle any electron burrowing and cell states are settled. In the hold stage boundaries are held high, so the output of the sub-exhibit can be utilized as inputs to the following stage. In the release stage barriers are imposed and cells are permitted down to an unpolarised state. Finally, in the fourth clock stage, the relax stage, cell barriers are brought down and cells stay in an unpolarised state. Figure 18.8 demonstrates every clock zone flag and shows the pipeline component. All cells in a certain zone are controlled by the same QCA clock signal. Cells in every zone perform a particular count and the condition of a zone is then settled. Thus, they can serve as data signs to the following zone and data moves in a pipelined manner.

18.5.1 Special cell arrangements and symmetric cells

During the design of a QCA circuit, the situation is likely to occur that QCA wires have to be crossed. In contrast to classic transistor technology where wires can only cross by inserting another layer, QCA wires can be crossed in the same layer. This works by introducing a QCA cell type where the four potential wells are not in the corners of the cell but in the middle of the edges, as shown in figure 18.9.

If several of these QCA cells are put together to form a wire, the adjustment of the succeeding cell is inverted to its predecessor and so on. The advantage of this type of QCA cell originates in its symmetric effect of Coulomb force on regular cells. Although the electrons do interact with electrons in neighboring regular QCA cells, through the symmetry they do not push the electrons in regular QCA cells to a particular adjustment. In the other direction electrons in a regular cell do not push the electrons in a symmetric cell into a particular potential well. This allows the construction of wire crossings, as shown in figure 18.10, of these QCA cells with

Figure 18.9. The symmetric adjustable QCA cell.

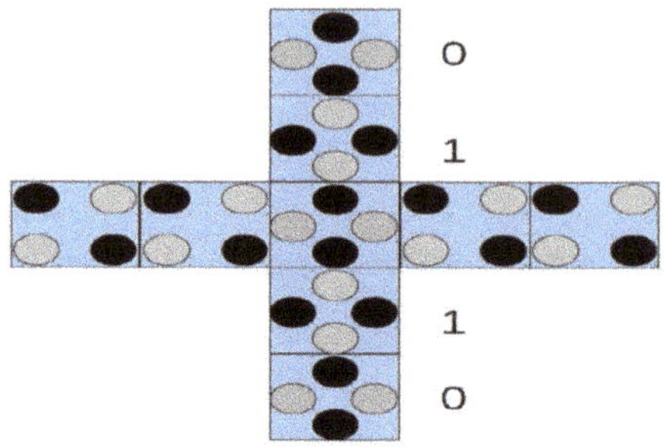

Figure 18.10. The QCA wire crossing.

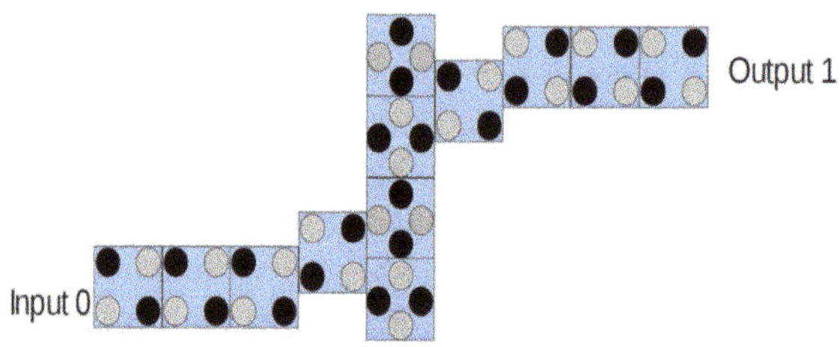

Figure 18.11. Connections between symmetric and regular QCA cells.

regular ones. A crossing is built using a continuous wire of special cells, building a gap across a wire of regular cells.

Of course it has to be possible to connect symmetric QCA cells to regular ones and vice versa. This works by putting a regular cell as a neighbor of two symmetric cells near the beginning or end of a wire of symmetric cells. One has to take care as to which two symmetric cells are chosen, as neighbors of this type are in the inverted arrangement depending on the necessary adjustment, the original or the inverted one, as shown in figure 18.11.

18.5.2 NOT gate clock zones

Since the NOT gate has a slightly critical zone of arranged QCA cells, it is important to put the cells in the proper clock zones to avoid randomly flipped adjustments of electrons near the forking wire.

The clock zone of the input should end at the beginning of the 'fork'. The complete fork itself, i.e. the U-shaped wire, should be in the subsequent clock zone

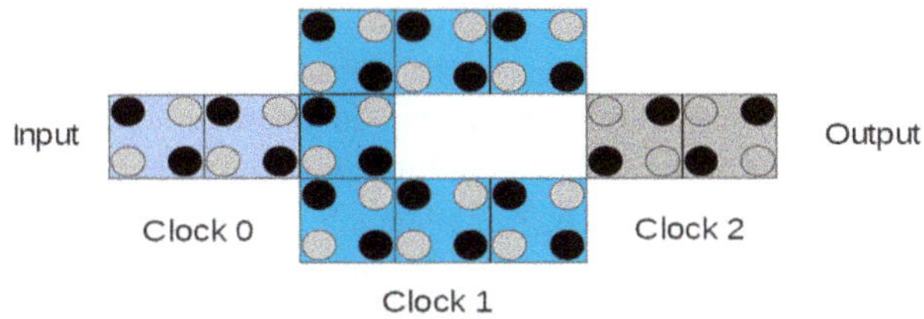

Figure 18.12. QCA NOT gate clock zones.

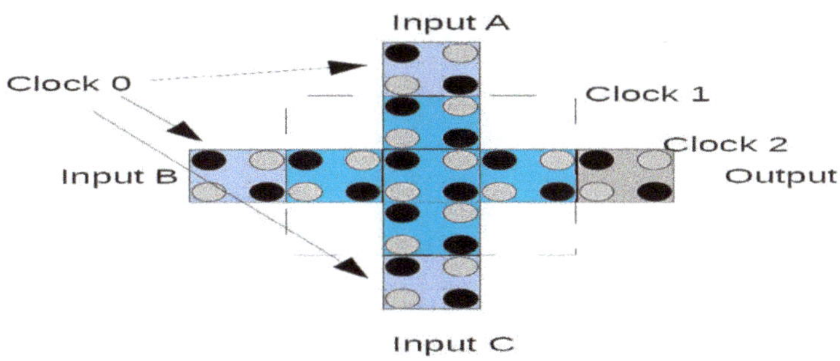

Figure 18.13. QCA majority voter clock zones.

of the input and the output should be in the subsequent clock zone of the fork. The QCA NOT gate clock zones are shown in figure 18.12.

18.5.3 Majority voter clock zones

For the majority voter it is important that all cells are in the same clock zone. Putting some cells in different clock zones can lead to incorrect results. The center cell and the three input cells plus the output cell have to be in the same clock zone. The QCA majority voter clock zones are presented in figure 18.13.

18.6 Summary

This chapter presents the basic components of quantum-dot cellular automata. We have described the QCA circuit's overall cost in terms of the cell area, delay, kink energy, and power. This chapter explains the QCA clock and its zones, QCA wires and their crossing, and the connection between symmetric and regular QCA cells. The ideas in this chapter will help the reader understand the different implementations of QCA logic.

Further reading

[1] Nano-Arch Online 2012 Quantum-dot cellular automata (QCA) *University of Erlagen* http://www.cs3.tf.uni-erlangen.de/Research/KOMINA/QCA_slides.pdf (Accessed: 6 December 2018)

[2] Gladshtein M 2011 Quantum-dot cellular automata serial decimal adder *IEEE Trans. Nanotechnol.* **10** 1377–82

[3] Liu M and Lent C S 2005 Power dissipation in clocked quantum-dot cellular automata circuits *63rd Device Research Conf. Digest* vol 1 (Piscataway, NJ: IEEE) pp 123–4

[4] Liu M 2006 Robustness and power dissipation in quantum-dot cellular automata *PhD Thesis* University of Notre Dame https://www3.nd.edu/~lent/pdf/nd/Robustness_and_Power_Dissipation_in_Quantum-Dot_Cellular_Automata.pdf (Accessed: 20 March 2020)

[5] Momenzadeh M, Huang J, Tahoori M B and Lombardi F 2005 Characterization, test, and logic synthesis of and-or-inverter (AOI) gate design for QCA implementation *IEEE Trans. Comput.-Aided Des. Integr. Circuits Syst.* **24** 1881–93

[6] Tahoori M B, Huang J, Momenzadeh M and Lombardi F 2004 Testing of quantum cellular automata *IEEE Trans. Nanotechnol.* **3** 432–42

[7] Timler J and Lent C S 2002 Power gain and dissipation in quantum-dot cellular automata *J. Appl. Phys.* **91** 823–31

Chapter 19

QCA adder and subtractor

A quantum dot is a nano-crystal made of semiconductor material that exhibits quantum mechanical properties. The flow of information in this quantum cellular automata (QCA) cell is based on the interaction of switching cells. It encodes the binary information as the electronic configuration of a cell. In this chapter the adder and subtractor circuits are discussed, using the QCA implementation of the exclusive-OR (Ex-OR) gate. This model has the potential for combinational circuits that are compatible with QCA gates within the nanoscale.

The basic logic gates, i.e. Ex-OR and exclusive-NOR (Ex-NOR), are used in the design of digital circuits. These gates have special functions and applications. They are particularly useful in arithmetic operations as well as in code generators. The Ex-OR and Ex-NOR gates are usually found as two-input gates. No multiple-input Ex-OR/Ex-NOR gates are available since they are too complex to fabricate using current hardware. The design has a complexity of cells and co-planar cross-overs or multiple layers that need to be implemented.

19.1 The Ex-OR gate

It is possible to implement all combinational and sequential logic functions by properly arranging cells so that the polarization of one cell can adjust the polarization of a nearby cell. A schematic representation of an Ex-OR gate in terms of a QCA majority gate and the truth table of this gate are shown in figure 19.1 and table 19.1, respectively.

A representation of an Ex-OR gate using a QCA designer tool is shown in figure 19.2.

19.2 The QCA half-adder and half-subtractor

The half-adder is implemented by using the combination of an Ex-OR gate and an AND gate. The half-adder circuits could be implemented using different logics. The

doi:10.1088/978-0-7503-2747-3ch19

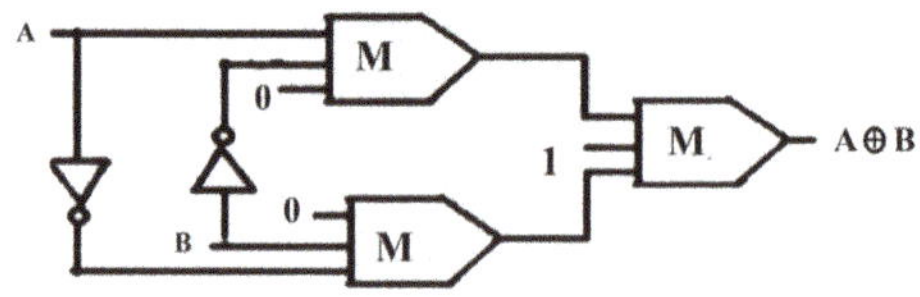

Figure 19.1. Schematic of the Ex-OR gate.

Table 19.1. The truth table of the Ex-OR gate.

A	B	$A \oplus B$
0	0	0
0	1	1
1	0	1
1	1	1

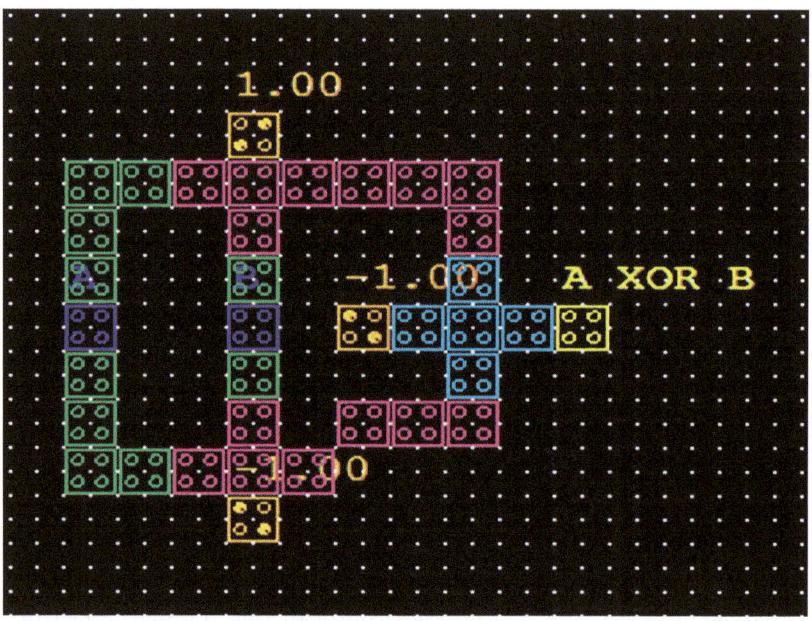

Figure 19.2. The QCA Ex-OR gate.

schematic of a half-adder is shown in figure 19.3 with its truth table shown in table 19.2, and the QCA implementation of figure 19.3 is presented in figure 19.4.

The half-adder circuit might be used to make the subtractor circuit with the addition of one NOT gate. The schematic diagram of a subtractor circuit is shown in figure 19.5 with its truth table in table 19.3.

Figure 19.6 presents the QCA implementation of the half-subtractor circuit.

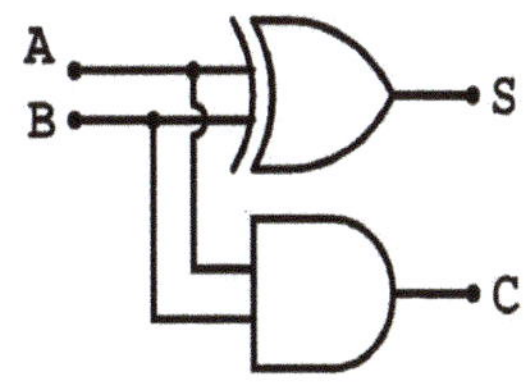

Figure 19.3. The design of a half-adder circuit.

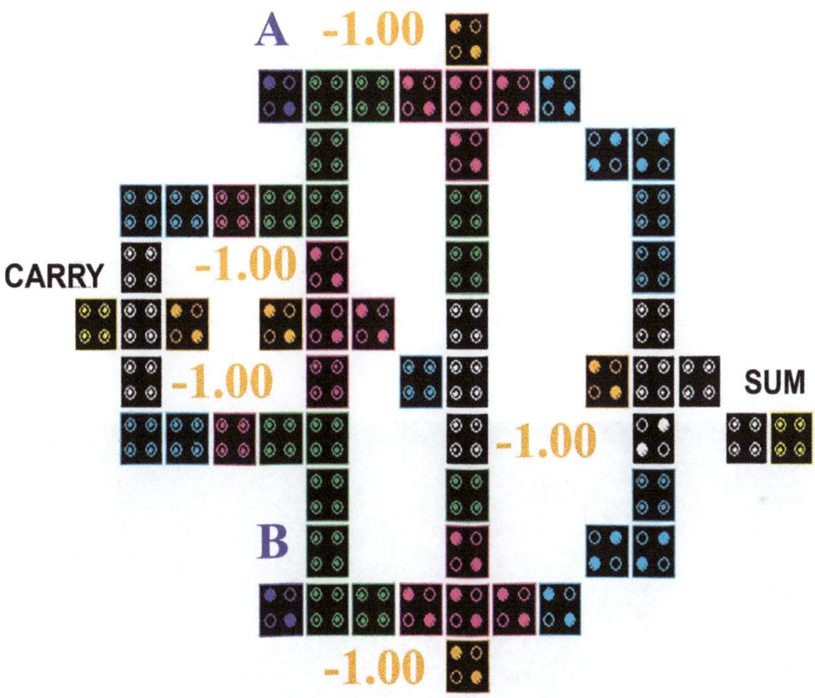

Figure 19.4. The design of a QCA half-adder circuit.

Table 19.2. The truth table of the half-adder.

Input		Output	
A	*B*	Sum	Carry
0	0	0	0
0	1	1	0
1	0	1	0
1	1	0	1

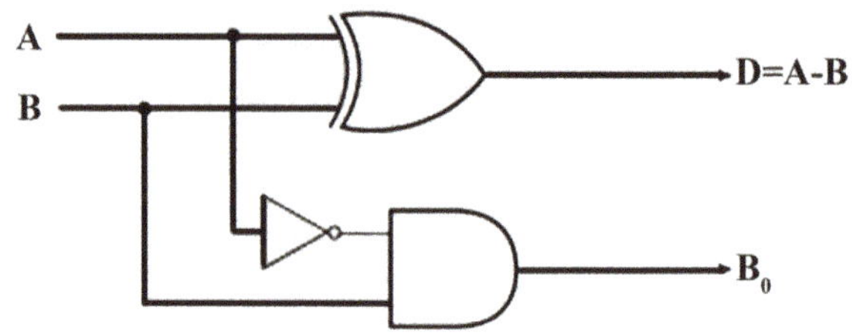

Figure 19.5. The half-subtractor.

Table 19.3. The truth table of the half-subtractor.

Input		Output	
A	*B*	Difference	Borrow
0	0	0	0
0	1	1	0
1	0	1	0
1	1	0	1

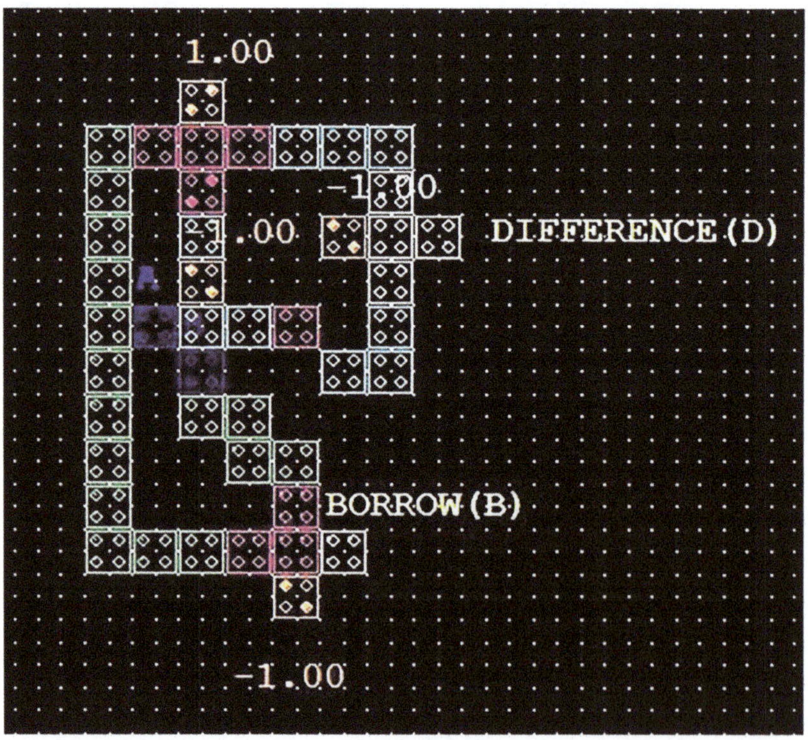

Figure 19.6. The QCA half-subtractor.

19.3 The QCA full-adder and full-subtractor

Imagine a full-adder with three inputs (*A*, *B*, and *C*). The *C* input will act as the carry bit which is transferred from the previous stage. The implementation function of a full-adder is

$$\text{Sum} = A \oplus B \oplus C, \tag{19.1}$$

$$\text{Carry} = AB + AC + BC. \tag{19.2}$$

As can be seen from equations (19.1) and (19.2), each full-adder has two outputs (sum and carry) and the sum output is the result of the Ex-OR of the circuit inputs. According to equation (19.2) it should also be noted that carry can be defined as an output of the three-input majority gate (with *A*, *B*, and *C* as its inputs). Hence, we can design an optimal full-adder in QCA technology with this simple idea. The implementation of this full-adder is shown in figure 19.7. This figure shows that the space occupied and the number of cells used in the proposed circuit are smaller than in the circuits using only three-input and five-input majority gates.

As can be seen from figure 19.7, the design can be generalized to higher bit number full-adder design using the co-planar method. This is the advantage of this full-adder.

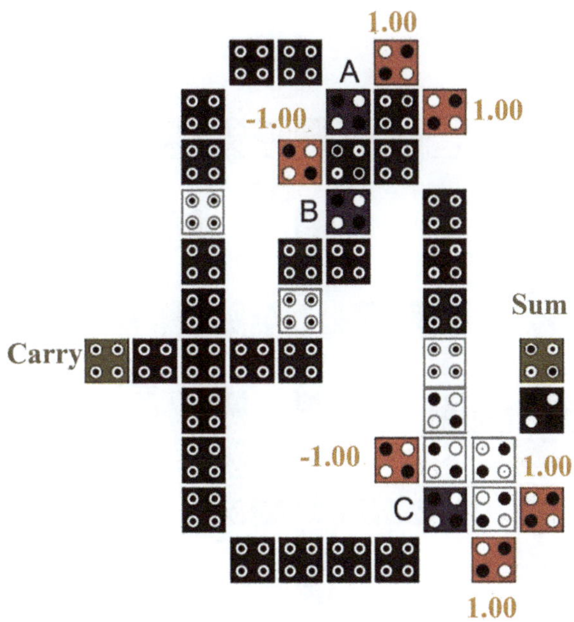

Figure 19.7. Implementation of the full-adder in QCA technology. Reproduced with permission from [9]. Copyright 2019 Springer.

19.3.1 Implementation of the full-adder and full-subtractor

A full-subtractor circuit is composed of three inputs like the full-adder. The sub and borrow outputs of the full-subtractor are as follows:

$$\text{Sub} = A \oplus B \oplus C, \tag{19.3}$$

$$\text{Borrow} = AB + AC + BC. \tag{19.4}$$

Since the sub-output of the full-subtractor is the same as sum in the full-adder (see equations (19.1) and (19.2)), implementing full-adder and full-subtractor circuits in one integrated structure is reasonable. Again it should be mentioned that from equations (19.3) and (19.4), to generate carry and borrow we need a majority gate. Since both circuits have the same inputs and use the majority and similar Ex-OR gates, it is possible to design an integrated circuit for both operations. Therefore, a new full-adder and full-subtractor structure with the minimum number of quantum cells will be introduced.

Figure 19.8 shows the implementation of the proposed circuit. As is obvious in the figure, this circuit makes use of all the gates in the adder. Although the inverter gate and one majority gate are added to the circuit, it is more efficient in comparison to other circuits in terms of the number of cells and area. As can be seen, the full-adder and full-subtractor has fewer cells and a smaller area with good delay performance.

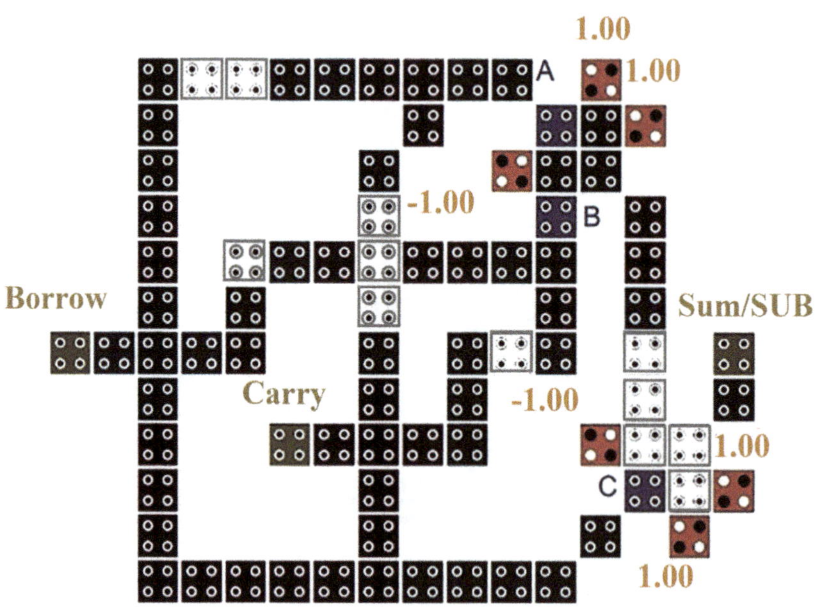

Figure 19.8. Implementation of the full-adder and full-subtractor in QCA technology. Reproduced with permission from [9]. Copyright 2019 Springer.

19.4 Summary

This chapter presents the QCA adder and subtractor circuits. The half-adder and half-subtractor circuits are designed from the Ex-OR gate and the schematic and QCA implementation of the circuits are described. Moreover, the full-adder and half-subtractor are presented together in the QCA implementation.

Further reading

[1] Ahmad F, Bhat G M and Ahmad P Z 2014 Novel adder circuits based on quantum-dot cellular automata (QCA) *Circuits Syst.* **5** 142–52

[2] Hayati M and Rezaei A 2015 Design of novel efficient adder and subtractor for quantum-dot cellular automata *Int. J. Circuit Theory Appl.* **43** 1446–54

[3] Karthigai Lakshmi S and Athisha G 2010 Efficient design of logical structures and functions using nanotechnology based quantum dot cellular automata design *Int. J. Comp. Appl.* **3** 35–42

[4] Polisetti S and Santhosh K 2008 QCA based multiplexing of 16 arithmetic and logical subsystem—a paradigm for nano computing *3rd Annual IEEE Int. Conf. on Nano/Micro Engineering Molecular System, Hainan Island, China* (Piscataway, NJ: IEEE)

[5] Rani T S, Karmakar S, Metri A A and Sharan P 2015 Comparative study of half adder and subtractor circuits based on quantum-dot cellular automata (QCA) *Int. Conf. on Advances in Computer and Communication Engineering* vol 3 pp 61–7

[6] Shahidinejad A and Selamat A 2012 Design of first adder/subtractor using quantum-dot cellular automata *Adv. Mater. Res.* **403** 3392–7

[7] Tougaw P D and Lent C S 1999 Logical devices implementation using quantum dot cellular automata *J. Appl. Phys.* **75** 1818

[8] Walus K, Dysart T J, Jullien G A and Budiman R A 2004 QCAdesigner: a rapid design and simulation tool for quantum-dot cellular automata *IEEE Trans. Nanotechnol.* **3** 26–31

[9] Zoka S and Gholami M 2019 A novel efficient full adder–subtractor in QCA nanotechnology *Int. Nano Lett.* **9** 51–4

IOP Publishing

Quantum Computing

A pathway to quantum logic design

Hafiz Md Hasan Babu

Chapter 20

The QCA multiplier and divider

In this chapter the QCA multiplier and divider circuits are discussed with schematic diagrams and QCA layouts. For the multiplier circuit, the filter network is explained and illustrated and finally the QCA implementation is presented. For the divider circuit, non-restoring division techniques are explained using a schematic and the QCA layout.

20.1 The QCA multiplier

The multiplier has important functions in signal processing and various other applications. With advancements in technology, high speed, low power consumption, and regularity of layout and hence low area, or even combinations can applied in one multiplier. This makes them suitable for various high speed, low power, and compact implementations. In general, the multiplication method uses add and shift algorithms.

If we want to perform multiplication of two numbers, we need to start with the filter networks, i.e. the finite impulse response (FIR) filter, and the output of the network is defined by

$$y_i = \sum_{k=0}^{N-1} b_k x_{i-k}. \tag{20.1}$$

Let Z^{-1} is the one cycle delay operator such that $Z^{-1}x_i = x_{i-1}$. Z^o is defined by the unit operator and Z^{-n} is defined by $Z^{-n} = Z^{-1}Z^{-n+1}$. The characteristics are the following:

- $Z^{-n}x_i = x_{i-n}$.
- $Z^{-1}F(x) = F(Z^{-1}x)$.
- IF C is the time invariant, $Z^{-1}C = C$.

$$
\begin{aligned}
y_i &= \sum_{k=0}^{N-1} b_k x_{i-k} \\
&= \sum_{k=0}^{N-1} b_k Z^{-k} x_i \\
&= \left(\sum_{k=0}^{N-1} b_k Z^{-k}\right) x_i.
\end{aligned}
\tag{20.2}
$$

Equation (20.2) can be implemented by the network presented in figure 20.1. The circles in the figure with b_i represent multiplication by the constant written inside them and $\oplus$ entails the addition of two inputs. The data x_i, b_i and y_i are words of arbitrary size.

In the pipeline design both the upper and lower signal lines take the same additional delay units. It is assumed that $Z^{-\frac{1}{4}}$ comprises the possibility and application of the $Z^{-\frac{1}{2}}$ delay element to each section with upper and lower lines. Equation (20.3) proves that figure 20.2 provides the correct filter output result with $N/2$ cycle delays.

The pipelined FIR filter outputs are the following:

$$
\begin{aligned}
&= Z^{-\frac{1}{2}}\left(b_{N-1}Z^{-\frac{3}{2}(N-1)} + Z^{-\frac{1}{2}}\left(b_{N-2}Z^{-\frac{3}{2}(N-2)} + \cdots + Z^{-\frac{1}{2}}(b_0 Z^0)\right)\right)x_i \\
&= Z^{-\frac{N}{2}} b_{N-1} Z^{-(N-1)} x_i + Z^{-\frac{N}{2}b_{N-2}} Z^{-(N-2)} x_i + \cdots + Z^{-\frac{N}{2}} b_0 x_i \\
&\quad \times Z^{-\frac{N}{2}}\left(\sum_{k=0}^{N-1} b_k Z^{-k}\right)x_i \\
&= Z^{-\frac{N}{2}} y_i.
\end{aligned}
\tag{20.3}
$$

20.1.1 Multiplication networks

The above relations of a channel system can be connected for the duplication procedure. Let us consider an example where an unsigned number framework is imagined; figure 20.3 shows the bit item matrix for an unsigned multiplication. For

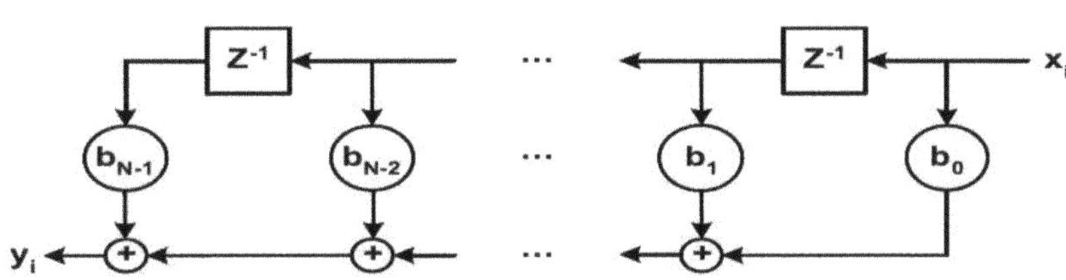

Figure 20.1. The FIR filter network.

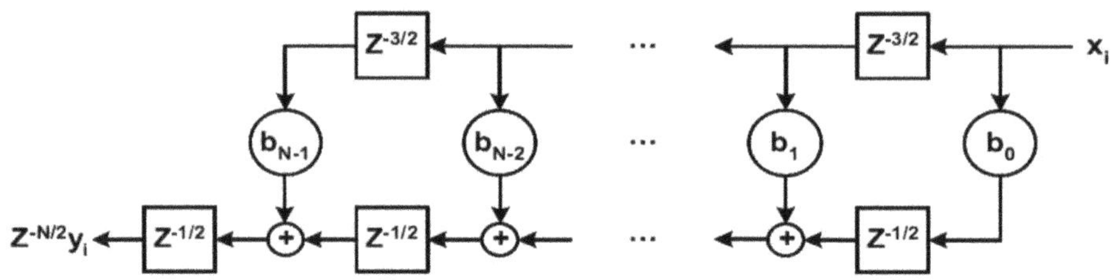

Figure 20.2. The pipelined FIR filter network.

						a_{N-1}	a_{N-2}	...	a_1	a_0
					X	b_{N-1}	b_{N-2}	...	b_1	b_0
						$a_{N-1}b_0$	$a_{N-2}b_0$	...	a_1b_0	a_0b_0
					$a_{N-1}b_1$	$a_{N-2}b_1$	...	a_1b_1	a_0b_1	
				⋮	⋮	...	⋮	⋮		
		$a_{N-1}b_{N-2}$	$a_{N-2}b_{N-2}$	...	a_1b_{N-2}	a_0b_{N-2}				
	$a_{N-1}b_{N-1}$	$a_{N-2}b_{N-1}$	...	a_1b_{N-1}	a_0b_{N-1}					
p_{2N-1}	p_{2N-2}	p_{2N-3}	...	p_N	p_{N-1}	p_{N-2}	...	p_2	p_1	p_0

Figure 20.3. The bit product matrix for unsigned multiplication.

this situation only single digit calculations are utilized in the network, so all additions and multiplications utilize binary calculations. A one-digit multiplication is addressed by a logical AND and a one-digit addition presents to a full-adder. The fundamental difference between the FIR filter and the multiplication network is the treatment of the carry out of the full adder. The channel organizes an inside use carry flow stream, yet the multiplication arrangement needs distinct signal streams, so the system for multiplication ought to be changed in a similar manner to the filter channel.

Let $(a_i; b_i)$ be the multiplicand and multiplier pair and p_i be the product sum of bit position i. Bits a_i and p_i correspond to the words x_i and y_i of the filter example. The position i is considered as the input applied at time i. Define the sum and carry-out of the full-adder at the ith time and jth location as (s_{ij}, c_{ij}) when $0 \leqslant i \leqslant 2N - 1$ and $0 \leqslant j \leqslant N - 1$, where j is numbered from right to left. Assume that the sum generation takes at least $Z^{-\frac{1}{2}}$ and the carry generation takes at least $Z^{-\frac{1}{4}}$. Although figures 20.1 and 20.2 ignored the zeroth full-adder, the derivation includes that adder. The implementation can be performed in two ways:

$$\begin{aligned}(s_{ij}, c_{ij}) &= \mathrm{Add}\left(b_j Z^{-\frac{7}{4}j} a_i,\ Z^{-\frac{3}{4}} s_{i(j-1)},\ Z^{-\frac{1}{4}} c_{i(j+1)}\right)\\ &= \mathrm{Add}\left(b_j a_{i-\frac{7}{4}j},\ s_{\left(i-\frac{3}{4}\right)(j-1)},\ c_{\left(i-\frac{1}{4}\right)(j+1)}\right)\end{aligned} \tag{20.4}$$

$$(s_{ij}, c_{ij}) = \text{Add}\left(b_j Z^{-\frac{3}{2}j} a_i, Z^{-\frac{1}{2}} s_{i(j-1), Z^{-1}c_{ij}}\right)$$
$$= \text{Add}\left(b_j a_{i-\frac{3}{2}}, s_{\left(i-\frac{1}{2}\right)(j-1)}, c_{(i-1)j}\right). \tag{20.5}$$

Equation (20.4) sends the carry-out in a backward direction with minimum delay. Equation (20.5) instead uses a feedback loop to the adder itself using a one clock delay unit. Both handle the carry-out correctly, carrying it to the higher bit calculation. We call them a carry shift multiplication (CSM) and a carry delay multiplication (CDM), respectively. Figures 20.4 and 20.5 present the network diagrams based on equations (20.4) and (20.5). From the stream direction of the information in and out, they are called right-to-left (RtL) systems. The calculated yield line of a full-adder is the carry-out. The CSM configuration has the advantage of the least delay for the convey shift, while the CDM configuration minimises the latency of the yield.

In this realization, the minimum latency from the first input to first output is either $3N/4$ or $N/2$ cycles. From figure 20.1 the redirected output to the right side, which is the same side as the input in figure 20.5 and is present for the redirection graph and for a pipeline design, can be redrawn as shown in figure 20.6 using the following equation:

$$Z^{-\frac{1}{2}} y_i = Z^{-\frac{1}{2}}\left(\sum_{k=0}^{N-1} b_k Z^{-k}\right) x_i$$
$$= Z^{-\frac{1}{2}}\left(\sum_{k=0}^{N-1} b_k Z^{-\frac{k}{2}} Z^{-\frac{k}{2}}\right) x_i \tag{20.6}$$
$$= Z^{-\frac{1}{2}}\left(\sum_{k=0}^{N-1} Z^{-\frac{k}{2}} b_k Z^{-\frac{k}{2}}\right) x_i.$$

Finally, figure 20.6 is a network design which is comparable to figure 20.2. The main difference is that there exists a much smaller latency from the first input to the first output. Figure 20.7 presents the block diagram of a redirected pipelined FIR filter network which can easily be derived from figure 20.6.

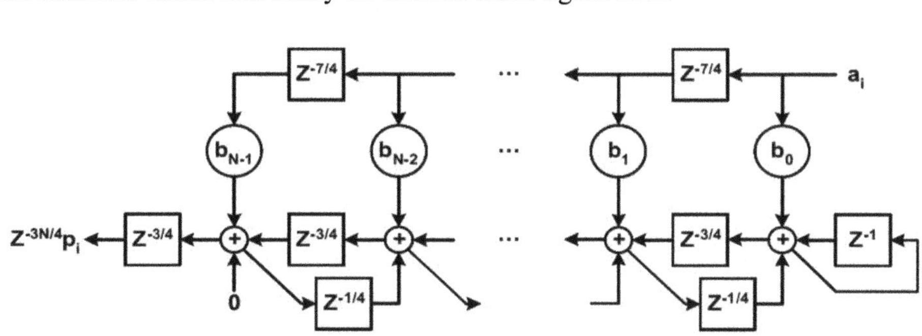

Figure 20.4. The RtL carry shift multiplication network.

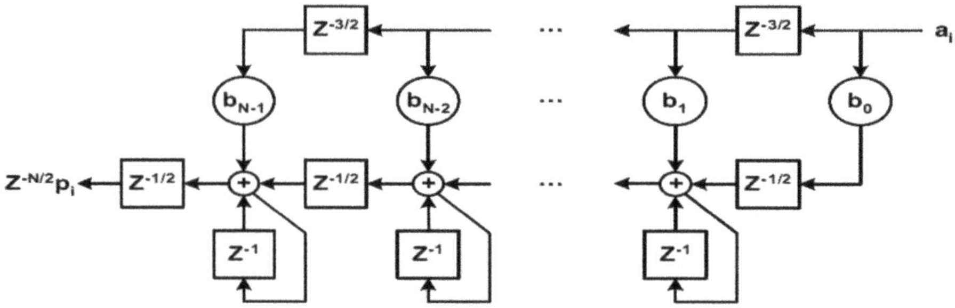

Figure 20.5. The RtL carry delay multiplication network.

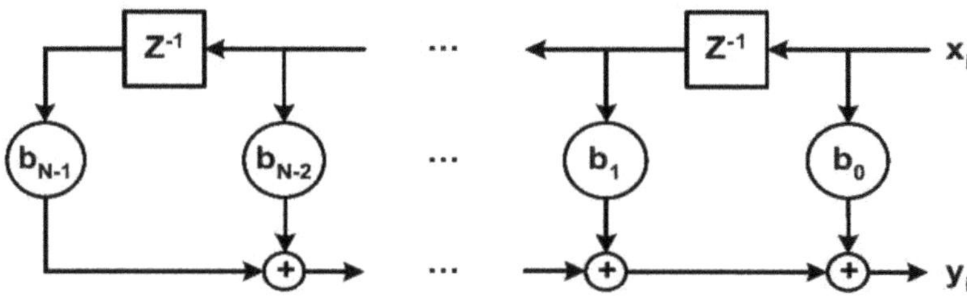

Figure 20.6. The redirected FIR filter network.

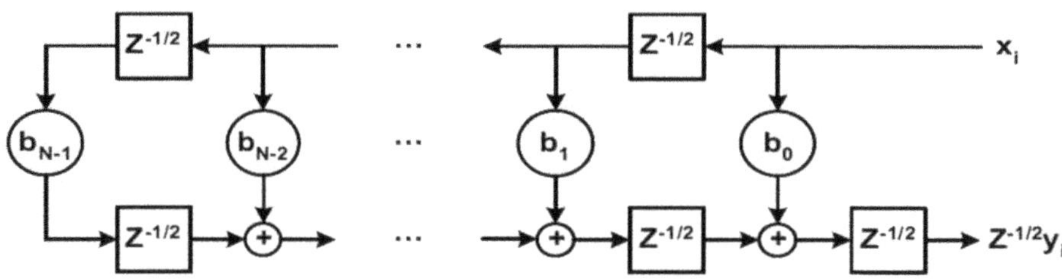

Figure 20.7. The redirected pipelined FIR filter network.

Based on figure 20.6, the multiplication networks are represented by equations (20.7) and (20.8) depending on the carry-out handling. Figures 20.8 and 20.9 show the respective network implementations. Similarly, the networks are called right-to-right (RtR) networks and they are distinguished by the carry flow as either CSM or CDM networks.The CSM design has the minimum delay of the carry shift and the CDM design has the minimum latency to the output.

$$\begin{aligned}(s_{ij}, c_{ij}) &= \mathrm{Add}\left(b_j Z^{-\frac{1}{4}j} a_i,\ Z^{-\frac{3}{4}} s_{i(j+1)},\ Z^{-\frac{1}{4}} c_{i(j-1)}\right)\\ &= \mathrm{Add}\left(b_j a_{i-\frac{1}{4}j},\ s_{\left(i-\frac{3}{4}\right)(j+1)},\ c_{\left(i-\frac{1}{4}\right)(j-1)}\right)\end{aligned} \tag{20.7}$$

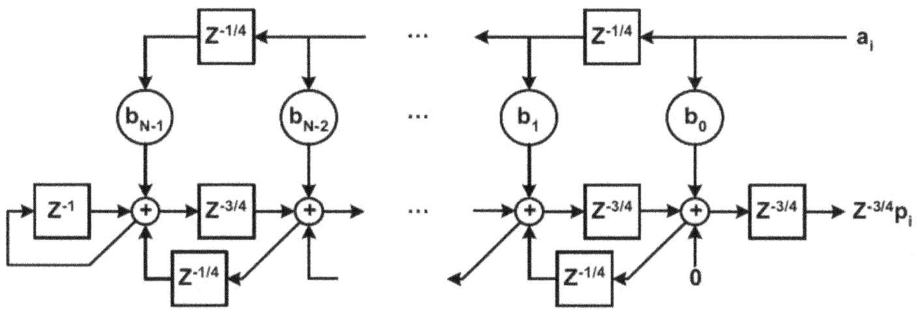

Figure 20.8. The RtR carry shift multiplication network.

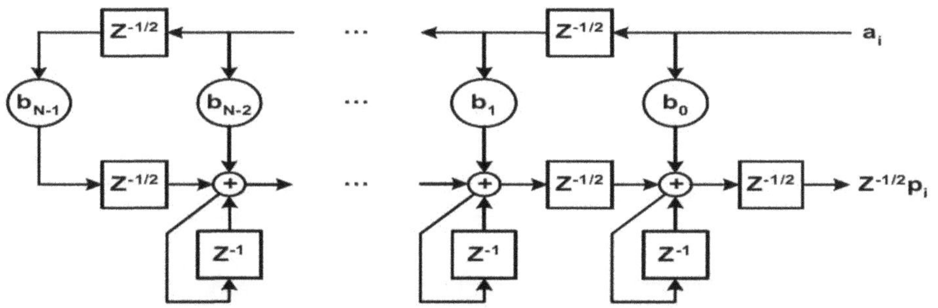

Figure 20.9. The RtR carry delay multiplication network.

$$
\begin{aligned}
(s_{ij},\, c_{ij}) &= \mathrm{Add}\left(b_j Z^{-\frac{1}{2}j} a_i,\ Z^{-\frac{1}{2}} s_{i(j+1)},\ Z^{-1} c_{ij}\right) \\
&= \mathrm{Add}\left(b_j a_{i-\frac{1}{2}},\ s_{\left(i-\frac{1}{2}\right)(j+1)},\ c_{(i-1)j}\right).
\end{aligned}
\tag{20.8}
$$

20.1.2 QCA multiplication networks

The QCA circuit has a four phase clocks and the circuit areas are divided into four clock zones. One clock zone delay is denoted by the D^{-1} operator which corresponds to a quarter cycle. That is, $D^{-4} = Z^{-1}$. Based on the QCA circuit characteristics, one clock zone delays a quarter clock, so that the delay matches one D^{-1} operation. Assume a logical AND operation delays by one D^{-1} operation and one full-adder sum and carry-out computation from the input by D^{-2} and D^{-1} operations, respectively. Wires also cause some delay based on the wire length. After incorporating these characteristics the filter networks are redrawn as in figures 20.10 and 20.11. The delay amounts in the upper and lower signal flows in either case are chosen to make one clock cycle difference between the adjacent paths.

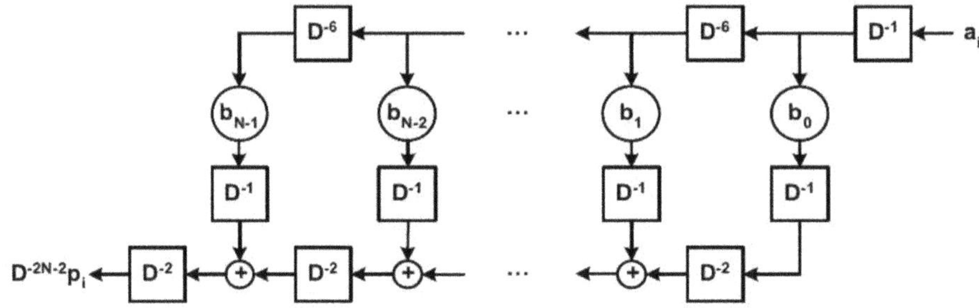

Figure 20.10. The FIR filter network for QCA.

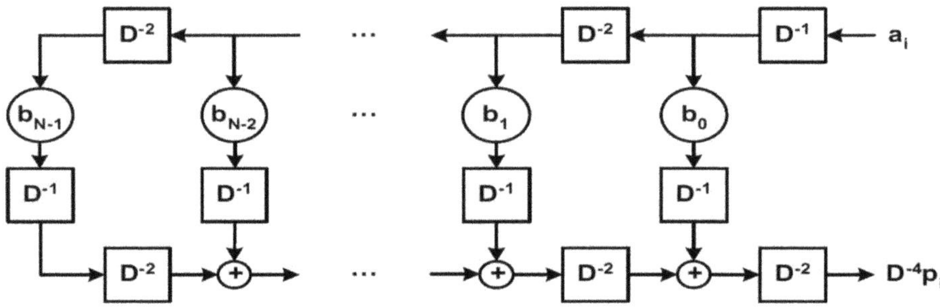

Figure 20.11. The redirected FIR filter network for QCA.

From the filter network examples, the multiplier networks for QCA are drawn. Based on equations (20.4), (20.5) and (20.7)–(20.12) are re-organized for QCA multiplication. The former figures are revised by the carry flow routes. The serial multiplier can be realised with four distinct possibilities, as shown in figures 20.12–20.15. The multiplier block diagram and the modified multiplier block diagram are presented in figures 20.16 and 20.17, respectively.

$$\begin{aligned}(s_{ij}, c_{ij}) &= \mathrm{Add}\left(b_j D^{-7j-2} a_i,\ D^{-3} s_{i(j-1)},\ D^{-1} c_{i(j+1)}\right)\\ &= \mathrm{Add}\left(b_j a_{i-7j-2},\ s_{(i-3)(j-1)},\ c_{(i-1)(j+1)}\right)\end{aligned} \tag{20.9}$$

$$\begin{aligned}(s_{ij}, c_{ij}) &= \mathrm{Add}\left(b_j D^{-6j-2} a_i,\ D^{-2} s_{i(j-1)},\ D^{-4} c_{ij}\right)\\ &= \mathrm{Add}\left(b_j a_{i-6j-2},\ s_{(i-2)(j-1)},\ c_{(i-4)j}\right)\end{aligned} \tag{20.10}$$

$$\begin{aligned}(s_{ij}, c_{ij}) &= \mathrm{Add}\left(b_j D^{-j-2} a_i,\ D^{-3} s_{i(j+1)},\ D^{-1} c_{i(j-1)}\right)\\ &= \mathrm{Add}\left(b_j a_{i-j-2},\ s_{(i-3)(j+1)},\ c_{(i-1)(j-1)}\right)\end{aligned} \tag{20.11}$$

$$\begin{aligned}(s_{ij}, c_{ij}) &= \mathrm{Add}\left(b_j D^{-2j-2} a_i,\ D^{-2} s_{i(j+1)},\ D^{-4} c_{ij}\right)\\ &= \mathrm{Add}\left(b_j a_{i-2j-2},\ s_{(i-2)(j+1)},\ c_{(i-4)j}\right).\end{aligned} \tag{20.12}$$

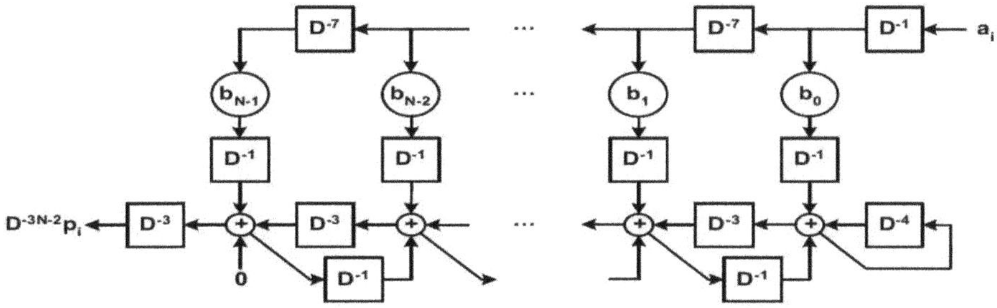

Figure 20.12. The RtL CSM network for QCA.

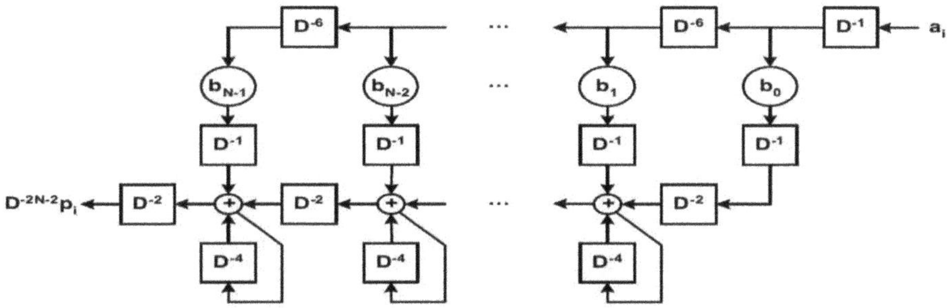

Figure 20.13. The RtL CDM network for QCA.

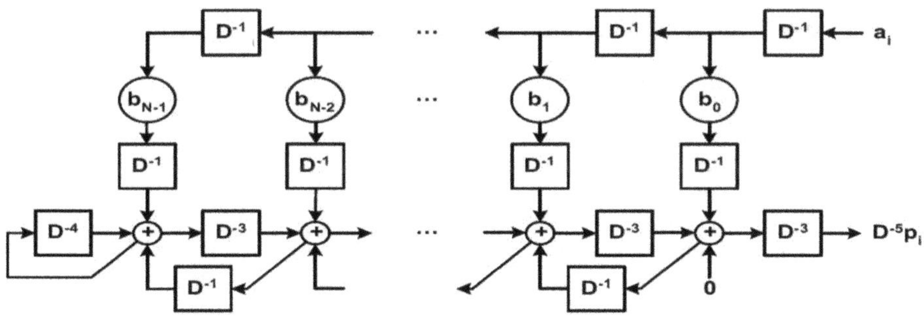

Figure 20.14. The RtR CSM network for QCA.

20.1.3 Multiplier design

The multiplication track is spread over a structure like figure 20.5, except multiplying $Z^{-\frac{1}{2}}$ to all the horizontal stream delay units. This design intrinsically experiences a lengthy delay and has an redundant right-most full-adder. An RtR structure is nominated for QCA execution since it is known to reduces the inactivity of the first input and output. The final proposal has two substitutes, as depicted in figures 20.12 and 20.13. Figure 20.14 presents the general block diagrams of QCA multipliers and figure 20.15 shows the block diagrams of the improved designs for QCA circuits.

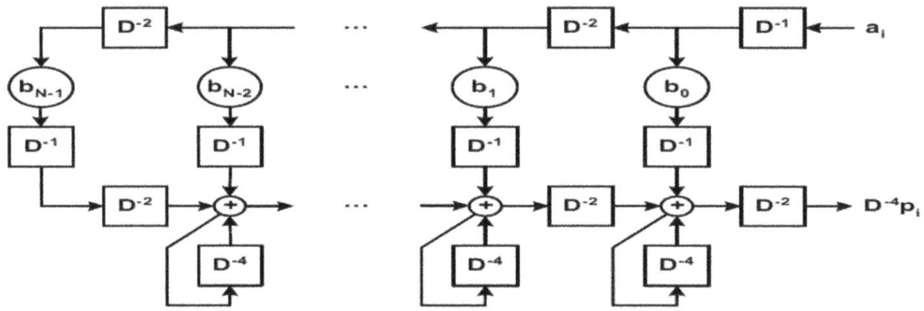

Figure 20.15. The RtR CDM network for QCA.

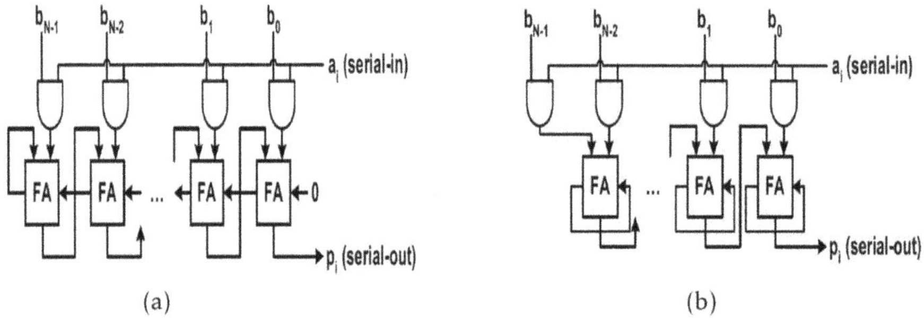

Figure 20.16. Multiplier block diagrams. (a) RtR CSM. (b) RtR CDM.

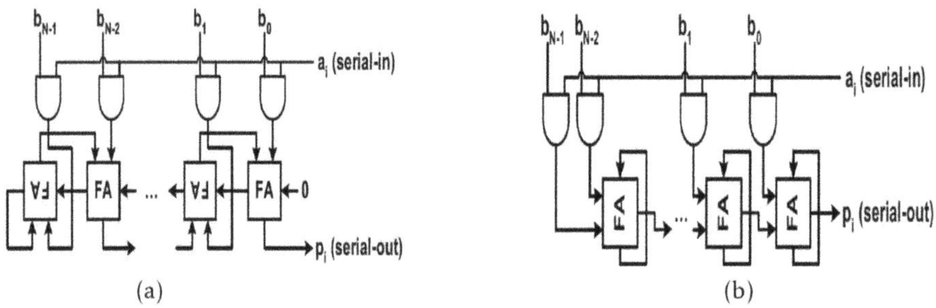

Figure 20.17. Modified multiplier block diagrams. (a) RtR CSM. (b) RtR CDM.

20.1.4 QCA implementation

Figure 20.18 describes the bit product matrix for four-bit multiplication.

Full-adders are used for the carry shift multiplier and bit-serial addresses are used for the carry delay multiplier. Those addresses are made using the 'carry flow adder (CFA)' which is a layout optimized ripple carry adder. The bit-serial adder is similar to the full-adder except that the carry-in and carry-out are connected internally with a one clock delay. Figures 20.19 and 20.20 show the schematics and layouts of both adders.

				a_3	a_2	a_1	a_0
X				b_3	b_2	b_1	b_0
				a_3b_0	a_2b_0	a_1b_0	a_0b_0
			a_3b_1	a_2b_1	a_1b_1	a_0b_1	
		a_3b_2	a_2b_2	a_1b_2	a_0b_2		
	a_3b_3	a_2b_3	a_1b_3	a_0b_3			
p_7	p_6	p_5	p_4	p_3	p_2	p_1	p_0

Figure 20.18. Bit product matrix for four-bit multiplication.

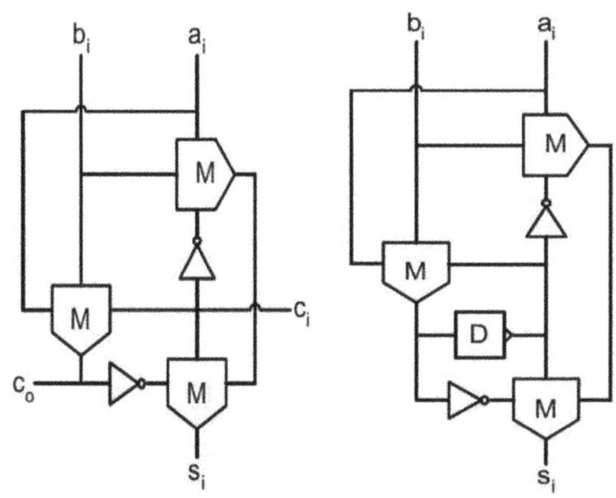

(a) (b)

Figure 20.19. One-bit adder schematic. (a) Full-adder. (b) Bit-serial adder.

Using the above mechanism, the four-bit RtR CSM and CDM are designed as shown in figure 20.17. These block diagrams are used to implement the multiplier circuits, as shown in figures 20.21 and 20.22. Multipliers for larger word sizes also can be implemented easily by adding additional bit slices, i.e. a 32-bit CSM and CDM are shown in figures 20.23 and 20.24.

20.2 The QCA divider

The divider is the most entangled and tedious circuit among all the math units. However, for the advancement of exact instrumentation, i.e. for spaceflight and radar innovations, the divider is vital. Numerous calculations are utilized for the execution of the divider, specifically re-establishing, non-re-establishing, Sweeney, Robertson and Tocher techniques, and the Newton iterative calculation. Re-establishing and non-reestablishing calculations contain on expansion and subtraction, which are appropriate for coordinated circuit usage. The generally utilized

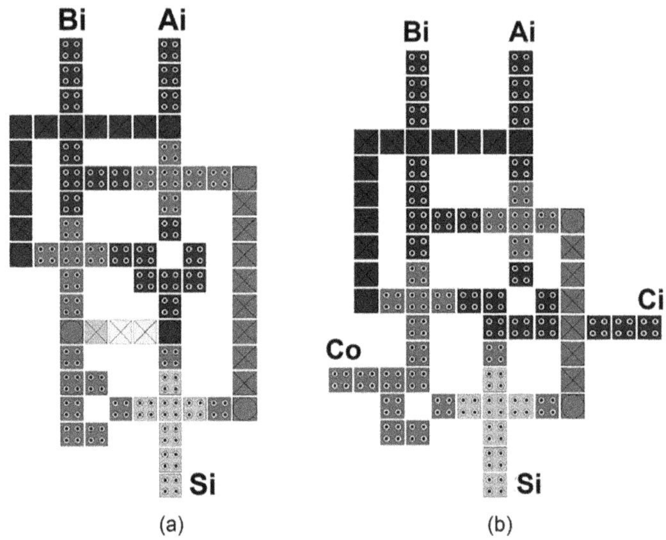

Figure 20.20. QCA one-bit adder layouts. (a) Full-adder. (b) Bit-serial adder. Reproduced with permission from [2]. Copyright 2007 IEEE.

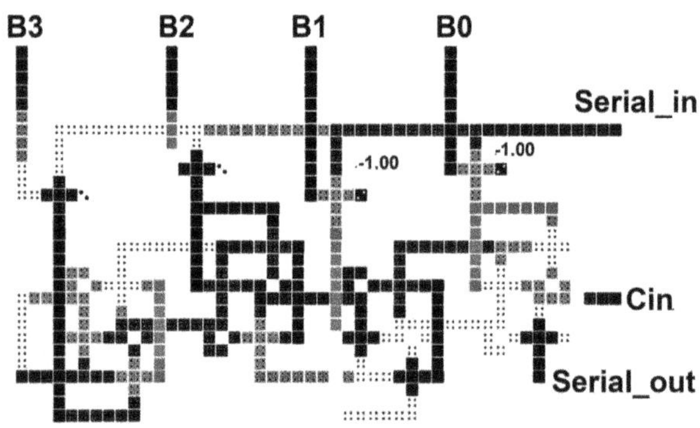

Figure 20.21. Layout of a four-bit QCA carry shift multiplier. Reproduced with permission from [2]. Copyright 2007 IEEE.

calculation in the divider is the non-re-establishing division, and Huanqing displayed the structure of a non-re-establishing binary array divider in QCA.

20.2.1 The non-restoring binary divider

To understand the non-restoring division, it is necessary to explain the restoring division algorithm. The following notation will be used in the divider section:

N : numerator or dividend

Y : denominator or divisor

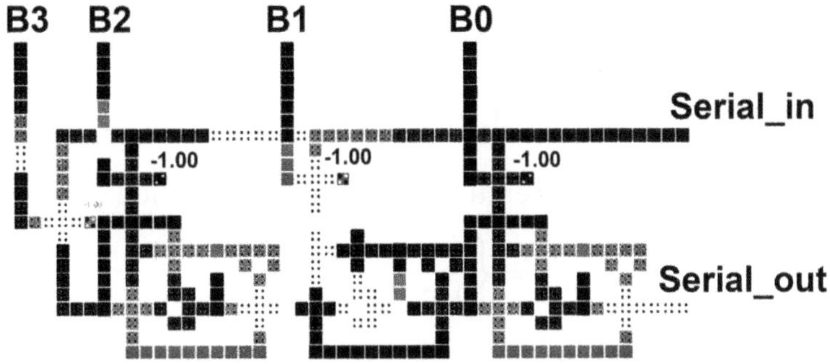

Figure 20.22. Layout of a four-bit QCA carry delay multiplier. Reproduced with permission from [2]. Copyright 2007 IEEE.

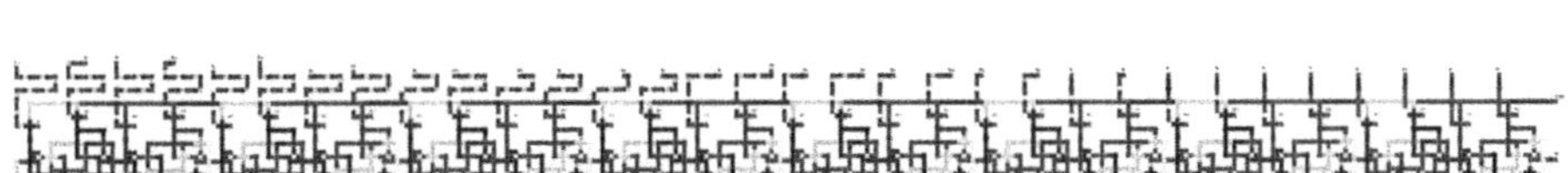

Figure 20.23. Layout of a 32-bit QCA carry shift multiplier. Reproduced with permission from [2]. Copyright 2007 IEEE.

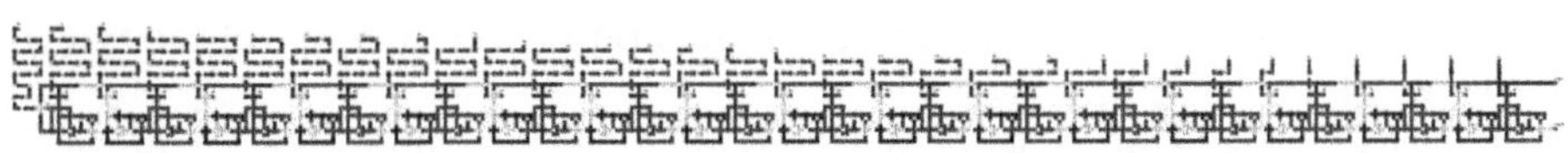

Figure 20.24. Layout of a 32-bit QCA carry delay multiplier. Reproduced with permission from [2]. Copyright 2007 IEEE.

R_i : partial remainder after ith iterations i number of iterations
n : number of bits
q : quotient set for the algorithm
Q : quotient for the division

The restoring algorithm is the basic method for division operation, and the details of the method are described by the following set of equations:

$$q_{i+1} = \begin{cases} 1, \text{ if } 2R_i > Y \\ 0, \text{ if } 2R_i < Y. \end{cases} \tag{20.13}$$

$$R_{i+1} = 2R_i - q_{i+1}.\ Y. \tag{20.14}$$

The partial remainder is going through a left shift of R_i and a subtraction: $R_{i+1} = 2R_i - 1$. If $R_i + 1$ is positive, then $q_{i+1} = 1$, otherwise $q_{i+1} = 0$ and a restoring addition is needed as shown in equation (20.15). The addition is used to restore the proper partial remainder, $R_{i+1} = R_{i+1} + Y = 2R_i$, as shown in equation (20.17).

To overcome the problems of restoring division, undertake an alternative method of binary division, non-restoring division also known as the addsubtraction alternate algorithm, and the relevant divider is called the non-restoring division (NRD). In this algorithm if a negative partial remainder is produced at the end of a certain cycle, no restoration is executed.

The non-restoring division operation is performed by the following formulas:

$$q_{i+1} = \begin{cases} 1, \text{ if } R_i > 0 \\ 0, \text{ if } R_i < 0 \end{cases} \tag{20.15}$$

$$R_{i+1} = \begin{cases} 2R_i - Y, \text{ if } R_i > 0 \\ 2R_i + Y, \text{ if } R_i < 0 \end{cases} \tag{20.16}$$

$$r = \begin{cases} 2^{-n}.\ R_n, & \text{if } R_n > 0 \\ 2^{-n}.\ (R_n + Y) & \text{if } R_n < 0, \end{cases} \tag{20.17}$$

where $i = 0, 1, \cdots, n - 1$ is the recursion index. The initial partial remainder R_0 equals the dividend, and r is the final remainder.

In this manner, from equations (20.15)–(20.17) above, it is anything but difficult to see that each time the negative partial remainder portion shows up, the non-re-establishing calculation makes one expansion stride not precisely as the re-establishing strategy. Thus there will be a critical reduction of the quantity of clock cycles over the whole division process.

In binary non-restoring division, partial remainders are obtained by a subtraction or an addition between the dividend and the successively right-shifted versions of the divisor. The quotient bit is determined by the sign of the partial remainder which also decides whether to add or subtract the shifted divisor in the next cycle.

A two-dimensional array of pipelined two's-complement adder/subtractor cells (CAS cells) can be used to implement the non-restoring algorithm. Fundamentally the array consists of columns of carry propagate adders with one controlled input bit P. Figures 20.25(a) and 20.25(b) illustrate a array divider with a two-bit divisor $(0.\ y_1y_2)$ and a four-bit dividend $(0.\ x_1x_2x_3x_4)$. The first bit of both the divisor and dividend are their signs, it is used to represent a positive number. Each logic cell is made up of a full-adder and an Ex-OR gate. The Ex-OR gate provides the divisor input to the full-adder. The control signal P determines the function of the CAS cell, namely whether an addition or a subtraction is to be executed. The two-bit divisor and the double-length four-bit dividend are imported at the top and right edges of the array, respectively. At first, a a three-bit quotient is produced at the left side of the array. Then every quotient bit is transported to the next row as the control signal P and the low carry of the right-most cell. The three-bit final reminder appears at the bottom of the array.

In this section, the presented QCA multiplication circuit depends on pipeline engineering. The sequential parallel multiplier is a conceivable answer for this

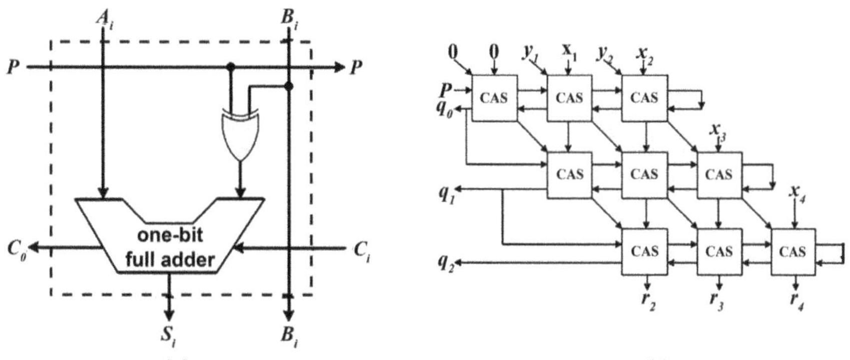

Figure 20.25. The binary array divider. (a) CAS cell. (b) Four-bit dividend, two-bit divisor.

requirement, and the standard QCA pipeline configuration has the advantage of reducing wire delays.

20.2.2 Divider implementation

The non-restoring array divider is composed of CAS cells that have one full-adder and one Ex-OR gate. Figure 20.26 shows the CAS cell layout. The Ex-OR gate is demonstrated in the dashed box and it can be designed using only four majority gates and one inverter; its delay is a full clock cycle. The rest of the cell is a one-bit full-adder; it can be constructed using only three majority gates and two inverters. The full-adder takes three inputs, (Ex-$OR_O UT$, A, and C_i) and gives two outputs [$S_i = A \oplus B \oplus C_i$ and $C_o = MAJ(A, B, C_i)$]. Both the outputs have a latency of $1\frac{1}{4}$ clock cycles. The different colors in the layout are used to distinguish different clock zones.

An n-bit divider is formed by combining n^2 CAS cells into the regular array shown in figure 20.27; only $n = 3$ is presented here for simplicity. In the figure the total number of QCA cells is 3742 and the area of the divider layout is 6.22 μm^2. Compare this to the eight-bit NRD implemented by 45 nm CMOS technology which has an area of 2015.64 μm^2. It is very easy to see that the QCA NRD is much more area efficient than the normal divider because the area of the eight-bit QCA NRD is 56.98 μm^2, which also shows the advantage of the QCA device over traditional transistors.

20.3 Summary

QCA multiplier and divider circuits are designed with detailed calculations. For the multiplier, a four-bit and 32-bit carry shift multiplier and a four-bit and 32-bit carry delay multiplier are described. For the divider, a non-re-establishing binary array divider is discussed using QCA technology. It is built using CAS cell blocks.

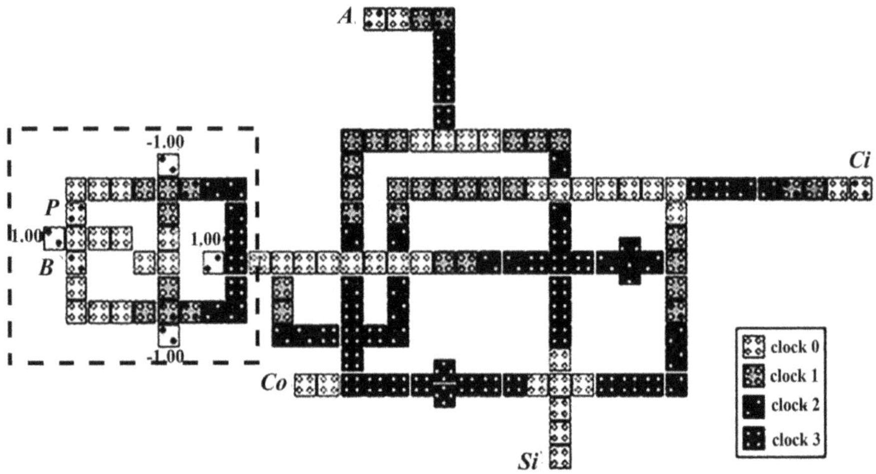

Figure 20.26. The QCA layout of the CAS cell.

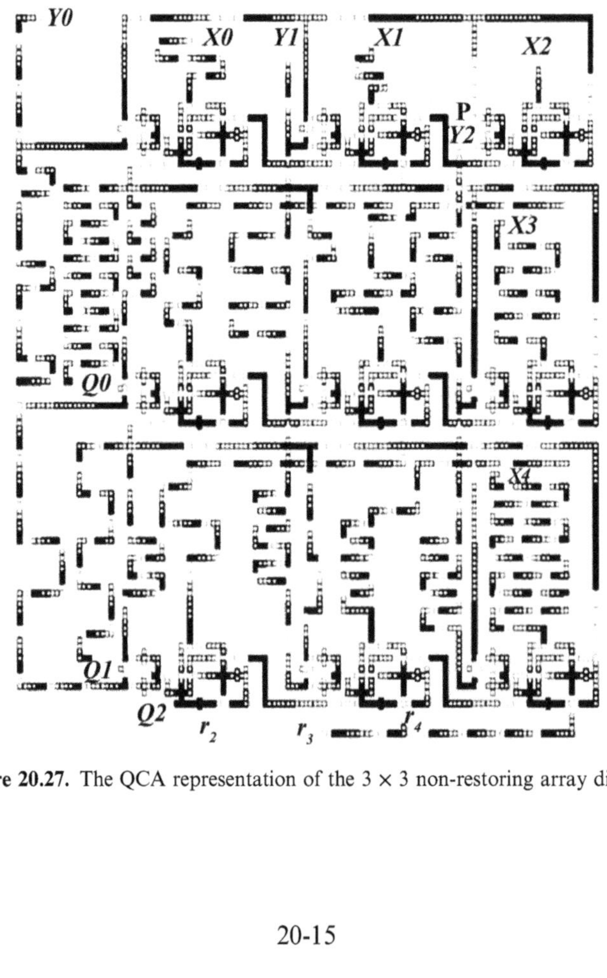

Figure 20.27. The QCA representation of the 3 × 3 non-restoring array divider.

Further reading

[1] Cappa M and Hamacher V C 1973 An augmented iterative array for high-speed binary division *IEEE Trans. Comput.* **C-22** 172–5

[2] Cho H *et al* 2007 Serial parallel multiplier design in quantum-dot cellular automata *18th IEEE Symp. on Computer Arithmetic (ARITH'07)* (Piscataway, NJ: IEEE) pp 7–15

[3] Cho H and Swartzlander E E 2006 Modular design of conditional sum adders using quantum-dot cellular automata *Sixth IEEE Conf. on Nanotechnology, 2006* vol 1 (Piscataway, NJ: IEEE) pp 363–6

[4] Cohen D 1987 A mathematical approach to computational network design *Systolic Signal Processing Systems* (New York: Marcel Dekker, Inc) pp 1–29

[5] Cui H, Cai L, Yang X, Feng C and Qin T 2014 Design of non-restoring binary array divider in quantum-dot cellular automata *IET Micro Nano Lett.* **9** 464–7

[6] Shin S-H, Jeon J-C and Yoo K-Y 2013 Wire-crossing technique on quantum-dot cellular automata *Second Int. Conf. on Next Generation Computer and Information Technology* vol 27 pp 52–7

[7] Walus K, Dysart T J, Jullien G A and Budiman R A 2004 QCAdesigner: a rapid design and simulation tool for quantum-dot cellular automata *IEEE Trans. Nanotechnol.* **3** 26–31

[8] Wang Y and Lieberman M 2004 Thermodynamic behavior of molecular-scale quantum-dot cellular automata (QCA) wires and logic devices *IEEE Trans. Nanotechnol.* **3** 368–76

Chapter 21

QCA asynchronous and synchronous counters

In this chapter, quantum-dot cellular automata (QCA) asynchronous and synchronous counters are described. In the asynchronous counter section, a dual-edge triggered J-K flip-flop is presented with its schematic and QCA implementation. In the synchronous counter section J-K-type and T-type flip-flops are explained with graphic symbols, schematics, and QCA layouts.

21.1 The asynchronous counter

In this section a falling-edge triggered J-K flip-flop is designed based on QCA implementation. The dual-edge triggered J-K flip-flop is designed using QCA by arranging special clock zones exquisitely and serially. The dual-edge triggered flip-flop requires half of the clock compared to the falling-edge triggered J-K flip-flop and the clock power consumption can be reduced significantly. Based on the dual-edge triggered J-K flip-flop an asynchronous backward counter design is explored.

21.1.1 The dual-edge triggered J-K flip-flop

A dual-edge triggered J-K flip-flop is designed with a dual-edge triggered structure. The QCA cells are quantum dots 18 nm in width and 5 nm in height. They are placed on a grid with a cell center-to-center distance of 20 nm. Figure 21.1 shows the layout of the dual-edge triggered structure. Its input is a clock pulse (CP) and the output will only be high when the CP is at its rising and falling edge. The output is added as the level input for the level-triggered J-K flip-flop shown in figure 21.2.

It is necessary to take some precaution for the CP, because it is different to the QCA clock zone signal. It is just an input of a flip-flop while the QCA clock zone signal controls signal flow. Here the different shades of gray represent the different clock zones (clock zones 0, 1, 2, and 3). The truth table of the dual-edge triggered structure is given in table 21.1.

doi:10.1088/978-0-7503-2747-3ch21

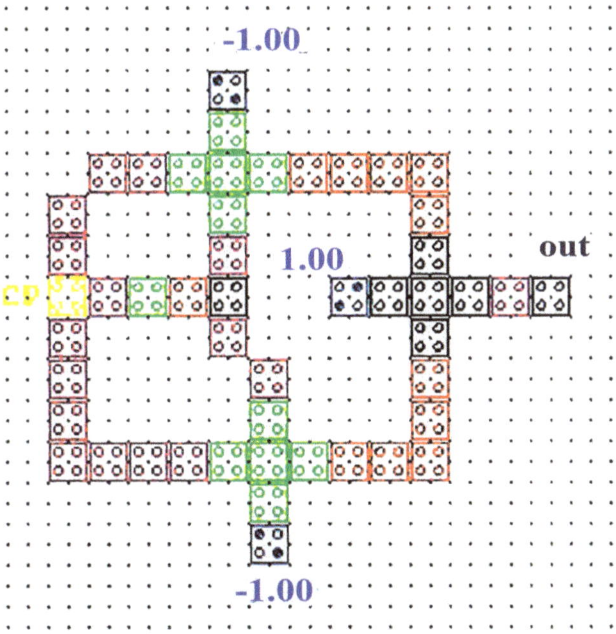

Figure 21.1. Layout of the dual-edge triggered structure.

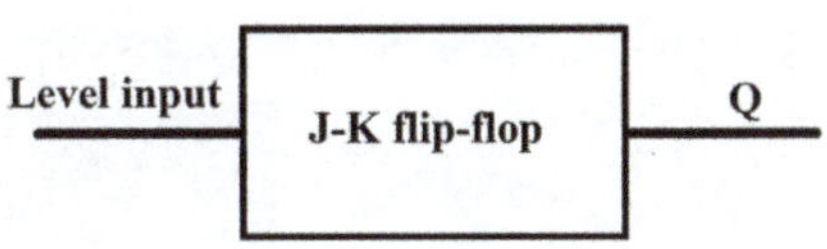

Figure 21.2. Level-triggered J-K flip-flop description.

Table 21.1. The truth table of figure 21.1.

CP	$\overline{CP}$	CP_{old}	$\overline{CP_{old}}$	AND 1 = $\overline{CP} \cup CP_{old}$	AND 2 = $CP \cup \overline{CP_{old}}$	OR
1	0	1	0	0	0	0
1 $n \rightarrow 0$	1	1	0	1	0	1
0	1	0	1	0	0	0
0 $n \rightarrow 1$	0	0	1	0	1	1

21.1.2 The design of dual-edge triggered J-K flip-flop

A dual-edge triggered J-K flip-flop is constructed using the aforementioned dual-edge triggered structure. The dual-edge triggered J-K flip-flop comprises QCA logic gates, such as AND gates, OR gates, and inverters. A schematic diagram of the dual-edge triggered J-K flip-flop is shown in figure 21.3, in which the dual-edge triggered structure is presented in the dotted box. The state transfer equation of the dual-edge triggered J-K flip-flop is as

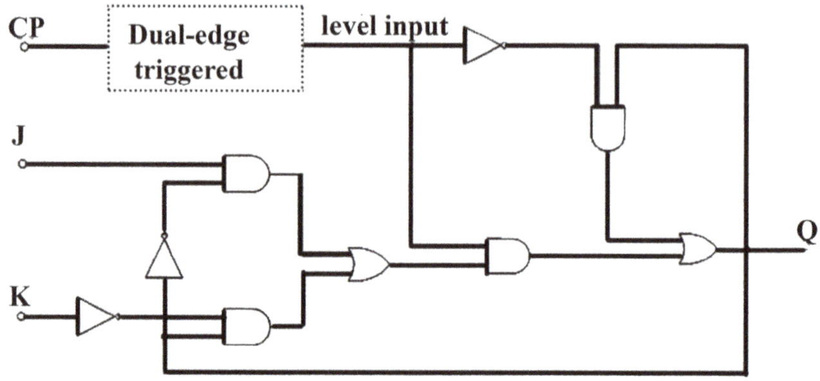

Figure 21.3. Schematic diagram of the dual-edge triggered J-K flip-flop.

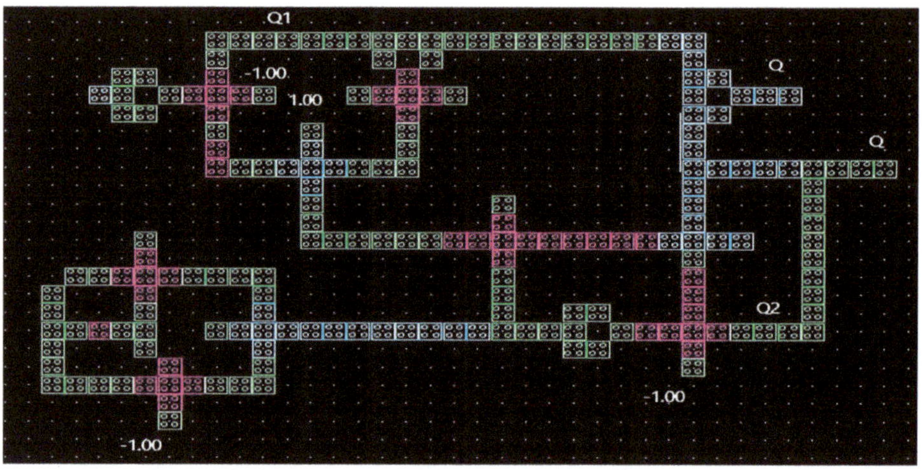

Figure 21.4. Layout of the dual-edge triggered J–K flip-flop

$$Q^{n+1} = \overline{\text{level input}}\ Q^n + (\text{level input})(J\overline{Q^n} + \bar{K}Q^n). \tag{21.1}$$

The layout of the dual-edge triggered J-K flip-flop is shown in figure 21.4. The three yellow boxes denote the J, K, and CP inputs, respectively. From the layout it can be observed that there is a clock cycle delay at the output of $(J\bar{Q}^n + \bar{K}Q^n)$ to ensure timing and delay matching. This makes sure that the CP can capture the values of J, K at the front edge when the CP is falling or rising. When the edge of the CP is applied, the level input becomes 1. Then the level-triggered J-K flip-flop carries through a different operation according to J and K. The dual-edge triggered structure and the level-triggered J-K flip-flop compose a dual-edge triggered J-K flip-flop. The detailed functional description of this dual-edge triggered J-K flip-flop is as follows:

1. If $J = 0$, $K = 0$, and level input = 0, the output of the dual-edge triggered J-K flip-flop does not change. If the falling or rising edge of the clock pulse is applied, level input = 1 and, according to equation (21.1), the output at the end of the edge is $Q^{n+1} = Q^n$. This is called a holding state.
2. If $J = 0$, $K = 1$, and level input = 0, the output of the dual-edge triggered J-K flip-flop does not change. If the falling or rising edge of the clock pulse is applied, level input = 1 and, according to equation (21.1), the output at the end of the edge is $Q^{n+1} = 0$. This is called set 0 in the J-K flip-flop.
3. If $J = 1$, $K = 0$, and level input = 0, the output of the J-K flip-flop does not change. If the falling or rising edge of the clock pulse is applied, level input = 1 and, according to equation (21.1), the output at the end of the edge is $Q^{n+1} = 1$. This is called set 1 in J-K flip-flop.
4. If $J = 1$, $K = 1$, and level input = 0, the output of the J-K flip-flop does not change. If the falling edge of the CP is applied, level input = 1 and, according to equation (21.1), the output at the end of the edge is $Q^{n+1} = Q^n$. At this time the J-K flip-flop will act as a one-bit counter.

21.1.3 The asynchronous backward counter

Any sequential logic circuits can be designed using the J-K flip-flop and additional combinational circuits. Counters are widely used sequential circuits in digital systems. In this section the QCA implementation of a mod-8 asynchronous backward counter is described using the dual-edge triggered J-K flip-flop.

The equation of the n-bit asynchronous backward counter is

$$
\begin{aligned}
J_i &= K_i = 1,\ i = 1 \\
J_i &= K_i = \overline{Q_{i-1}},\ i = 1, 2, 3, \cdots, n \\
\mathrm{CP}_i &= \left\{ \frac{\mathrm{CP},\ i = 1, 2}{Q_{i-1,\ i=3,\ 4,\ \cdots,\ n}} \right\}.
\end{aligned}
\tag{21.2}
$$

A counter circuit has n flip-flops and it includes 2^n possible states. A mod-8 counter can count from 0 to 7. On the other hand the backward counter, which is from 7 to 0, that is the state $Q_3Q_2Q_1$, is 111 → 110 → 101 → 100 → 011→ 010 → 001 → 000 as presented in equation (21.2). Large modular sizes of counters can be implemented by simply adding additional bit slices of dual-edge triggered J-K flip-flops. The schematic diagram of a three-bit asynchronous backward counter is shown in figure 21.5(a) and its QCA implementation is shown in figure 21.5(b). In this design, three dual-edge triggered J-K flip-flops are applied. The first two dual-edge triggered J-K flip-flops are clocked by the CP, and the last dual-edge triggered J-K flip-flop is clocked by the inversion of Q_2. It can be seen from the figure that the inputs of J and K of the first dual-edge triggered J-K flip-flop are logic 1 and the output Q_1 will change state for every CP's rising and falling edges. Then the CP is delayed the same number of clock cycles as the first dual-edge triggered J-K flip-flop

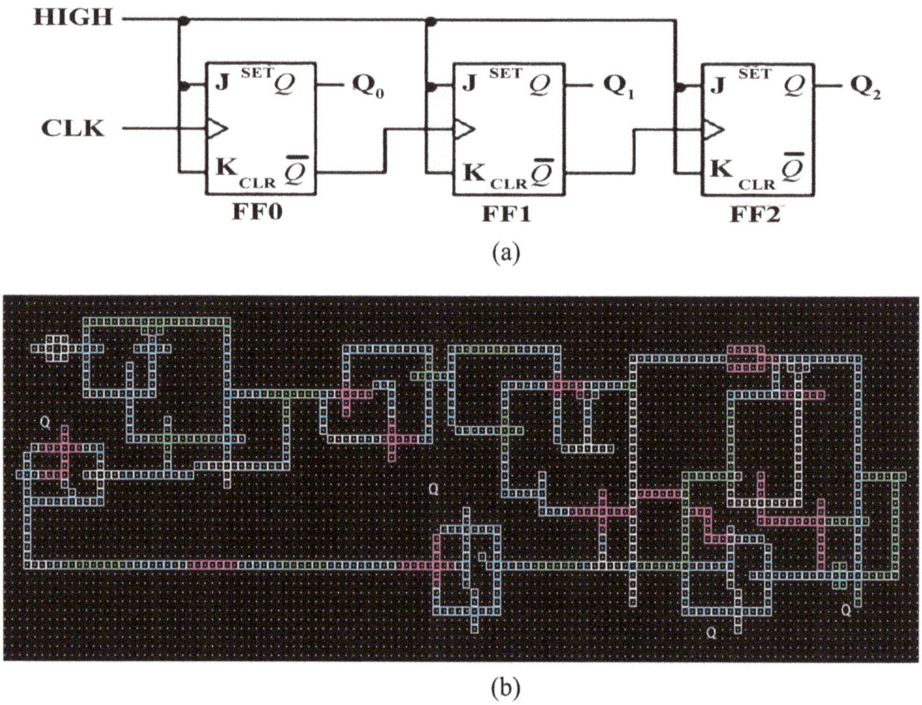

Figure 21.5. The QCA three-bit asynchronous backward counter. (a) Block diagram. (b) QCA layout.

is delayed. Then the delayed CP and Q_1 are simultaneously imported to the second dual-edge triggered J-K flip. The Q_2 will change its state depending on the Q_1 value. When $Q_1 = 0$, Q_2 will change its state, and when $Q_1 = 1$, Q_2 will hold its state. The mechanism of the last dual-edge triggered J-K flip-flop is the same as the previous flip-flops. The only difference is that it is driven by Q_2. Similarly, a QCA encoder with multi-feedback is presented in figure 21.6.

21.2 The synchronous counter

The synchronous counter circuit can be constructed using any types of flip-flops. In QCA technology, synchronous counters have been implemented based on the J-K-type flip-flop (JK-FF) and T-type flip-flop (T-FF). However, to date there has been no implementation of synchronous counters using the D-FF (D-type flip-flop) in QCA technology. Here, efficient QCA synchronous counters are designed based on D-FFs. A QCA falling-edge triggered cascaded design is discussed in this section which employs well-optimized level-sensitive D-FFs in parallel with an 'edge-to-level' converter to realize synchronous counters with different bit sizes.

An innovative QCA design of a level-sensitive D-FF is explained here with the goal of building a high performance constructive model for implementing different sequential circuits in QCA technology, in particular QCA counters. The D-FF is a memory element with two inputs (*D* and CLK) and an output (*Q*) whose graphic symbol is shown in figure 21.7(a). It implements the following functionality:

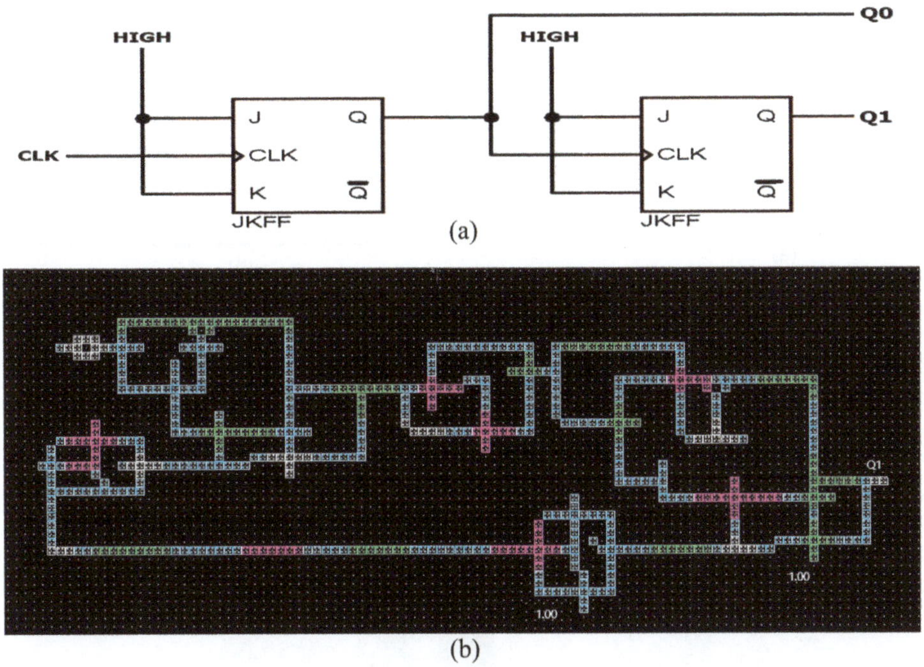

Figure 21.6. The QCA encoder with multi-feedback. (a) Block diagram. (b) QCA layout.

transparent (Q follows input D) and hold (Q remains unchanged). The logic function of the level-sensitive D-FF design can be expressed using equation (21.3).

The operation of this D-FF is shown in table 21.2. According to this table when the clock (CLK) signal is activated by 1, the value of data input (D) is stored in the output (Q) and when the CLK signal is deactivated by 0, the output is not changed. Due to the fact that majority gates are key components in designing QCA circuits, the corresponding design equation can also be represented based on majority voter gates as shown in equation (21.4).

$$Q_t = \text{CLK}.\ D + \overline{\text{CLK}}.\ Q_{t-1} \tag{21.3}$$

$$Q_t = \text{Maj}\big(\text{Maj}(\text{CLK},\ D,\ X0'),\ \text{Maj}(\overline{\text{CLK}},\ Q_{t-1},\ X0')\big). \tag{21.4}$$

The corresponding schematic diagram of the level-sensitive D-FF is illustrated in figure 21.7(b). According to this figure, the design requires three majority gates connected together in two successive gate levels. Figure 21.7(c) depicts the QCA layout of a D-FF with regular clock zones. It is implemented using the 4 × 4 USE grid with square dimensions of 5 × 5 QCA cells. As shown in figure 21.7(c), when the CLK signal is in 'high', the input bit D is transferred into the output Q and stored in the closed loop. On the other hand, when the CLK signal is in 'low', the stored value is preserved in the loop due to a QCA pipelined process. It consists of 74 cells with an area of 0.1 μm^2 and a latency of 1.5 QCA clocking cycles from input to output.

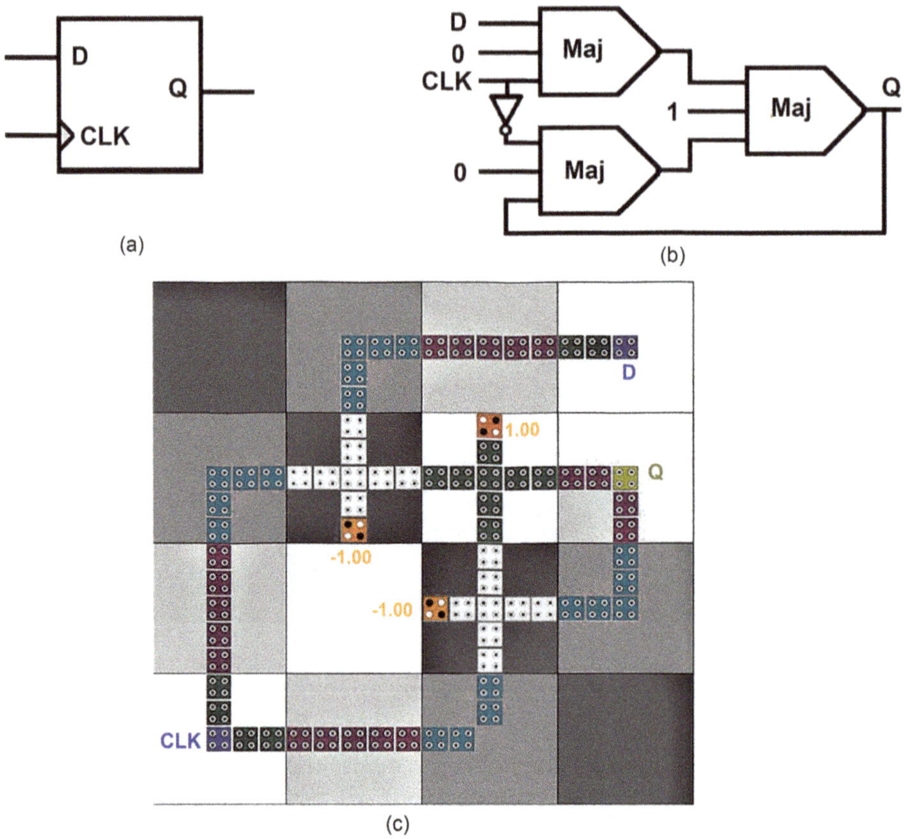

Figure 21.7. The QCA sensitive D-flip-flop. (a) Graphic symbol. (b) Schematic diagram. (c) QCA implementation. Reproduced with permission from [1]. Copyright 2017 Elsevier.

Table 21.2. The operation table of the level-sensitive D-FF.

$CLK_{in}(t-1)$	$CLK_{in}(t)$	CLK_{out}
0	0	0
0	1	0
1	0	1
1	1	0

21.2.1 QCA synchronous counters

In this subsection a well-optimized QCA design of an n-bit synchronous counter is described. The counter design comprises n D-FFs and other simple combinational circuits to generate 2^n ascending count states. The level-sensitive D-FF is susceptible to noise if the period of the positive level is long in the clock signal. In this case, the

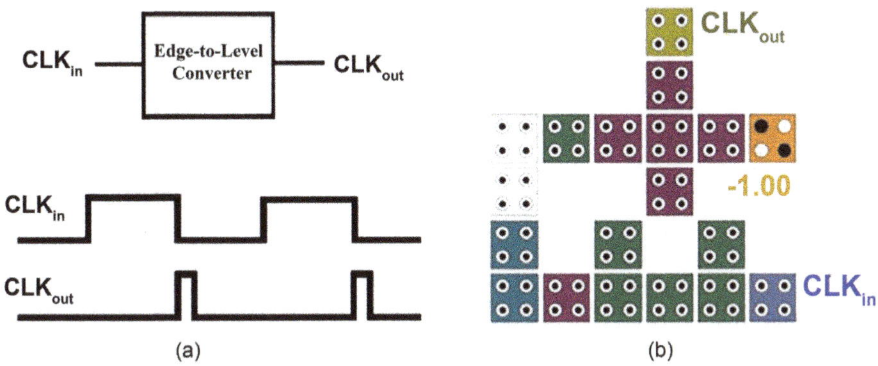

Figure 21.8. The falling-edge converter. (a) Signals. (b) QCA layout. Reproduced with permission from [1]. Copyright 2017 Elsevier.

Table 21.3. The operation table of the falling-edge converter.

CLK	D	Q_t	State
0	0	Q_{t-1}	Hold
0	1	Q_{t-1}	
1	0	0	Transparent
1	1	1	

output of the D-FF would have a race-round condition. In order to avoid this unstable phenomenon, an edge-to-level converter is essential to enable the edge detecting technique in the D-FF. The edge-to-level converter is used to create a narrow pulse with each falling transition of input signal. The input and output signals of a falling-edge converter are shown in figure 21.8(a). The edge-to-level converter is constructed in QCA technology and its layout is shown in figure 21.8(b). This converter utilizes the inherent capability of QCA clocking zones. Apparently, an AND logic operation between the inversion value of the current clock signal and its delayed version (by one QCA clock cycle) are used to generate a narrow pulse at each negative transition (high-to-low) of the input signal. This converter operates based on the operation table demonstrated in table 21.3.

The sensitive D-FF can be used to design a one-bit counter by attaching a falling-edge converter at the clock terminal, as shown in figure 21.8(a). We can see from figure 21.8(a) that a one-bit counter can be constructed using a simple and robust D-FF, feeding back the inversed output value directly to the input D at each falling-edge of the clock signal (CLK). Moreover, an AND gate is placed at the converter output to add the control mechanism to the counter through the count enable (CE) terminal. As a result, the output Q_0 changes to its opposite value on each CLK falling-edge when the CE signal is actively high. The corresponding QCA layout is explored in figure 21.9(b). The QCA layout is achieved by employing the sensitive

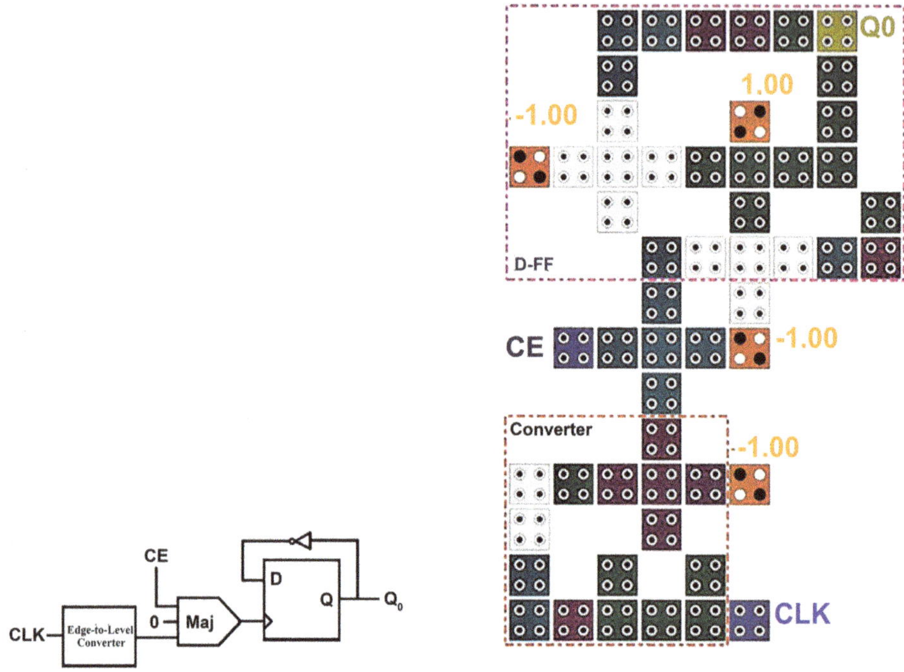

(a) (b)

Figure 21.9. The one-bit counter. (a) Block diagram. (b) QCA layout. Reproduced with permission from [1]. Copyright 2017 Elsevier.

D-FF and falling-edge converter as shown by the dashed rectangular boxes in the figure.

The one-bit counter in figure 21.9 can be expanded for building *n*-bit synchronous counters. As an example, a three-bit synchronous counter is designed, as shown in figure 21.10, by employing the same building block that is depicted by a solid red rectangular box. As illustrated in figure 21.10(a), this counter is implemented by cascading three level-sensitive D-FFs. It is worth mentioning that to create a QCA falling-edge design, an edge-to-level converter is connected to the clock terminal of each D-FF. Count enable signals are evaluated as $CE_0 = 1$,

$CE_1 = Q_0$ and $CE_2 = Q_0Q_1$. As a result, the three outputs of D-FFs ($Q_2Q_1Q_0$) count through 000001010011100101110111 (from decimal 07, returning 0 again). The corresponding efficient QCA layout is shown in figure 21.10(a). The QCA layout is achieved by employing three level-sensitive D-FFs that are simultaneously clocked by a falling-edge converter. The counter is a high-speed QCA structure where the input-to-output delay is only two QCA clocking cycles.

Generally, an *n*-bit synchronous counter can be implemented, as shown in figure 21.11, in a single-layer by involving a cascade structure of *n* level-sensitive D-FFs that are clocked in parallel using a falling-edge converter and controlled by a logical AND of the previous outputs.

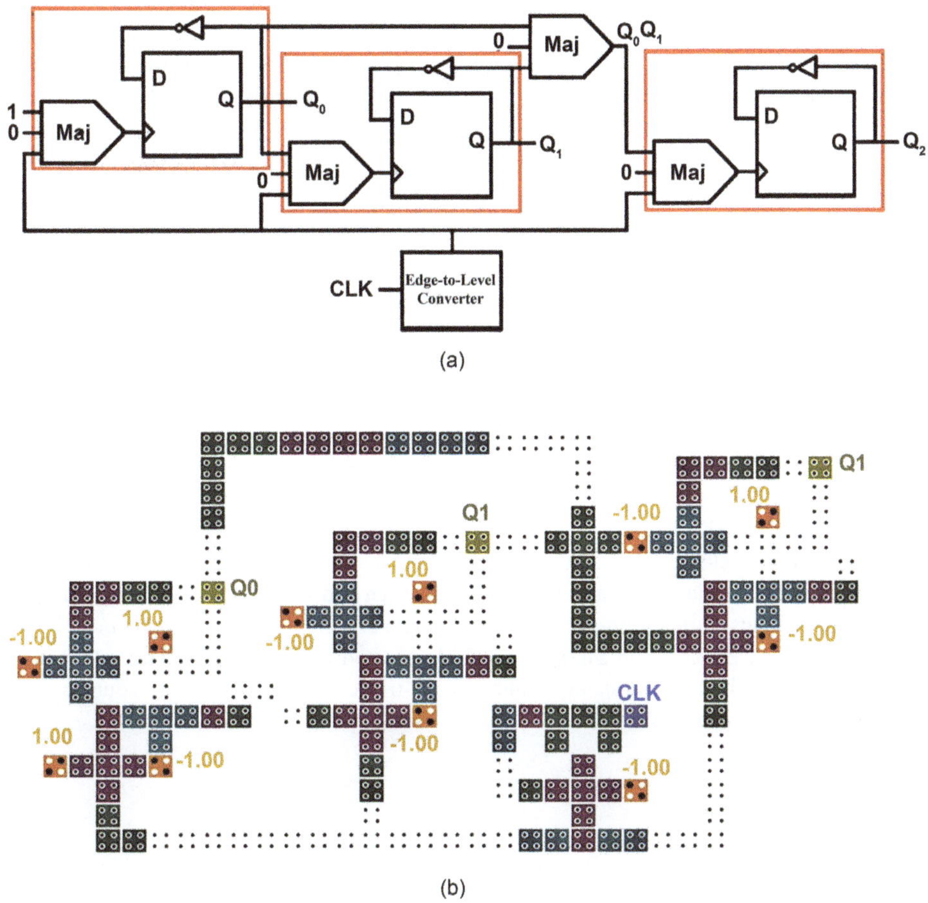

Figure 21.10. The three-bit counter. (a) Block diagram. (b) QCA layout. Reproduced with permission from [1]. Copyright 2017 Elsevier.

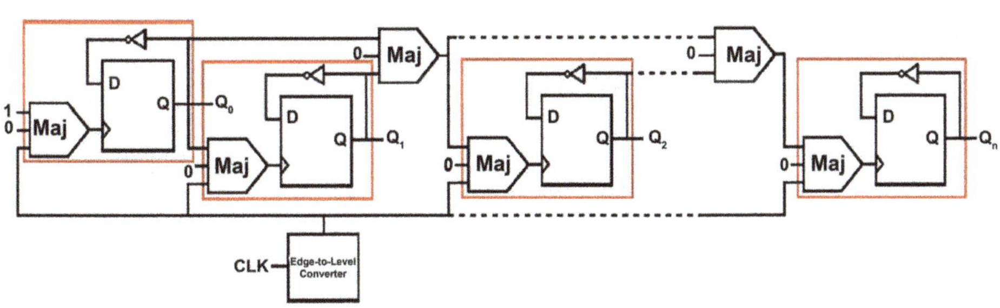

Figure 21.11. The *n*-bit synchronous counter.

21.3 Summary

In this chapter, asynchronous and synchronous counters are explained with block diagrams and QCA implementations. A three-bit asynchronous backward counter and two-bit counter circuits are implemented in QCA. For the synchronous counter, one-bit and three-bit counter circuits are designed with block diagrams and QCA implementations.

Further reading

[1] Abutaleb M M 2017 Robust and efficient quantum-dot cellular automata synchronous counters *Microelectron. J.* **61** 6–14

[2] Angizi S, Sayedsalehi S, Roohi A, Bagherzadeh N and Navi K 2015 Design and verification of new *n*-bit quantum-dot synchronous counters using majority function-based J-K flip-flops *J. Circuits Syst. Comput.* **24** 15501

[3] Nelson V P and Nagle H T 1995 *Digital Logic Circuit Analysis and Design* (Englewood Cliffs, NJ: Prentice Hall)

[4] Sheikhfaal S, Navi K, Angizi S and Navin A H 2015 Designing high speed sequential circuits by quantum-dot cellular automata: memory cell and counter study *Quantum Matter* **4** 190–7

[5] Wu C-B, Xie G-J, Xiang Y-L and Lv H-J 2014 Design and simulation of dual-edge triggered sequential circuits in quantum-dot cellular automata *J. Comput. Theor. Nanosci.* **11** 1620–6

[6] Yang X, Cai L, Zhao X and Zhang N 2010 Design and simulation of sequential circuits in quantum-dot cellular automata: falling edge-triggered flip-flop and counter study *Microelectron. J.* **41** 56–63

Chapter 22

The QCA decoder and encoder

In modern computing, the encoder and decoder are used to convert data from one form to another. The encoder and decoder are frequently used in communications systems, namely telecommunication, networking, and end-to-end transfer of data. They are also used for the easy transmission of data, which are replaced with codes and then transmitted. At the end of the receiver, the coded data are collected from the code and then processed for display. The encoder and decoder have a wide variety of application domains, such as the speed synchronization of multiple motors in industry, war zones, flying robots with night vision cameras, robotic vehicles with metal detectors, radiofrequency based home automation systems, and automatic health monitoring systems.

22.1 The QCA decoder

Ex-OR and decoder are the most valuable functions in the Boolean circuit family and have been of central concern in the QCA field for quite a while. One form of QCA Ex-OR gate is investigated in this section which will be utilized to design a decoder circuit. The truth table of the Ex-OR gate is given in table 22.1. The investigation covers the way that the general circuit can be streamlined if the significant yields as $\bar{A}B$ and $A\bar{B}$ can be adjusted. Along these lines, one single settled cell is utilized with polarization 1 in the larger part functions to play out the AND activities in both cases, i.e. $\bar{A}B$ and $A\bar{B}$. Figures 22.1(a) and 22.1(b) suggest this concept.

Two majority voters (MVs) generating $\bar{A}B$ and $A\bar{B}$ are ORed together to obtain the output of the Ex-OR gate. The three-input majority function implementation of an Ex-OR gate is

$$\begin{aligned} Y &= M(F1(\bar{A}, O, B), 1, F2(A, 0, \bar{B})) \\ &= M(\bar{A}B, 1, A\bar{B}) \\ &= \bar{A}B + A\bar{B}. \end{aligned} \tag{22.1}$$

doi:10.1088/978-0-7503-2747-3ch22

Table 22.1. The truth table of the Ex-OR logic.

Inputs		Output
A	B	$Y = A \oplus B$
0	0	0
0	1	1
1	0	1
1	1	0

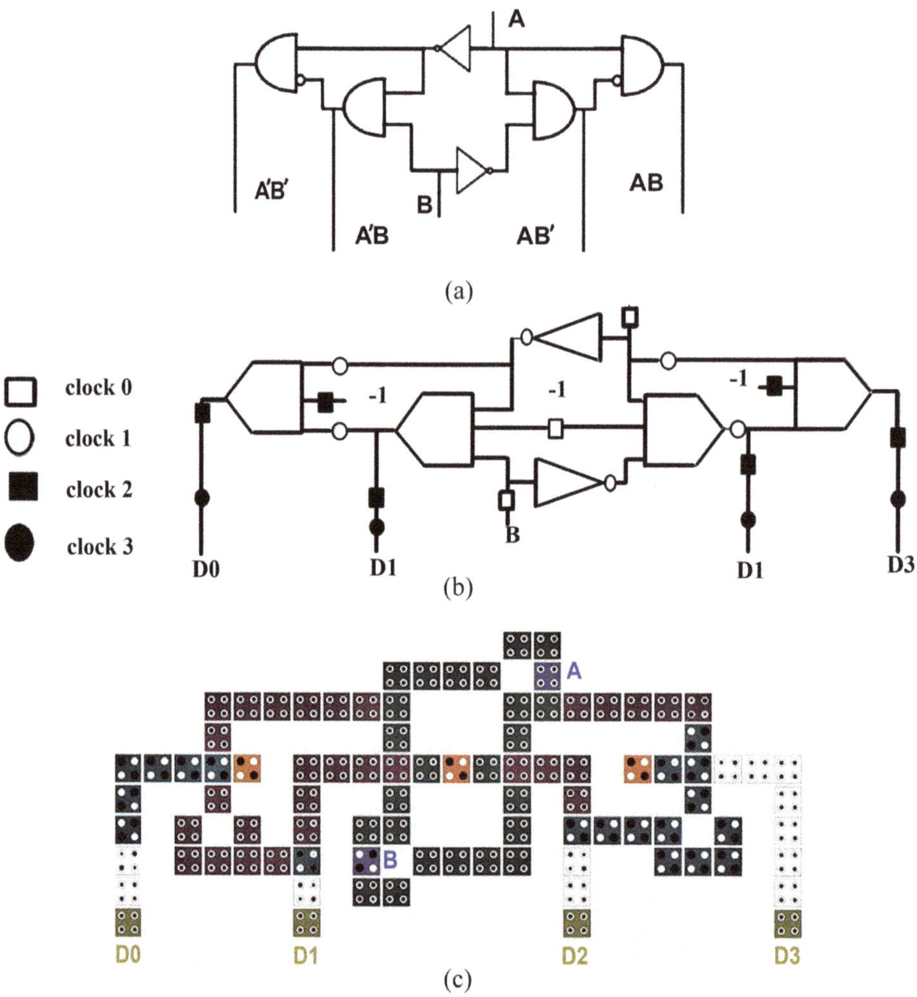

Figure 22.1. The QCA 2-to-4 decoder. (a) Schematic for the 2-to-4 decoder circuit. (b) The three-input majority gate implementation of the 2-to-4 QCA decoder. (c) QCA layout. Reproduced with permission from [2]. Copyright 2016 Elsevier.

22.1.1 The QCA 2-to-4 decoder

The Ex-OR gate is extended to design a 2-to-4 decoder. A 2-to-4 decoder consists of those two previous components as $\bar{A}B$ and $A\bar{B}$ along with two others, $\bar{A}B$ and $A\bar{B}$. The procedure for designing decoders results in a reduced cell count. The following equations implement the QCA decoder circuit.

A. Calculation for $\bar{A}\bar{B}$

$$\begin{aligned} P &= \overline{(\bar{A}B)} \cdot \bar{A} \\ &= (\bar{\bar{A}} + \bar{B}) \cdot \bar{A} \\ &= (A + \bar{B}) \cdot \bar{A} \\ &= A \cdot \bar{A} + \bar{A} \cdot \bar{B} \\ &= \bar{A} \cdot \bar{B}. \end{aligned} \tag{22.2}$$

B. Calculation for AB

$$\begin{aligned} P &= \overline{(\bar{B}A)} \cdot A \\ &= (\bar{\bar{B}} + \bar{A}) \cdot A \\ &= (B + \bar{A}) \cdot A \\ &= A \cdot \bar{A} + A \cdot B \\ &= A \cdot B. \end{aligned} \tag{22.3}$$

Table 22.2 and figure 22.2(a) show the truth table and the schematic diagram of the 2-to-4 decoder. An entire four clock zones or one clock cycle are required to synchronize the full circuit. In the plan the outputs of any majority gate act as the input to the next gate at the same clock sector. Thus the interval of the whole circuit has been condensed to a large degree.

The majority functions relevant to this implementation are given in equations (22.4) and (22.5). The majority gate representation and the QCA layout of the circuit design are given in figures 22.1(b) and 22.1(c), respectively.

Table 22.2. The truth table for the 2-to-4 decoder.

Inputs		Output			
A	B	$\bar{A} \cdot \bar{B}$	$\bar{A} \cdot B$	$A \cdot \bar{B}$	AB
0	0	1	0	0	0
0	1	0	1	0	0
1	0	0	0	1	0
1	1	0	0	0	1

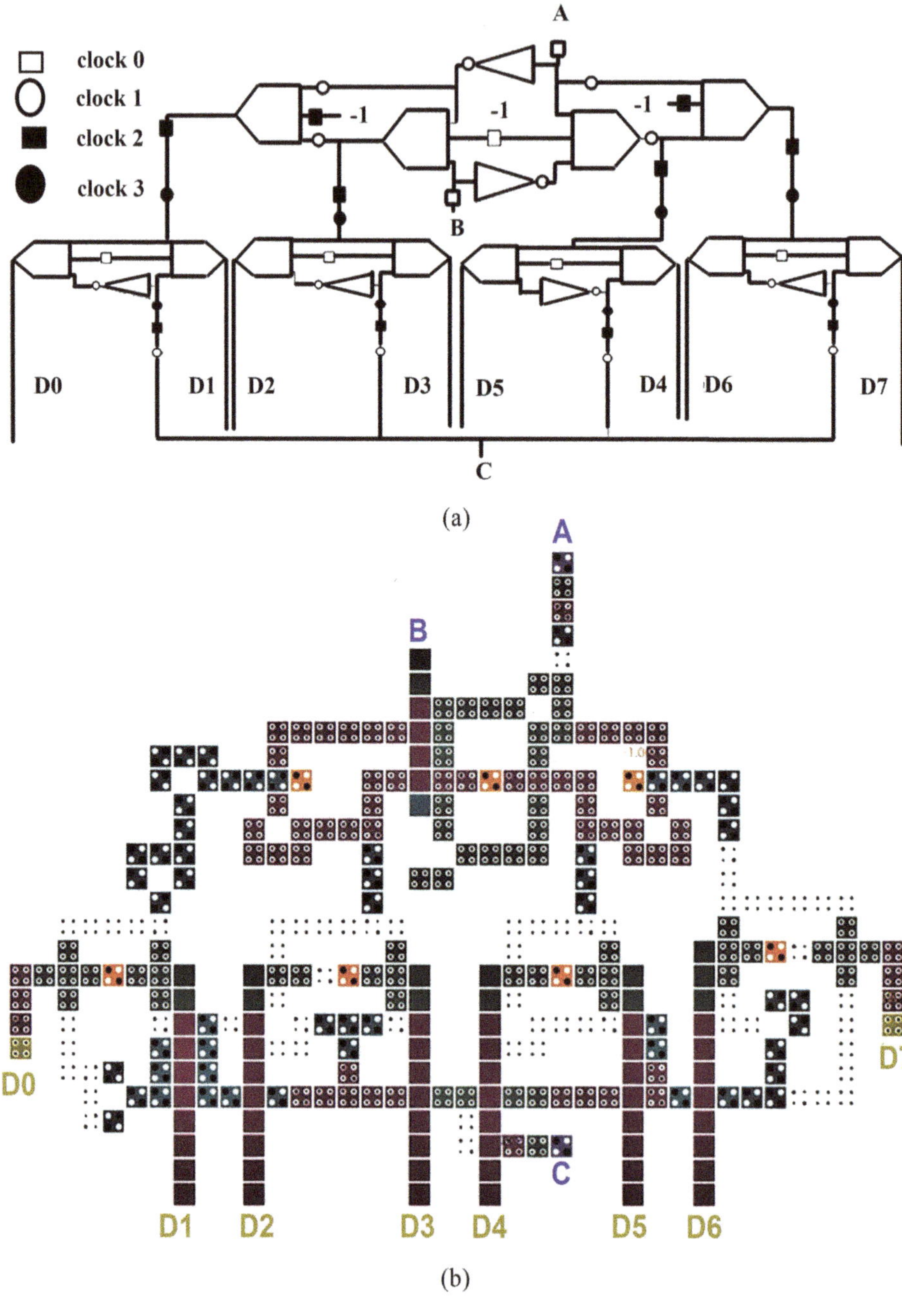

Figure 22.2. The 3-to-8 QCA decoder. (a) A three-input majority gate implementation of a 3-to-8 QCA decoder. (b) QCA layout. Reproduced with permission from [2].Copyright 2016 Elsevier.

$$\begin{aligned} R &= F(\bar{A}, 0, \overline{(\bar{A}B)}) \\ &= \bar{A} \cdot (A + B) \\ &= \bar{A}\bar{B}. \end{aligned} \tag{22.4}$$

$$\begin{aligned} S &= F(A, 0, \overline{(\bar{B}A)}) \\ &= A \cdot (A + \bar{B}) \\ &= AB. \end{aligned} \tag{22.5}$$

22.1.2 The QCA 3-to-8 decoder

The 2-to-4 decoder module is used to actualize the 3-to-8 decoder. Furthermore, the acquired parts from the 2-to-4 decoder modules are used to accomplish the yields of the 3-to-8 decoder. The third contribution as C has been brought into the non-rearranged and modified structures through QCA wires to perform AND tasks with each of the recently arrived segments to determine the comparing yields. Here the wire intersections are required to expand the data sources and yields of the circuit outside the circuit. Now, it adds up to eight clock zones (two clock cycles) which is maintained according to synchronization of the tasks. Comparative tasks are helpful in planning the 4-to-16 partner as well. For this situation and onwards crosses, it cannot be used. The related majority gate execution and the QCA design of the 3-to-8 decoder circuit are given in figures 22.2(a) and 22.2(b), respectively. Furthermore, the equivalent of the 4-to-16 decoder is given in figure 22.3, and the programmable logic array (PLA)-based decoder circuit is illustrated in figure 22.4.

22.2 The QCA encoder

The encoder generates a high-performance convolutional code, i.e. a turbo code, which closely approaches the channel capacity. The turbo code is a kind of error correction code (ECC) with a good performance of checking and correcting errors. It is widely used in mobile communications and satellite communications to protect information against distortion. It can also be applied to fault-tolerant computing. That is, the use of ECC leads to a scheme in which computation takes place in the encoded space, whereby errors are corrected locally, and encoding and decoding are only necessary at the beginning and end, respectively, of the computation.

22.2.1 The QCA turbo encoder design

The turbo encoder implements the convolution computation, feedback, and interleaver, as shown in figure 22.5.

1. *Convolution computation implementation.* In traditional circuits, the calculation components and registers of the resistor–capacitor (RC) encoder are isolated. The synchronization of the circuit is controlled by registers. However, in QCA the calculation components and interconnections generally present considerable delays that likewise assume a vital role in synchronization. The inborn shift registers and 'processing-in-wire' nature of QCA

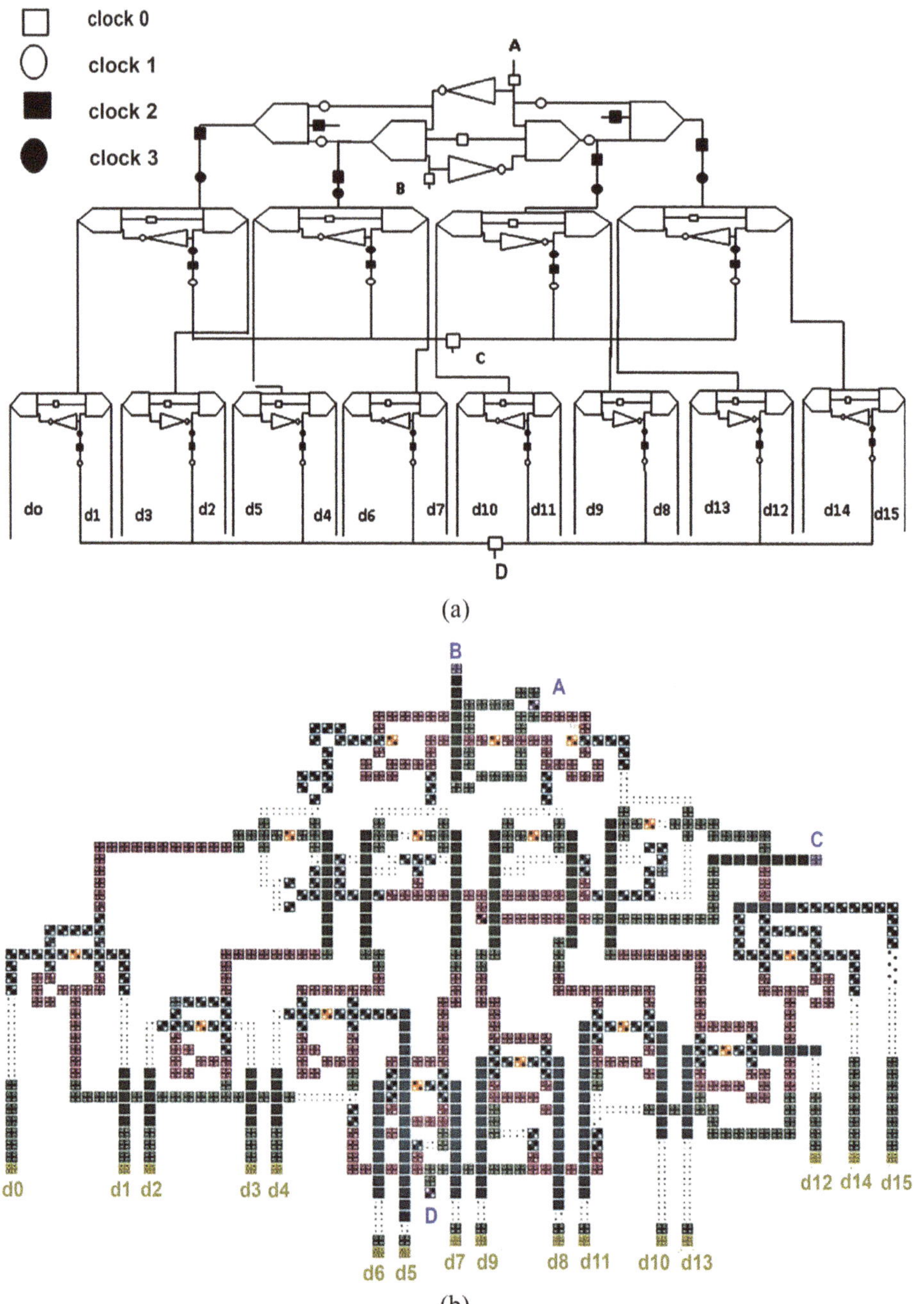

Figure 22.3. The QCA 4-to-16 decoder. (a) A three-input majority gate layout of the 4-to-16 QCA decoder. (b) QCA layout. Reproduced with permission from [2]. Copyright 2016 Elsevier.

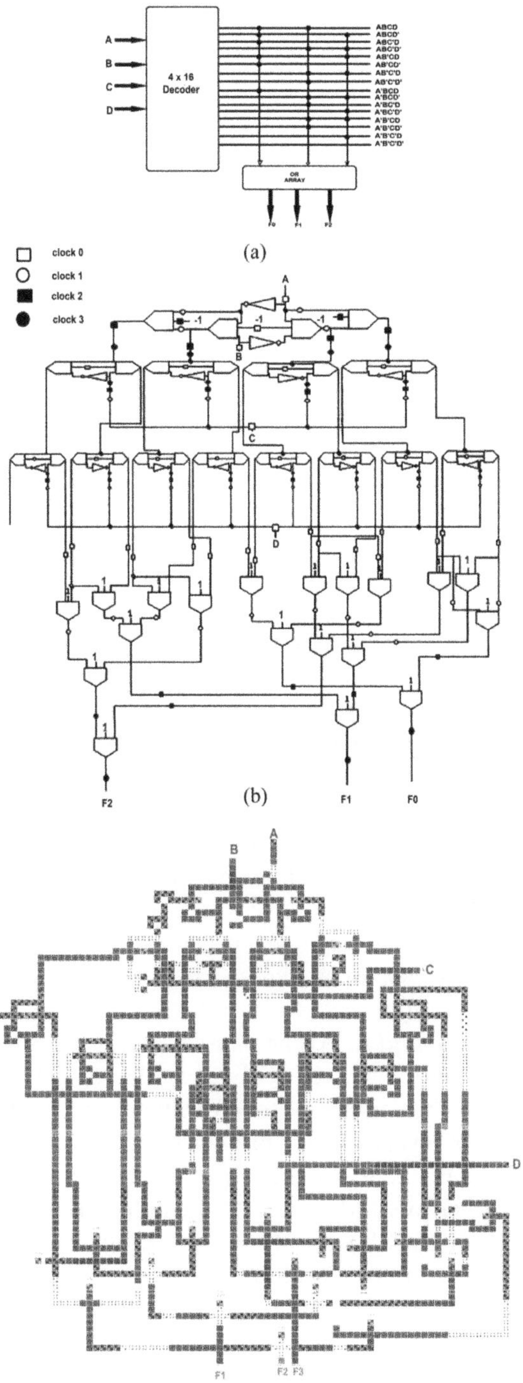

Figure 22.4. The QCA 4-to-16 PLA-based decoder. (a) Block diagram of the PLA-based adder. (b) Majority gate realization of the decoder-based PLA. (c) QCA layout. Reproduced with permission from [2]. Copyright 2016 Elsevier.

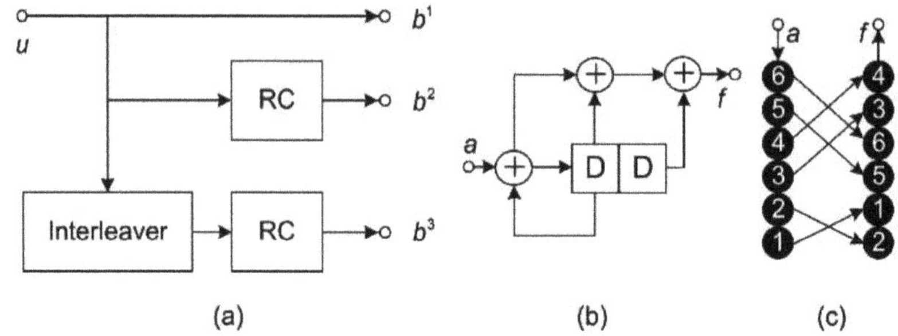

Figure 22.5. The turbo encoder: (a) block diagram of a turbo encoder, (b) block diagram of an RC encoder, and (c) model of an interleaver.

verifies that the RC encoder should be included in the calculation design with the register inserted.

2. *Feedback implementation.* Synchronization of feedback presents difficulty in executing an RC encoder. To meet this ongoing requirement, every novel information and comparing input bit should at the same time plot at the calculation unit. Consequently, the input must be implemented simultaneously. If this is the case, the numerous input circuits cannot offer sufficient delays to the current MUX based plan. Taking the case in figure 22.5(b), for example, the planning imperative of the feedback circle is one CLK delay ($1D = 1$ CLK delay). The present MUX cannot be utilized to make the information picking meaningful since it has expended something like one CLK delay. The delay expanded the number of circles by utilizing a time-scaling strategy. Thus there will be an extended interval between two nearby active sources of information. Even though this approach does not change the planning of a protest segment, it causes a chain reaction in the general framework. In this way, an latency sparing substitution of MUX is the center of feedback execution.
3. *Interleaver implementation.* At the parallel-to-serial converting stage the right shift-register of the interleaver is about to accept next group of transposed bits from the left shift-register. However, there are random states remaining in it due to the cyclic four-phase clocking mechanism. Therefore, the competition logjam between the random states and novel bits should be broken. A parallel-to-serial converter is employed to construct a serial shift/copy/shift-register (SCSR) structure for a programmable array of logic. It avoids competition by programming a suite of clock signals for the SCSR structure. However, this incurs high complexity of timing, in particular when many SCSR structures are used in a circuit. Hence, a simple parallel-to-serial converter is a major factor in achieving interleaver design.

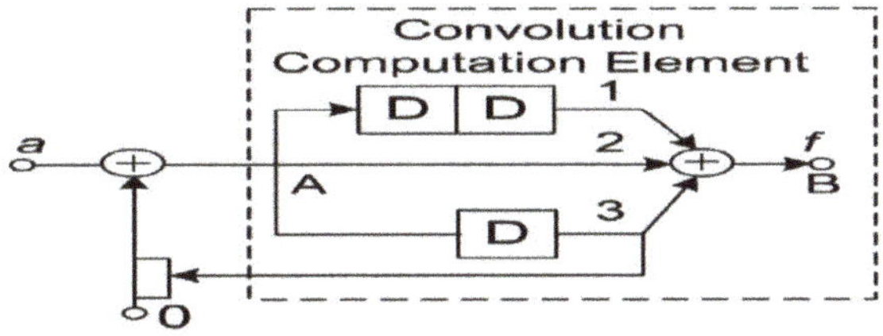

(a)

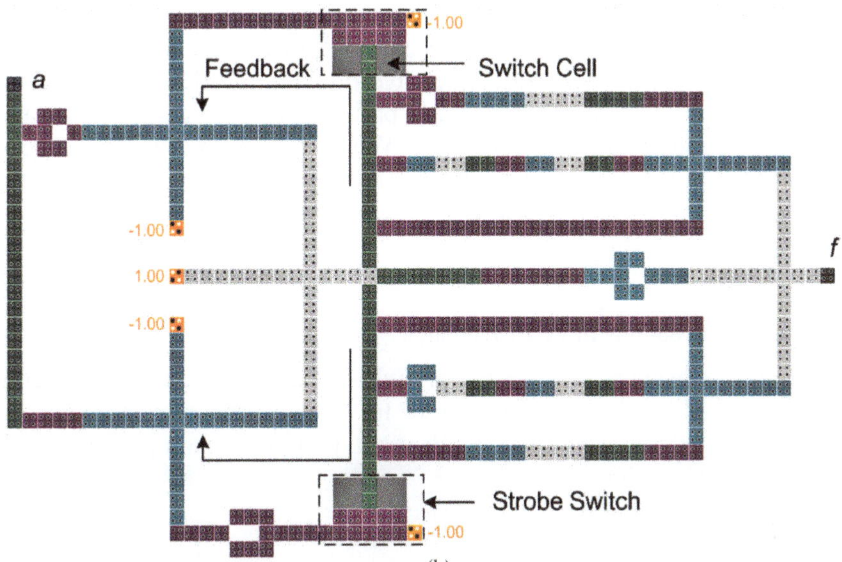

(b)

Figure 22.6. The QCA RC encoder with single-feedback. (a) Block diagram. (b) QCA layout. Reproduced with permission from [10]. Copyright 2015 IEEE.

22.2.2 The RC encoder with single-feedback

In this section, the design of the RC encoder in figure 22.6(a) is discussed. It is an extreme case of single-feedback as there is only one CLK delay for the loop. Therefore, if the design can cope with this case, it will be fit for other single-feedbacks.

Figure 22.6(a) demonstrates the changed block outline of the RC encoder in figure 22.5(b). It is linked by a convolution calculation element (set apart by a dashed square shape) and a solitary input circle. To perform the convolution calculation, we utilize a fan-out to offer parallel paths, rather than the sequential to parallel structure in an ordinary circuit. Every one of the paths is relegated with different quantities of delays. Thus every bit goes through them at different speeds. In this way the bits obtained at different times can meet at the calculation unit. Code words are obtained in progression.

Figure 22.6(b) demonstrates the QCA format of this RC encoder. The two arrows demonstrate the feedback paths that compare to the single-input in the square outline. The modulo-2 adder in the input circle works in the parallel frame rather than in the other plan since they expend more than one CLK, engendering delay. The strobe switches and switch cells are set apart by dashed square shapes and dark shadows, respectively.

In the feedback loop, the strobe switch is employed for data choosing. One end of the branch line is connected to a fixed 0 cell, while the main line is joined to the feedback path. The strobe switch alternates between 0 and the feedback signal. If there is no valid feedback signal, 0 is transmitted to the modulo-2 adder so as to protect the feed forward path from the interference caused by untimely feedback signals.

It is important to note that the delay in each path of the convolution computation element in figure 22.6(b) denotes the relative propagation delay from point A to point B. Given path 2 as the reference path, path 1 and path 3 consume two and one CLK delay more than path 2, respectively. When there are not sufficient delays to implement the three paths, the same number of delays can be supplemented to each path. This does not affect the final result except for increasing the corresponding delays.

22.2.3 The RC encoder with multi-feedback

The contrasted and single-feedback use of multi-inputs in QCA raises more complex time issues, caused by the interconnection. In addition, there are many cases that cannot be executed directly for a multi-input plot. For example, figure 22.7(a)

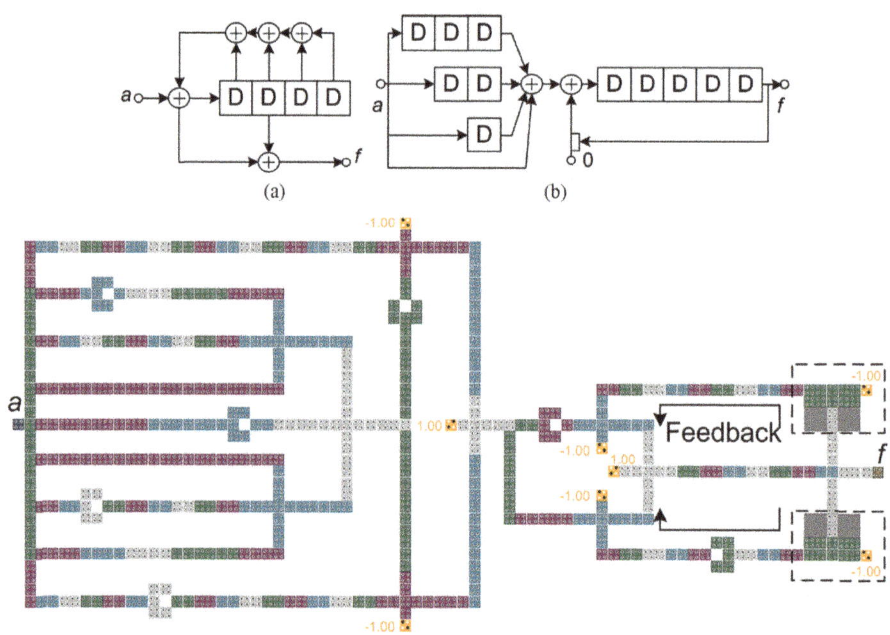

Figure 22.7. The QCA RC encoder with multi-feedback. (a) Classic block diagram. (b) Transformed block diagram. (c) QCA layout. Copyright 2015 IEEE. Reproduced with permission from [10].

demonstrates a square outline of a run-of-the-mill RC encoder with multi-feedback. Each register is associated with a feedback path. It is difficult to meet the constant prerequisite if the multi-input is directly built using QCA devices. To structure the RC encoder for multi-feedback, two questions arise: (i) can any multi-input be changed into an equivalent single-feedback or linked single-feedback and (i) how is this done? Taking the encoder in figure 22.7(a), for example, we can obtain the exchange function $H(D) = \frac{1+D^2}{1+D+D^2+D^3+D^4}$. The numerator and denominator of the transfer function indicate the feed forward and feedback, respectively. Obviously, the transfer function of the equivalent single-feedback must be of the type of $H(D) = \frac{f(D)}{1+D^n}$ and $n > 4$. Here, the single-feedback transformation is computed as

$$\begin{aligned} H(D) &= \frac{1 + D^2}{1 + D + D^2 + D^3 + D^4} \cdot \frac{(1 + D)}{(1 + D)} \\ &= \frac{1 + D + D^2 + D^3}{1 + D^5}. \end{aligned} \tag{22.6}$$

Thus the transformed block diagram and the corresponding layout are obtained, as shown in figures 22.7(b) and 22.7(c). In this way, a multi-feedback is transformed into the concatenation of a convolution computation element and a single-feedback loop. Similarly the transfer functions of the RC encoder with multi-feedback are denoted as $H(D) = \frac{1}{1+a_1D^1+a_2D^2+\cdots+a_nD^n}$, where $a_1, a_2, \cdots, a_n \in 0, 1$ and are not 0 at the same time.

Thus the first problem is equivalent to whether any transfer function $H(D) = \frac{1}{1+a_1D^1+a_2D^2+\cdots+a_nD^n}$ in the Galois field GF(16.6) can be transformed to the $H(D) = \frac{f(D)}{(1+D^p)}$ or $H(D) = \frac{f(D)}{(1+D^h)\cdots(1+D^l)}$ type.

Property 22.1. For any Galois field GF(q), q = prime power, x^{q^n} = product of all monic and irreducible polynomials over GF(q), whose degree divides n.

Since the above polynomials are Boolean arithmetic, they are in the field GF(2). In GF(2), x is a monic and irreducible polynomial of degree 1. Therefore, when $q = 2$, a similar property appears.

Property 22.2. For Galois field GF(2), $x^{2^n-1} - 1$ = product of all monic and irreducible polynomials (except the polynomial x) over GF(2), whose degree divides n. In GF(2), modulo-2 '–' is equal to modulo-2 '+'. Thus $x^{2^n-11} - 1$ is really equal to $x^{2^n-1} + 1$. Except for the polynomial x, the other monic and irreducible polynomials are all in the form of $1 + a_1x^1 + a_2x^2 + \cdots + a_nx^n$. Moreover, any product of these polynomials is still in this form in GF(2). Therefore, this theorem can settle the problem of multi-feedback discussed above.

According to the properties, a multi-feedback can be changed into a single-feedback when the denominator polynomial of its transfer function is irreducible. Otherwise, it can be transformed into concatenated single-feedbacks when the denominator polynomial is reducible. In the second case, the reducible polynomial should first be factorized into the product of irreducible polynomials. Then the multi-feedback can be further transformed into single-feedbacks according to the first case. The issues of the irreducibility test and the factorization of the polynomial in GF(2) are common mathematics issues and can be settled using calculators such as Magma. The relevant procedures are not discussed in detail here. To sum up, any multi-feedback of an RC encoder can be changed into the single-feedback type.

22.3 Summary

In this chapter, the decoder and encoder circuits are described using quantum dot cellular automata. 3-to-8 and 4-to-16 QCA decoder circuits are considered and presented explicitly with functional descriptions. For the encoder circuit, a QCA turbo encoder, and an RC encoder with single- and multi-feedback are described with block figures and QCA layouts.

Further reading

[1] WatElectronics 2019 Different types of encoder and decoder and its uses *WatElectronics.com* https://www.efxkits.us/different-types-encoder-decoder-applications/ (Accessed: 25 December 2018)

[2] De D, Purkayastha T and Chattopadhyay T 2016 Design of QCA based programmable logic array using decoder *Microelectron. J.* **55** 92–107

[3] Gladshtein M 2011 Quantum-dot cellular automata serial decimal adder *IEEE Trans. Nanotechnol.* **10** 1377–82

[4] Jagarlamudi H S, Saha M and Jagarlamudi P K 2011 Quantum dot cellular automata based effective design of combinational and sequential logical structures *World Acad. Sci. Eng. Technol., Int. J. Comput. Elect. Automat. Control. Inform. Eng.* **5** 1529–33

[5] Kianpour M and Sabbaghi-Nadooshan R 2012 A conventional design for CLB implementation of a FPGA in quantum-dot cellular automata (QCA) *Proc. of the 2012 IEEE/ACM Int. Symp. on Nanoscale Architectures* pp 36–42

[6] Kianpour M and Sabbaghi-Nadooshan R 2014 A novel quantum-dot cellular automata CLB of FPGA *J. Comput. Electron.* **13** 709–25

[7] Niemier M T and Kogge P M 1999 Logic in wire: using quantum dots to implement a microprocessor *ICECS'99. Proc. of ICECS'99. 6th IEEE Int. Conf. on Electronics, Circuits and Systems (Cat. No.99EX357)* vol 3 pp 1211–5

[8] Thompson R M 2011 A look at the special number field sieve using ideals as implementable in the magma computational algebra software system *Doctoral Dissertation*

[9] Vetteth A, Walus K, Dimitrov V S and Jullien G A 2002 Quantum dot cellular automata carry-look-ahead adder and barrel shifter *IEEE Emerging Telecommunications Technologies Conf.* pp 2–4

[10] Zhang M, Cai L, Yang X, Cui H and Feng C 2015 Design and simulation of turbo encoder in quantum-dot cellular automata *IEEE Trans. Nanotechnol.* **14** 820–8

Chapter 23

The QCA multiplexer and demultiplexer

The multiplexer (MUX) is the most common combinational circuit utilized in digital logic systems and it is an exceptionally valuable electronic circuit that is utilized in numerous applications, for example, signal directing, information interchange, and data transport control applications. An efficient quantum-dot cellular automata (QCA) MUX/demultiplexer (DEMUX) brings the importance of the current research networks. Multiplexers are utilized in different fields where various types of information need to be transmitted utilizing a solitary line, as the multiplexer allows the transmission of different kinds of information utilizing a solitary transmission line. It can be utilized to activate an immense amount of memory into the PC simultaneously.

23.1 The QCA 2-to-1 multiplexer

In this section, a new approach is described to implement the 2-to-1 multiplexer. Any Boolean function can be implemented using a MUX. It has 2^n inputs, one output, and $n = 1$ select line, which transfers one input to the output based on the value of the select lines. In addition, the MUX has been constructed using the rotating majority gate (RMG) which presents less complexity and enables powerful techniques in terms of implementing logic.

A schematic representation of the 2-to-1 MUX is shown in figure 23.1(a). It has two data inputs I_0 and I_1, one select line S_0, and one output (MUX). The block diagram is shown in figure 23.1(b). The output logic expression for the 2-to-1 MUX is

$$\text{Output} = I_0\overline{S_0} + I_1S_0. \tag{23.1}$$

From equation (23.1) it is clear that two AND gates, one OR gate, and an inverter are required to design the 2-to-1 MUX. The QCA equivalent block diagram is shown in figure 23.1(b). The layout of the 2-to-1 MUX design is presented in

doi:10.1088/978-0-7503-2747-3ch23

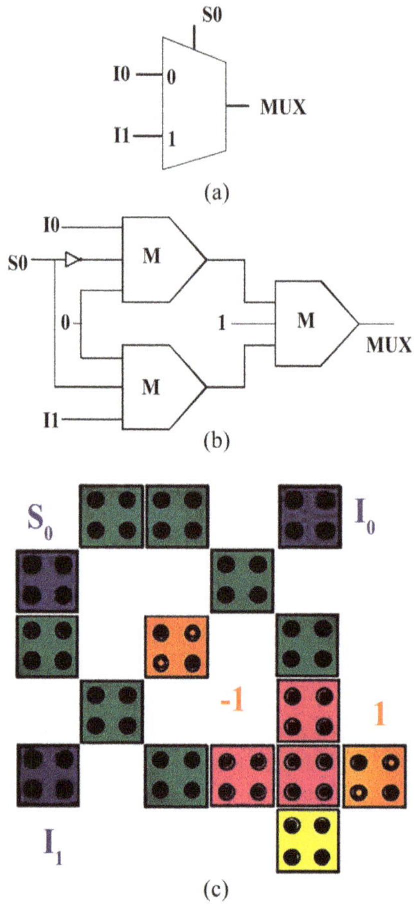

Figure 23.1. The 2-to-1 multiplexer. (a) Schematic symbol. (b) Majority logic diagram. (c) The QCA layout of the 2-to-1 MUX.

figure 23.1(c). The output (MUX) generates precise, highly polarized signals (shown inside the rectangles) leading to a high drivability for the circuit.

The implemented MUX has an area of 0.01 μm^2, a circuit complexity of 16 cells, and a latency of 0.5 clock delays. The QCA majority output logic expressions of the 2-to-1 MUX is

$$\text{Output} = M_j(M_j(I_0, \overline{S}_0, 0), M_j(I_1, S_0, 0), 1). \tag{23.2}$$

23.2 The QCA 4-to-1 multiplexer

A diagram of the 4-to-1 MUX is shown in figure 23.2(a). It is created using the concatenation of three 2-to-1 MUXs. The first two MUXs have a shared control signal link represented by S_0. The yields of first two MUXs are fed to the third

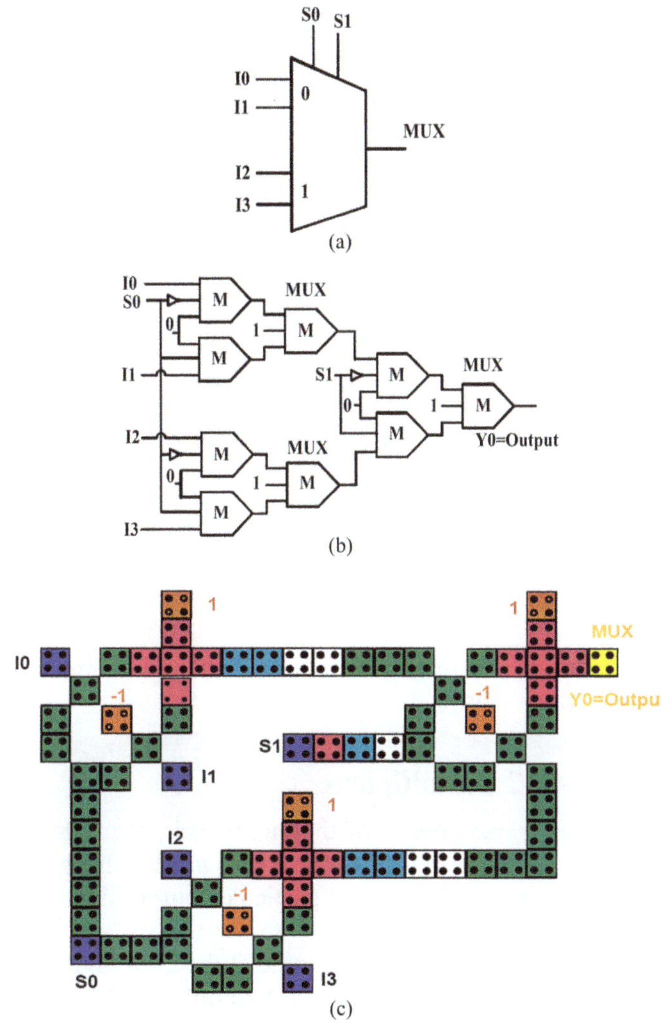

Figure 23.2. The 4-to-1 multiplexer. (a) Schematic symbol. (b) Majority logic diagram. (c) The QCA layout for the 4-to-1 MUX.

MUX, acting as an inputs where the regulator is provided by S_1. It has four data input lines I_0, I_1, I_2, and I_3 and a single output (Y_0). The block illustration of a 4-to-1 MUX is shown in figure 23.2(b). The output logic for the 4-to-1 MUX functions as

$$\begin{aligned} \text{Output} = {} & M_j(M_j(M_j(M_j(I_0, \overline{S_0}, 0), M_j(I_1, S_0, 0), 1), \overline{S_1}), \\ & \times M_j(M_j(M_j(I_2, \overline{S_0}, 0), M_j(I - 3, S_0, 0), 1), S_1)). \end{aligned} \tag{23.3}$$

From equation (23.3), it is clear that six AND gates, four OR gates, and three inverters are required to design a 4-to-1 MUX. The QCA layout of the design is

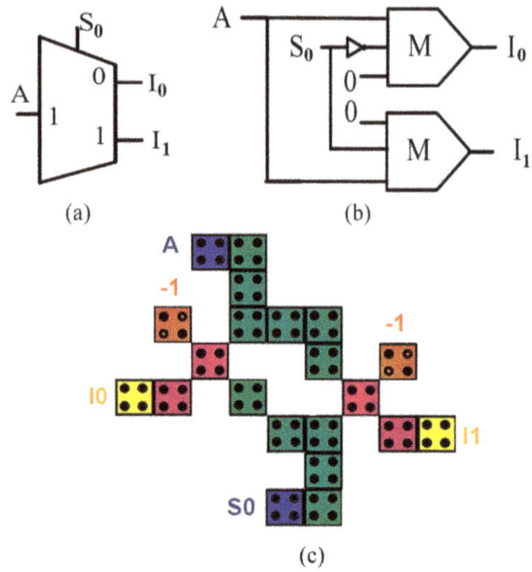

Figure 23.3. The 1-to-2 demultiplexer. (a) Schematic symbol. (b) Majority logic diagram. (c) QCA layout. Reproduced with permission from [1]. Copyright 2018 Elsevier.

shown in figure 23.2(c). It has an area of 0.11 μm^2, a circuit complexity of 79 cells, and a latency of 1.5 clock delays.

23.3 The QCA 1-to-2 demultiplexer

The DEMUX is a the reverse process of the multiplexer. It produces parallel lines from a serial data input line. It has 2^n outputs and n select lines which transfer a single data input line corresponding to the particular output data line based on the value of select lines.

A schematic representation and the truth table of the 2-to-1 DEMUX is shown in figure 23.3(a). It has a single data input line A, two outputs (I_0, I_1), and a select line S_0. The QCA equivalent block diagram is shown in figure 23.3(b). The output logic expression for the 2-to-1 MUX is

$$I_0 = A\overline{S_0} \tag{23.4}$$

$$I_1 = AS_0. \tag{23.5}$$

From equations (23.4) and (23.5), it is understandable that two AND gates and an inverter are required to construct the 2-to-1 DEMUX. The QCA layout of the 2-to-1 DEMUX design is shown in figure 23.3(c). The DEMUX has an area of 0.03 μm^2, a circuit complexity of 21 cells, and a latency of 0.5 clock delays. It is clear that when input $A = 1$ and control $S_0 = 0$, then $I_0 = 1$. Again, when input $A = 1$ and control $S_0 = 1$, then $I_1 = 1$. The QCA majority output logic expressions of the 2-to-1 DEMUX are

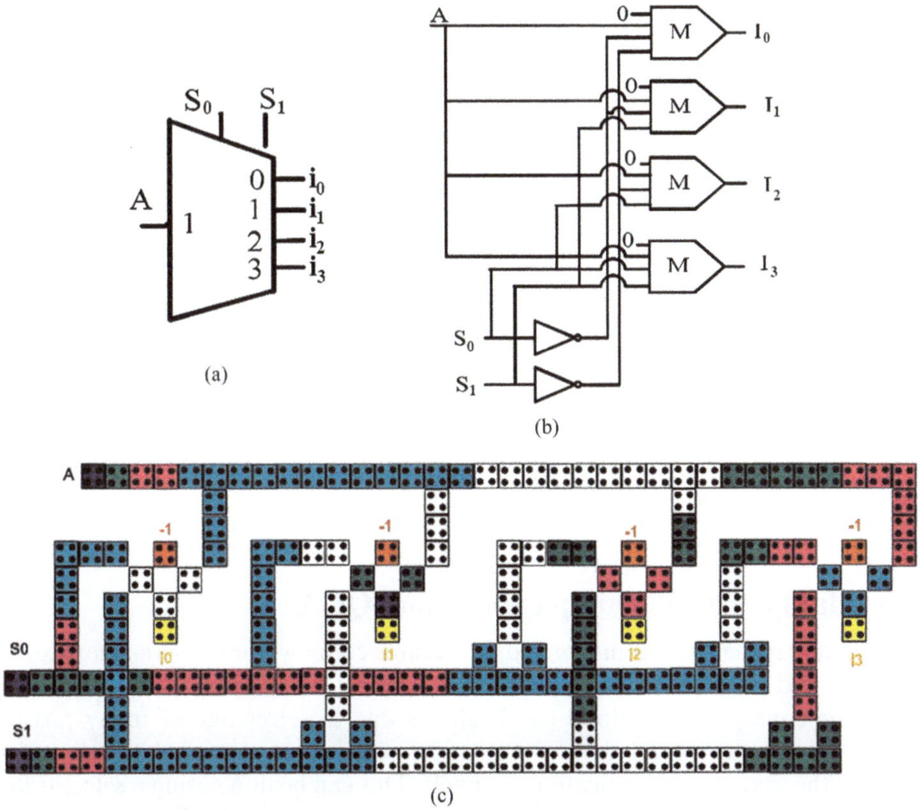

Figure 23.4. The 1-to-4 demultiplexer. (a) Schematic symbol. (b) Majority logic diagram. (c) QCA layout. Reproduced with permission from [1]. Copyright 2018 Elsevier.

$$I_0 = M_j(A,\ \overline{S_0},\ 0) \tag{23.6}$$

$$I_1 = M_j(A,\ S_0,\ 0). \tag{23.7}$$

23.4 The QCA 1-to-4 demultiplexer

The graphic illustration of the 1-to-4 DEMUX is shown in figure 23.4. It has a single data input line A, four outputs (I_0, I_1, I_2, I_3), and a select line S_0, S_1. The equivalent block diagram is shown in figure 23.4(b). The output logic expression for the 4-to-1 MUX is

$$I_0 = A\overline{S_0}.\ \overline{S_1} \tag{23.8}$$

$$I_1 = AS_0.\ \overline{S_1} \tag{23.9}$$

$$I_2 = A\overline{S_0}.\ S_1 \tag{23.10}$$

$$I_3 = AS_0 .\ S_1 . \tag{23.11}$$

From equations (23.8)–(23.11) it is clear that four three-input AND gates and two inverters are required to design the 1-to-4 DEMUX. The QCA equivalent block diagram is shown in figure 23.4(c). The DEMUX has an area of 0.18 μm^2, circuit complexity of 187 cells and latency of 1.75 clock delays.

The QCA majority output logic expression of the 1-to-4 DEMUX are

$$I_0 = M_j(A\overline{S_0}, \overline{S_1}, 0) \tag{23.12}$$

$$I_1 = M_j(AS_0, \overline{S_1}, 0) \tag{23.13}$$

$$I_2 = M_j(A\overline{S_0}, S_1, 0) \tag{23.14}$$

$$I_3 = M_j(AS_0, S_1, 0). \tag{23.15}$$

23.5 Multiplexing/demultiplexing using QCA

Multiplexing is the process in which data streams coming from different sources are combined and transmitted over a single data channel. In practice, a MUX is used to realize the concept of multiplexing. It allows one to select one of many possible sources. There are several data inputs and one of them is routed to the output (possibly the shared communication channel). This can be done using a selector line. The selected inputs determine which data input gets through. The demultiplexer performs the reverse process of multiplexing and routes the separated signals to their corresponding receiver. It allows one to select one of many possible destinations. A DEMUX has one data input and several data outputs. The data output can be selected using a selector line. Both the MUX and DEMUX are combined into a single device which has the capability to process outgoing and incoming signal lines, as shown in figure 23.5(a).

The idea of the proposed method is to design a novel ultra-high speed data multiplexing/demultiplexing circuit for computer network systems. The schematic logic symbol of the 2-to-1 MUX/DEMUX is shown in figure 23.5(a). It consists of a MUX, which has two inputs (I_0, I_1) and a single selector line S_0. The number of selector lines is equal to [$\log_2 n$], where n is the number of inputs. Conversely, a DEMUX is a device that takes a single input signal line and selects one of many data output lines, which is connected to the single input line (possibly the shared channel). The majority logic diagram of the 2-to-1 MUX/DEMUX is shown in figure 23.5(b). In this case, a logic value of 0 would connect I_0 to the output channel; a logic value of 1 would connect I_1 to the output channel. The QCA layout of the 2-to-1 MUX/DEMUX is shown in figure 23.5(c). MUXs are used in computer memory to maintain a vast amount of memory in computers, and also to decrease the number of copper lines necessary to connect the memory to the other parts of a computer.

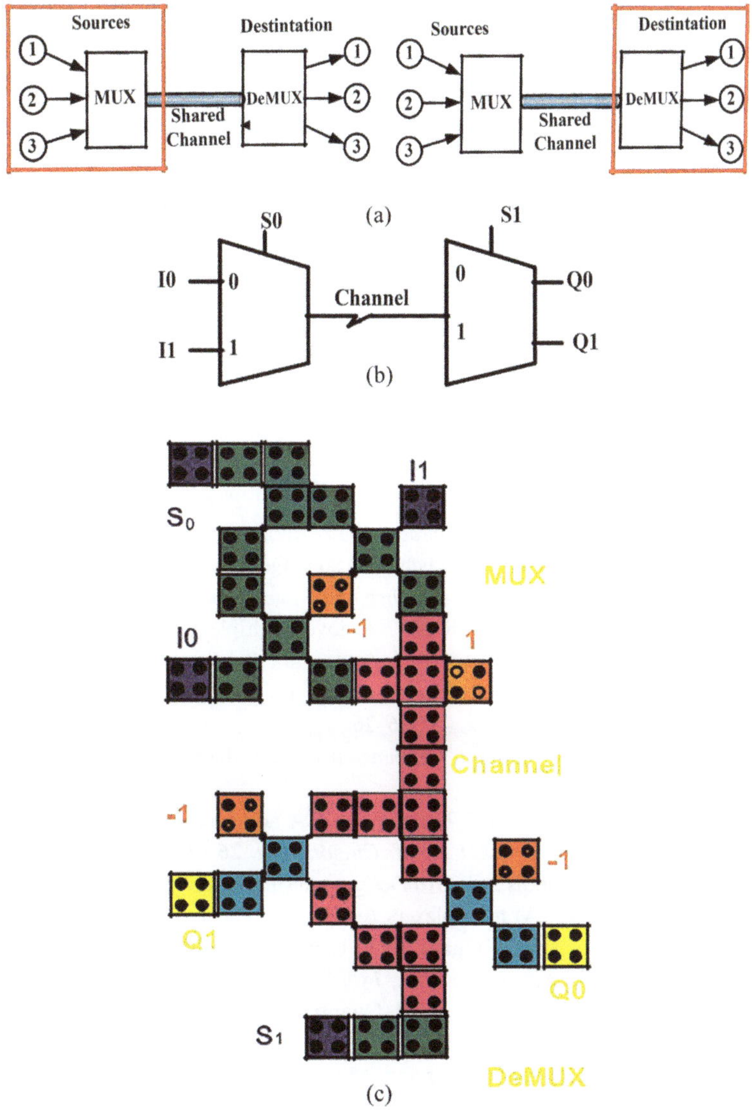

Figure 23.5. The 1-to-4 MUX/DEMUX using QCA. Reproduced with permission from [1]. Copyright 2018 Elsevier. (a) Schematic symbol. (b) Majority logic diagram. (c) QCA layout.

23.5.1 The effect of the selector line (S_0, S_1) on the 2-to-1 MUX/1-to-2 DEMUX

Here, S_0 and S_1 are the two selector lines at the transmitter and receiver ends of the MUX and DEMUX:

- If S_0 and S_1 are low, the top AND gate of the MUX and DEMUX is open and I_0 is copied to Q_0.

- If S_0 and S_1 is high, the bottom AND gate of the MUX and DEMUX is open and I_1 is copied to Q_1.
- Both the MUX and DEMUX are synchronized using the proper clocking signals.

23.6 Summary

This chapter presents an approach to deal with 2-to-1 MUX and 1-to-2 DEMUX circuits using a two-electron four-dot QCA. The 2-to-1 MUX has been used to execute the essential digital elements required for QCA. Any higher level MUX/DEMUX can be designed utilizing the 2-to-1 MUX and 1-to-2 DEMUX, respectively. In this way, utilizing the 2-to-1 MUX and 1-to-2 DEMUX, a 4-to-1 MUX and a 1-to-4 DEMUX are defined.

Further reading

[1] Ahmad F 2018 An optimal design of QCA based 2^n:1/1:2^n multiplexer/demultiplexer and its efficient digital logic realization *Microprocess. Microsyst.—Embedded Hardware Des.* **56** 64–75

[2] Amiri M A, Mahdavi M and Mirzakuchaki S 2008 QCA implementation of a MUX-based FPGA CLB *2008 Int. Conf. on Nanoscience and Nanotechnology* pp 141–4

[3] Asfestani M N and Heikalabad S R 2017 A novel multiplexer-based structure for random access memory cell in quantum-dot cellular automata *Physica* B **521** 162–7

[4] Askari M and Taghizadeh M 2011 Logic circuit design in nano-scale using quantum-dot cellular automata *Eur. J. Sci. Res.* **48** 516–26

[5] Das J C and De D 2016 Shannon's expansion theorem-based multiplexer synthesis using QCA *Nanomater. Energy* **5** 53–60

[6] Kim K, Wu K and Karri R 2007 The robust QCA adder designs using composable QCA building blocks *IEEE Trans. CAD Integr. Circuits Syst.* **26** 176–83

[7] Mardiris V A and Karafyllidis I G 2010 Design and simulation of modular 2^n to 1 quantum-dot cellular automata (QCA) multiplexers *Int. J. Circ. Theor. App.* **38** 771–85

[8] Momenzadeh M, Tahoori M B, Huang J and Lombard F 2004 Quantum cellular automata: new defects and faults for new devices *Proc. 18th Int. Parallel and Distributed Symp.* pp 207–14

[9] Mukhopadhyay D and Dutta P 2012 Quantum cellular automata based novel unit 2:1 multiplexer *Int. J. Com. App.* **43** 22–5

[10] Sabbaghi-Nadooshan R and Kianpour M 2014 A novel QCA implementation of MUX-based universal shift register *J. Comput. Electron.* **13** 198–210

[11] Tahoori M B, Momenzadeh M, Huang J and Lombardi F 2004 Defects and faults in quantum cellular automata at nano scale *Proc. 22nd IEEE VLSI Test Symp.* pp 291–6

[12] Teodosio T and Sousa L 2007 QCA-LG: a tool for the automatic layout generation of QCA combinational circuits *Norchip 2007* pp 1–5

IOP Publishing

Quantum Computing
A pathway to quantum logic design
Hafiz Md Hasan Babu

Chapter 24

The QCA RAM, ROM, and processor

In this chapter, the quantum-dot cellular automata (QCA) random access memory (RAM) cell, read only memory (ROM), and processor are discussed with graphical representations and QCA executions.

24.1 The RAM cell

Memory is one of the critical parts of any computerized framework. It is also of value to construct a fast and greatly upgraded QCA RAM.

Generally, there are two types of RAM cell designs in QCA, which are categorized as loop-based and line-based according to the operational manner of the QCA circuits. In the loop-based RAM cell structures, the storage mechanism is achieved using a loop that contains all four clocking zones. In contrast, line-based RAM cell structures use a QCA line to save the previous value of the output. There is an early effort to present a loop-based QCA RAM cell and its schematic diagram is illustrated in figure 24.1. The presented architecture is based on the D latch and a loop which is used to save the memory content. In the QCA implementation, it also takes two clocking cycles to transmit the input signal to the output.

However, a QCA random access memory without set and reset ability might be implemented. As is shown in figure 24.1(a), this design has a loop-based mechanism and is constructed based on an SR latch. When select = 1 and write/read = 0, the content of the RAM cell is not changed and can be read, but when select = 1 and write/read = 1 the new input will be transmitted to the output. In this structure a coplanar wire crossing method has been employed. Furthermore, this design has used only one cell in some of the clocking zones, which increases the noise sensitivity of the circuit, as shown in figure 24.1(a).

RAM cell engineering with set and reset capacity has been presented, and is shown in figure 24.1. This structure is made up of two 2-to-1 multiplexers. One of these multiplexers is used as the D latch by joining the yield flag to the main multiplexer's information input line. At the point when the select and compose/read signals are

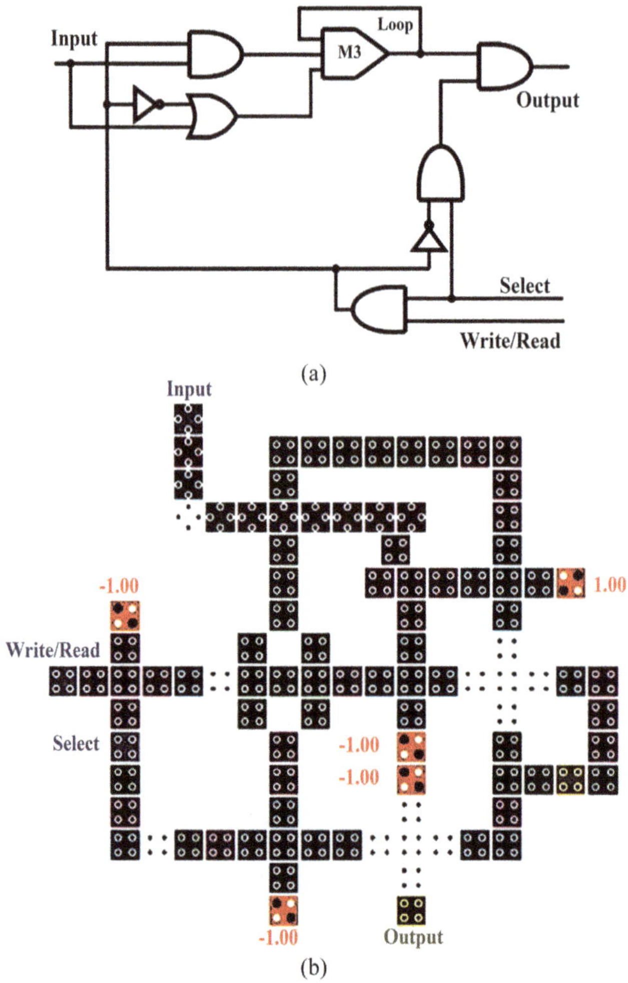

Figure 24.1. The RAM cell structure without set/reset. (a) Schematic symbol. (b) QCA implementation. Reproduced with permission from [1]. Copyright 2015 Elsevier.

equal to 1, new information will be embedded in the RAM cell, otherwise the reading task can be performed by setting the compose/read flag to 0. It utilizes a single multiplexed wire for set and reset tasks. This flag additionally decides the substance of the cell only when the select flag is deactivated by 0 and the compose/read flag is activated by 1. The inertness of this RAM circuit is 1.75 clock cycles for transmission of the info flag to yield the cell, as shown in figure 24.2(b). As is shown in figure 24.2, another design of the RAM cell is given an indistinguishable task from the principal configuration in figure 24.1(b). This structure is based on a D latch with a non-robust three-input majority gate structure, and the set and reset capacity of the RAM is deficient in this design.

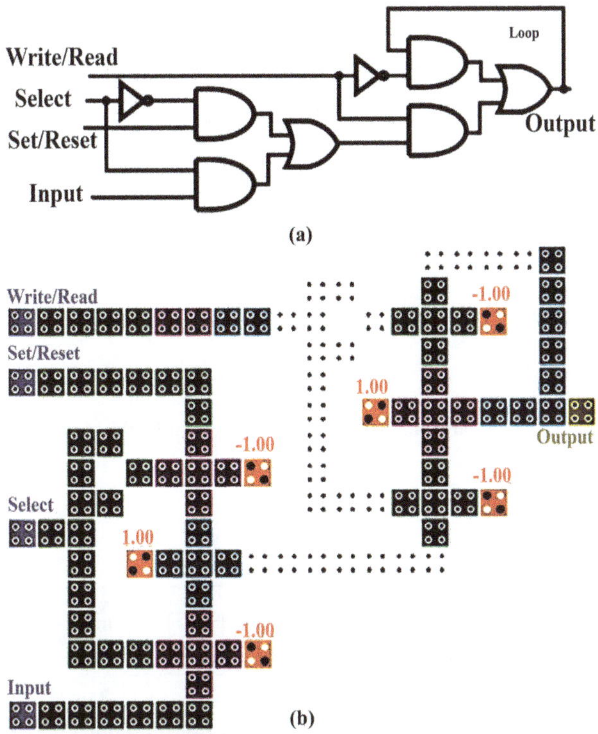

Figure 24.2. The RAM cell structure with set/reset. (a) Schematic symbol. (b) QCA implementation. Reproduced with permission from [1]. Copyright 2015 Elsevier.

24.2 The QCA ROM

Nanotechnology is contributing to the introduction of new digital media with higher capacity and performance, which allow digital data to be stored at the nano-scale. The QCA ROM is described with a simple structure. The simplified ROM can store 2.5 GBits of digital data per cm^2.

Typical ROMs consist of data bit cells, decoders, and tristate buffers. Each bit cell has a select input and an output line. When the select input line is active, binary information appears at the output line. Figure 24.3 shows the QCA structure of a ROM. The select input has been added to the cell using an AND operation. An array of bit cells can be selected based on the address given by a decoder.

The block diagram of a simple ROM is shown in figure 24.4. Each ROM consists of some bit cells. The block diagram of each bit cell and its corresponding QCA structure are shown in figure 24.5. Current semiconductor memories achieve random access by connecting the memory cells to the bit lines in parallel. These kinds of ROM cell outputs are connected to the bit lines, which behave similarly to wires or gates. As a result, when the bit cell is not selected, the output of that cell is 0 and when selected the output of the cell is 0 or 1 depending on the binary bit stored in the cell. Consequently, the bit value stored in the selected cell would come on to the output bit line of the ROM. This is a kind of serial OR that is very easy to implement

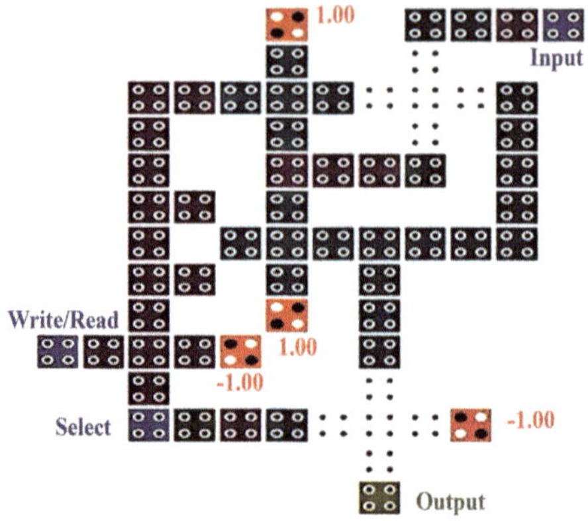

Figure 24.3. The QCA structure of a ROM.

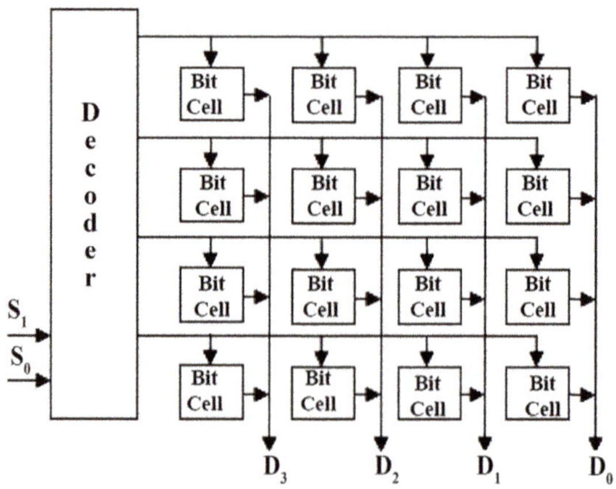

Figure 24.4. Block diagram of a typical 4 × 4 ROM. Reproduced with permission from [9]. Copyright 2008 IEEE.

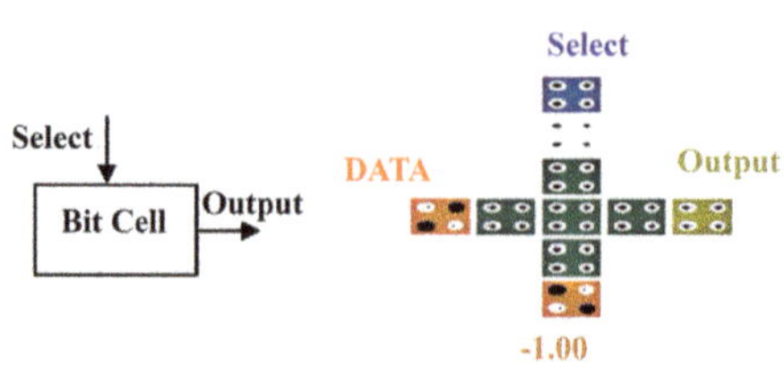

Figure 24.5. Block diagram of a ROM bit cell and its corresponding QCA structure. Reproduced with permission from [9]. Copyright 2008 IEEE.

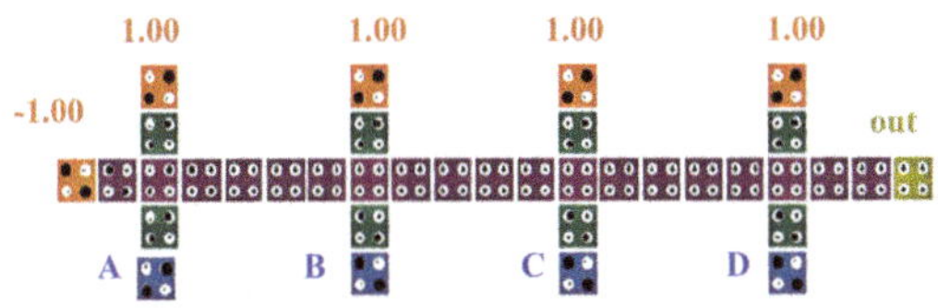

Figure 24.6. The QCA serial OR. Reproduced with permission from [9]. Copyright 2008 IEEE.

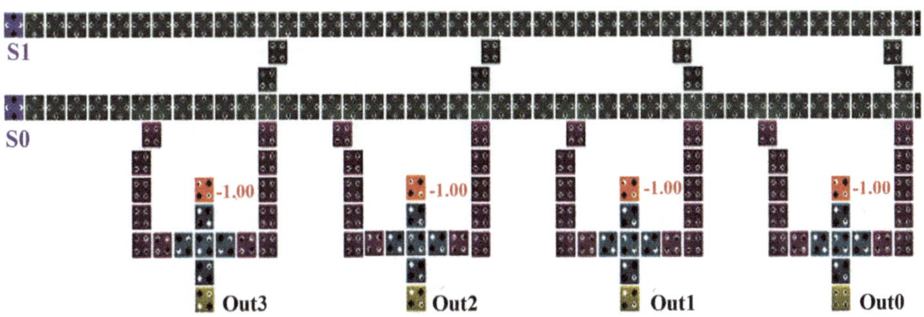

Figure 24.7. The QCA 2-to-4 decoder. Reproduced with permission from [9]. Copyright 2008 IEEE.

using QCA cells. Figure 24.6 shows how the serial OR can be carried out on multiple inputs with QCA technology. Figure 24.7 shows a 2-to-4 QCA decoder layout.

24.3 The QCA processor

A central processor unit (CPU), otherwise called a processor, is responsible for deciphering and executing program directions. As it is the principal part in PC frameworks, when another innovation emerges there is an effort to execute a processor design in it. The engineering of the one-bit subtract and branch if negative (SUBNEG) is executed utilizing carbon nanotube technology (CNT), which is one of the options for CMOS circuits. The latest processor executed in the QCA innovation depends on the accumulator design.

The SUBNEG architecture is used for designing the processor. As QCA are not based on transistors, some changes in the circuit are needed in order to make the SUBNEG viable for this technology. In figure 24.8 the combinational QCA processor design is presented. The circuit is composed of three main components: the instruction memory (IM), data memory (DM), and the ALU. The 'next instruction address' is a register that saves the known part of the next instruction address. For simplicity, we consider an architecture that contains a DM with 4-to-1-bit positions and an IM that holds four-bit to five-bit instructions.

QCA has the advantages of the inalienable pipelining innovation and registers not being required to store the state of the machine, since these data are stored by the clock zones. Nonetheless, this suggests the right state in the wire. Subsequently, we

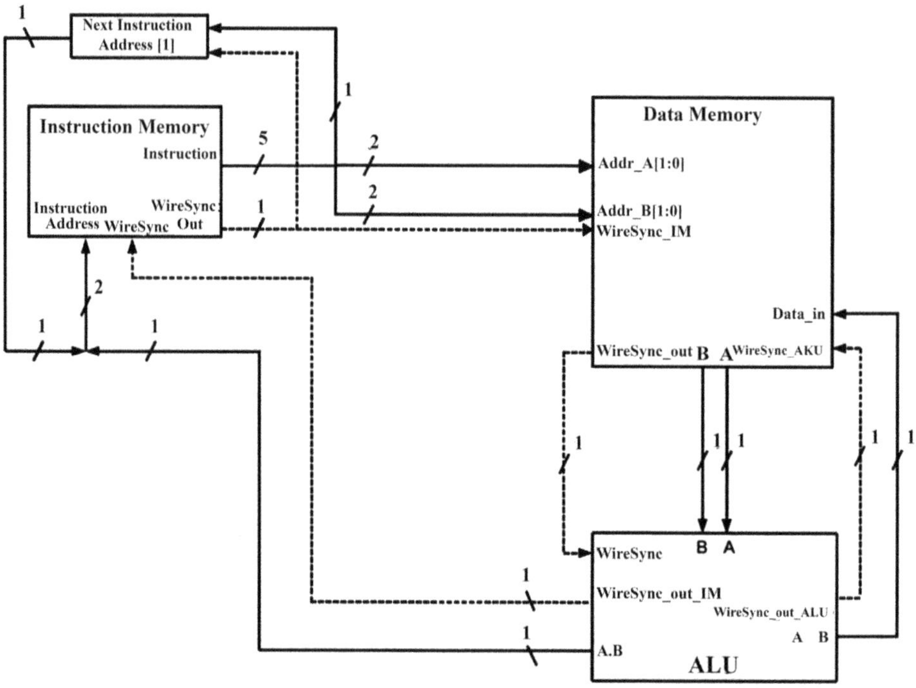

Figure 24.8. Combinational logic design of the processor architecture. Reproduced with permission from [4]. Copyright 2008 IEEE.

require a system to distinguish the right state in the circuit so as to perform composing accurately in DM and select the correct guidance in IM. In addition, in QCA there is no high impedance flag. So as to deal with this circumstance, an efficient strategy is devised: a wire in parallel to the processor circuit, conveying the polarization $P = +1$ if the related guidance in the processor should be executed. We call this instrument WireSync on the basis that this wire is in charge of synchronization of the the whole processor.

WireSync is illustrated in figure 24.9. Such a mechanism allows the execution of more than one instruction at the same time—each instruction with polarization $P = +1$ in WireSync will be executed. In the best scenario, where no data hazards are present, such as RAW, WAR, and WAW, it could be instructed that the running of the clock cycle will be one clock cycle difference. The SUBNEG architecture always has the RAW hazard: to go to the next instruction. Therefore, it is necessary to obtain the 'branching' result from the ALU. In addition, it will use only one instruction at time, i.e. there will be only one $P = +1$ in WireSync.

24.3.1 Instruction memory

The IM contains the instructions for the SUBNEG processor. This is a ROM type memory, i.e. the instructions are already stored. In the QCA circuit, fixed cell technology is used to represent these static data in the ROM and robust multiplexers

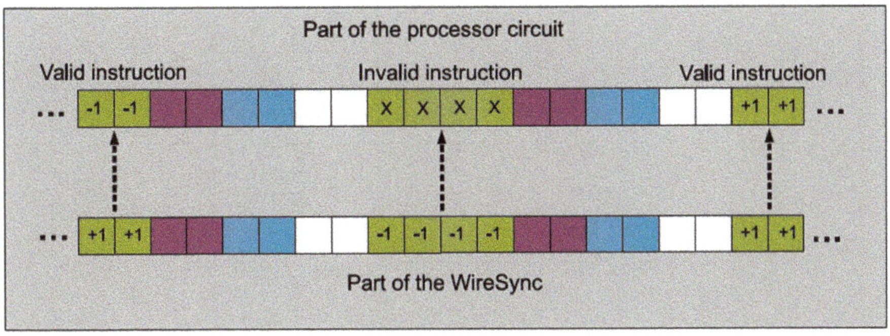

Figure 24.9. WireSync operation. Reproduced with permission from [4]. Copyright 2008 IEEE.

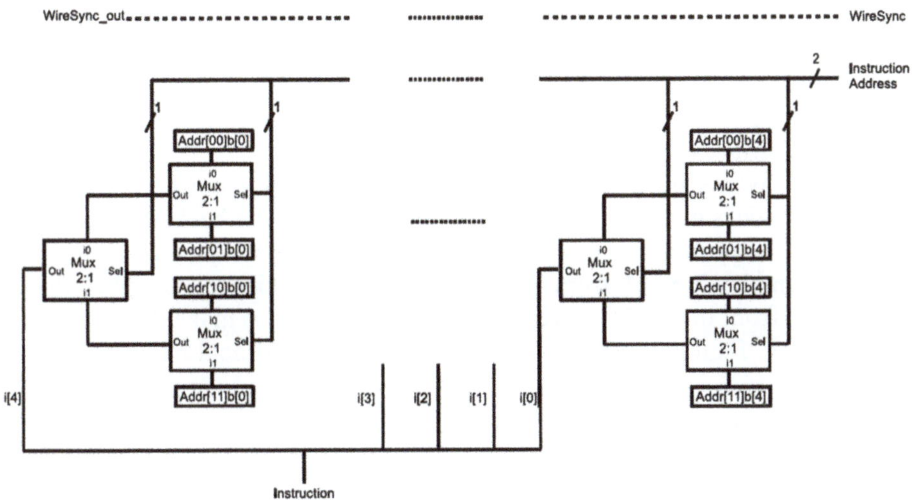

Figure 24.10. Combinational logic design of the instruction memory architecture.

are used to select the instruction. Figure 24.10 shows the IM as a combinational logic design. In figure 24.10 the instruction bits are a little *endian* (*i*[0] to *i*[4]) and have the following format: the least significant bit (*i*[0]) is the known part of the next instruction address, the two next bits (*i*[1] and *i*[2]) are the address for the second operand, and the last two bits (*i*[3] and *i*[4]) are the address for the first operand (A). Figure 24.11 illustrates the instruction format. As part of the IM, the bit *i*[0] is stored in a register, waiting for the other bit from the result from the ALU, to determine the next instruction address. The WireSync mechanism is applied to write the correct value to the register. The correct value from the instruction is followed by a parallel $P = +1$ in the WireSync from the IM. When *i*[0] reaches the special register for the next instruction address, $P = +1$ reaches it as well and the write in the register is executed.

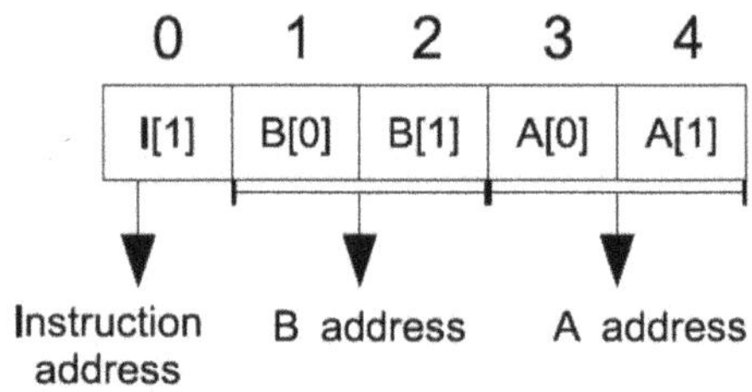

Figure 24.11. The SUBNEG instruction format.

24.3.2 Data memory

DM design is unpredictable since it performs two stages of activities. The initial step chooses A and B values for the ALU and stores the write back location, which is the enlist address B. The second step composes the subtraction (single Ex-OR) of the ALU in the location put away in the initial step. To store the write back address, in the first and second steps, the WireSync component is utilized. At the point when the privilege write back location comes to the write back address registers, a $P = +1$ in WireSync IM triggers composition of the right address.

The same awareness is applied to the write back of the ALU's subtraction outcome. A logic is made with the stored address in the write back address registers for each DM location in order to obtain a $P = +1$ in the conforming memory place output. When the $P = +1$ from the WireSync ALU arrives, it specifies that the correct value from the ALU arrives and a $P = +1$ is freed to make the correct write in the memory. Figure 24.12 establishes the DM combinational logic strategy.

24.3.3 The arithmetic logic unit

The ALU is the simplest component of the SUBNEG processor, because it performs only two operations. The first operation, a subtraction, consists of an Ex-OR and the negative test can be performed using a single logic equation $A{:}B$. The ALU unit works asynchronously, which implies that the WireSync has no use here unless it is following the right ALU result. The ALU logic design is shown in figure 24.13.

24.3.4 The integrated processor

The full QCA circuit with all the components discussed in the previous section is presented in figure 24.14. It represents a complete QCA circuit for the SUBNEG processor. The processor implementation is important for the robustness constraints, which define a maximum wire length of 12 cells, a minimum of two cells per clock zone, and a robust majority gate, which is largely used in the circuit. Multilayer crossing cells are used instead of coplanar crossings with rotated cells in order to make the circuit more robust.

The processor works as follows. First, the critical path of the QCA circuit is the minimum number of cycles in order to process the next instruction. Initially, the IM receives the two-bit instruction address, where the most significant bit comes from

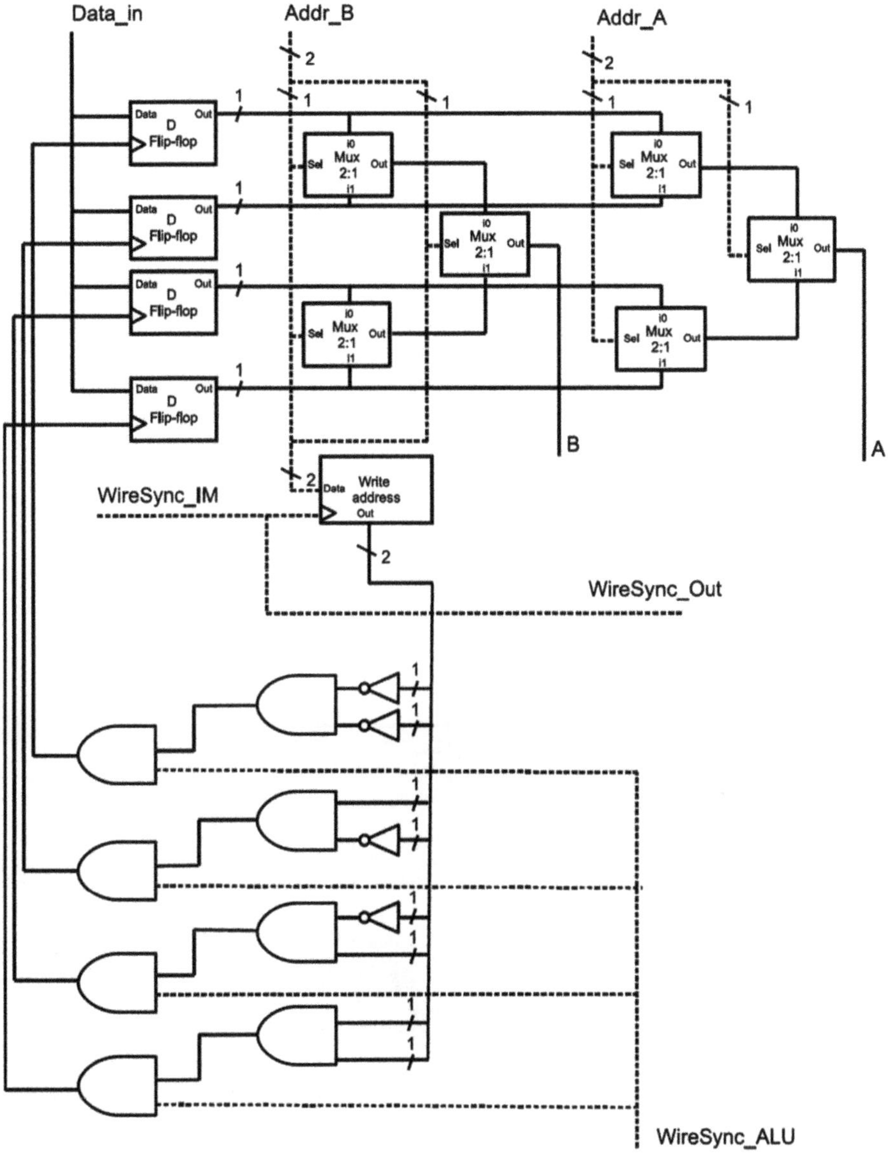

Figure 24.12. Combinational logic design of the data memory architecture. Reproduced with permission from [4]. Copyright 2008 IEEE.

the register that keeps the known bit from the previous instruction and the least significant bit comes from the ALU. It also receives a $P = +1$ from the WireSyncoutIM from IM. The instruction selection step, which is in the critical path, takes 11 full cycles to be completed, according to table 24.1, and passes the address of A and B, and the $P = +1$ of WireSync_out to the DM.

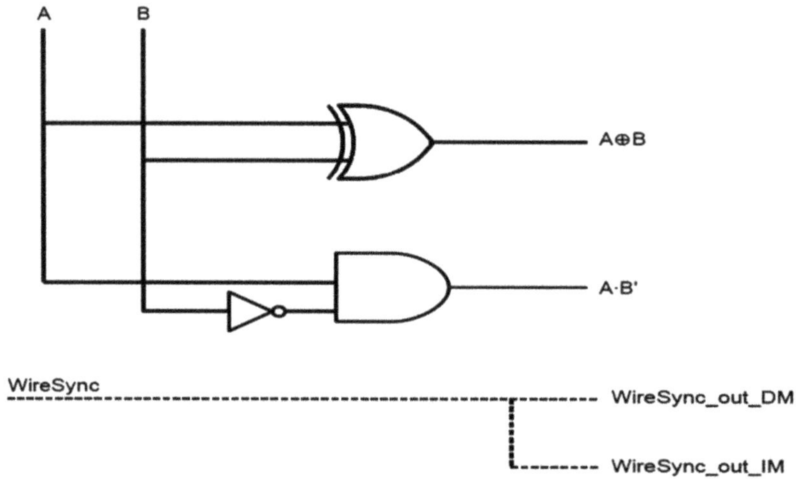

Figure 24.13. Combinational logic design of the ALU architecture.

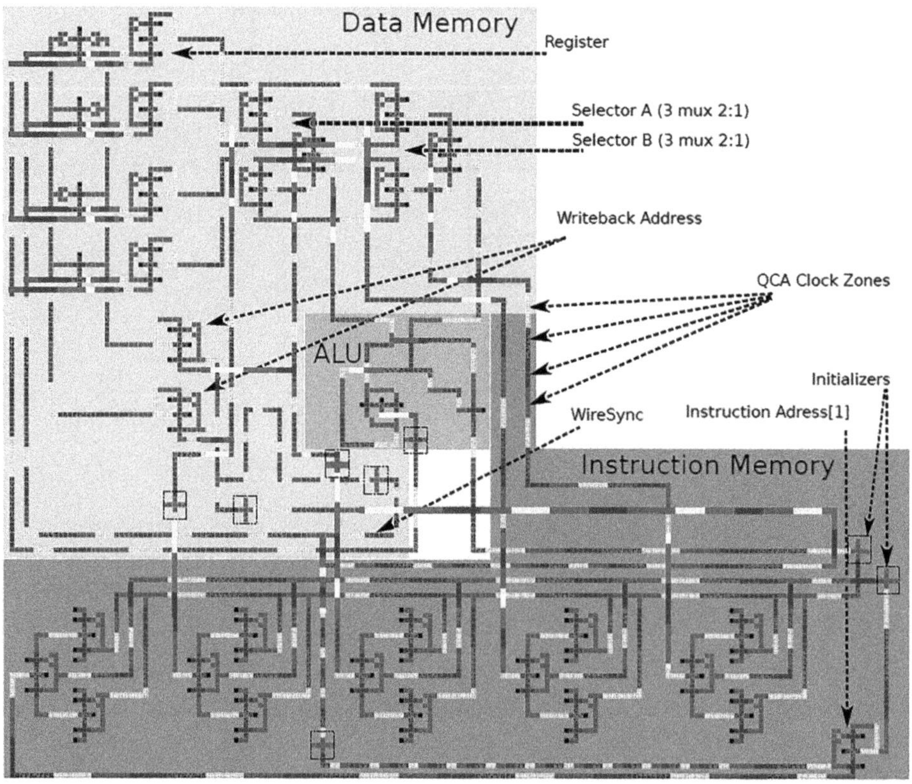

Figure 24.14. The SUBNEG processor.

Table 24.1. Inside components delays.

Component	Operation	Cycles
IM	Select instruction	11
IM	Save addr bit	7
DM	Select registers *A* and *B*	8
DM	Save register *B*	4
DM	Write value to register	12
ALU	Process registers *A* and *B*	3

Next, three operations occur simultaneously: bit[0] instruction storage, the data reading *A* (bit[4] and bit[3]) and *B* (bit[2] and bit[1]) from DM, and address *B* storage for the write back. They take seven, eight, and four cycles, respectively. As the data reading fits the critical path and it has the maximum value among the others, it is cetain that all the processes work at the right time. Then the *A* and *B* values and the $P = +1$ of WireSync are passed to the ALU which takes three cycles to make the operations. The ALU actions are in the critical path. Then the ALU re-passes the $P = +1$ to WireSyncoutDM and the subtraction result to the DM in order to make the write back, and the diverging result and the $P = +1$ in WireSyncoutIM to IM.

Finally, the write back is accomplished up to a maximum of 12 cycles (if the target register is '00') while the next instruction is selected, meaning that the write back is not in the critical path. It is important to mention that the write back finishes before the storage of address *B* occurs. Thus, no conflict is possible here even with two instructions sharing the same time window. Thus, the processor cycle ends with the consumption of 22 QCA cycles in the critical path.

24.4 Summary

In this chapter, the structure of a QCA RAM is demonstrated. There are two kinds of RAM cell engineering, specifically circle/loop-based and line-based. In the QCA ROM structure it is shown how an improved ROM may enable the storage of 2.5 Gbits of computerized information per cm^2. In addition, for processor planning the SUBNEG design was pursued and the circuit was constructed using IM, DM, and the ALU.

Further reading

[1] Angizi S, Sarmadi S, Sayedsalehi S and Navi K 2015 Design and evaluation of new majority gate-based RAM cell in quantum-dot cellular automata *Microelectron. J.* **46** 43–51

[2] Berzon D and Fountain T J 1999 A memory design in QCAs using the squares formalism *Proc. Ninth Great Lakes Symp. on VLSI* (Piscataway, NJ: IEEE) pp 166–9

[3] Dehkordi M A, Shamsabadi A S, Ghahfarokhi B S and Vafaei A 2011 Novel RAM cell designs based on inherent capabilities of quantum-dot cellular automata *Microelectron. J.* **42** 701–8

[4] Fazzion E, Fonseca O L, Nacif J A M, Neto O P V, Fernandes A O and Silva D S 2014 A quantum-dot cellular automata processor design *Proc. of the 27th Symp. on Integrated Circuits and Systems Design* (New York: ACM) p 29

[5] Hashemi S and Navi K 2012 New robust QCA D flip flop and memory structures *Microelectron. J.* **43** 929–40

[6] Hashemi S and Navi K 2012 New robust QCA D flip flop and memory structures *Microelectron. J.* **43** 929–40

[7] Kim K, Wu K and Karri R 2005 Towards designing robust QCA architectures in the presence of sneak noise paths *Proc. Conf. on Design, Automation and Test in Europe* vol 2 (IEEE Computer Society) pp 1214–9

[8] Lin A 2010 Carbon nanotube synthesis, device fabrication, and circuit design for digital logic applications *PhD Thesis* Stanford University

[9] Rahimi E and Nejad S M 2008 Quantum-dot cellular ROM: a nano-scale level approach to digital data storage *6th Int. Symp. on Communication Systems, Networks and Digital Signal Processing* (Piscataway, NJ: IEEE) pp 618–21

[10] Schulhof G, Walus K and Jullien G A 2007 Simulation of random cell displacements in QCA *ACM J. Emerg. Technol. Comput. Syst.* **3** 2

[11] Shamsabadi A S, Ghahfarokhi B S, Zamanifar K and Movahhedinia N 2009 Applying inherent capabilities of quantum-dot cellular automata to design: D flip-flop case study *J. Syst. Archit.—Embedded Sys. Des.* **55** 180–7

[12] Shulaker M M, Hills G, Patil N, Wei H, Chen H-Y, Wong H-S P and Mitra S 2013 Carbon nanotube computer *Nature* **501** 526

[13] Baris T and Hong B 2008 Improving line-based QCA memory cell design through dual phase clocking *IEEE Trans. Very Large Scale Integr. Syst.* **16** 1648–56

[14] Vankamamidi V, Ottavi M and Lombardi F 2005 A line-based parallel memory for QCA implementation *IEEE Trans. Nanotechnol.* **4** 690–8

[15] Vankamamidi V, Ottavi M and Lombardi F 2008 A serial memory by quantum-dot cellular automata (QCA) *IEEE Trans. Comput.* **57** 606–18

[16] Vetteth A, Walus K, Dimitrov V S and Jullien G A 2003 Quantum-dot cellular automata of flip-flops *ATIPS Lab.* **2500** 1–5

[17] Walus K, Vetteth A, Jullien G A and Dimitrov V S 2003 RAM design using quantum-dot cellular automata *NanoTechnology Conf.* vol 2 pp 160–3

[18] Walus K, Mazur M, Schulhof G and Jullien G A 2005 Simple 4-bit processor based on quantum-dot cellular automata (QCA) *16th IEEE Int. Conf. on Application-Specific Systems, Architecture Processors* (Piscataway, NJ: IEEE) pp 288–93

[19] Yang X, Cai L and Zhao X 2010 Low power dual-edge triggered flip-flop structure in quantum dot cellular automata *Electron. Lett.* **46** 825–6

[20] Yang X, Cai L, Zhao X and Zhang N 2010 Design and simulation of sequential circuits in quantum-dot cellular automata: falling edge-triggered flip-flop and counter study *Microelectron. J.* **41** 56–63

www.ingramcontent.com/pod-product-compliance
Ingram Content Group UK Ltd.
Pitfield, Milton Keynes, MK11 3LW, UK
UKHW051858100926
8556IPUK00005B/687